The Divided Skies

THE
DIVIDED
SKIES

Establishing
Segregated
Flight Training
at Tuskegee, Alabama,
1934–1942

Robert J. Jakeman

The University of Alabama Press
Tuscaloosa and London

First Paperback Printing 1996

designed by zig zeigler

∞
The paper on which this book is printed meets the
minimum requirements of American National Standard
for Information Science-Permanence of Paper for Printed
Library Materials, ANSI Z39.48-1984.

Library of Congress Cataloging-in-Publication Data

Jakeman, Robert J., 1948–
The divided skies : establishing segregated flight training at
Tuskegee, Alabama, 1934–1942 / Robert J. Jakeman.
p. cm.
Includes bibliographical references and index.
ISBN 0-8173-0859-8
1. Flight training—Alabama—Tuskegee—History. 2. Tuskegee Institute (Ala.)—History. 3. Afro-Americans in aeronautics—Alabama—History. I. Title.
TL713.T87J35 1992
358.4′137′08996073—dc20
                                                    91-11459

British Library Cataloguing-in-Publication Data available

 # Contents

	Preface	vii
1.	Aviation and Tuskegee Institute: The Early Years	1
2.	Black Americans and the Military	33
3.	The Black Public Becomes Air-minded	53
4.	Civil Rights Emerges as a National Issue	69
5.	The Aviation Legislation of 1939	88
6.	The Civilian Pilot Training Program at Tuskegee	112
7.	Tuskegee Emerges as the Center of Black Aviation	132
8.	The Campaign for Air Corps Participation Broadens	158
9.	Fruits of the Campaign: The Ninety-ninth Pursuit Squadron	183
10.	The Reaction to the Ninety-ninth	216
11.	Military Flight Training Begins	240
12.	Making the Dream a Reality: The Air Corps' First Black Pilots	270
13.	Conclusion	306
	Notes	317
	Bibliography	391
	Index	403

 # Preface

In March 1942, at Tuskegee Army Air Field in rural Alabama, five men received the silver wings of Army Air Forces pilots.* Like thousands before them, they had completed standard Army flight training, a demanding course of study that included extensive classroom instruction and many hours of flying time. Outward similarities notwithstanding, their graduation marked an important milestone in the history of military aviation in the United States— they were the first African Americans to qualify as military pilots in any branch of the nation's armed forces. Until these five entered flight training the previous summer, blacks had been systematically excluded from the Air Corps and from the aviation program of the U.S. Navy. By the end of World War II, however, almost 1,000 African Americans had won their wings at Tuskegee Army Air Field. The Navy's policy of racial exclusion in aviation continued throughout World War II; it was not until 1948 that the first black American received the gold wings of a Navy pilot.

Roughly half of the men who graduated from flight training at

---

*The United States Air Force was established as an independent branch of service in 1947. Prior to 1947, ground-based air units were part of the United States Army. The Signal Corps was responsible for army aviation until World War I when the Air Service was established; in 1926, the Air Service was redesignated the Air Corps. On the eve of World War II, the army initiated a major expansion and reorganization of its air arm. By March 1942 the War Department abandoned the Air Corps and organized three co-equal commands—the Army Ground Forces, the Army Service Forces, and the Army Air Forces. Although nominally part of the U.S. Army, the Army Air Forces was virtually an independent service within the War Department. See Charles A. Ravenstein, *Organization of the Air Force* (Maxwell Air Force Base, Alabama: Albert F. Simpson Historical Research Center, 1982).

Tuskegee flew combat missions as fighter pilots in the European and Mediterranean theaters. These black pilots—known today as the Tuskegee Airmen—compiled a respectable combat record. They flew over 15,000 sorties and destroyed over 100 German aircraft, including several of the jet fighters that appeared near the end of the war. Among their decorations were some 150 Distinguished Flying Crosses and hundreds of Air Medals. When the war ended, an all-black bomber group was in training for combat in the Pacific.

The story of the Tuskegee Airmen is indeed a compelling one. It has been the subject of numerous television documentaries, newspaper feature articles, and museum exhibitions. A recent theater production in New England dramatized the exploits of the Tuskegee Airmen. Such coverage in the popular media has raised public awareness of the role black airmen played during World War II. Similarly, most studies of black Americans during the World War II era, as well as those of black participation in the armed forces, include an account of the men who trained at Tuskegee and fought in the skies over the Mediterranean and Europe.

Although the wartime exploits of the Tuskegee Airmen have received considerable popular and scholarly attention in recent years, the story of the people and events that culminated in the establishment of a segregated flight training program at Tuskegee has been largely ignored. This important prologue to the better-known story of black pilots of World War II is the subject of this study. It begins by recounting Tuskegee Institute's first tentative efforts to enter the field of aviation during the mid-1930s, and concludes with the graduation of the first class of black pilots in early 1942.

Chapter one outlines Tuskegee's early involvement with aviation and the school's first attempt to add aviation courses to its curriculum. By spring of 1936 the institute had all but finalized plans to hire an instructor and initiate aviation instruction the following fall. The plan suddenly collapsed, however, and over three years elapsed before Tuskegee students finally had the opportunity to learn to fly.

The establishment of an aviation course at Tuskegee in 1939 was, in part, the result of a national crusade by African Americans for broader participation in the nation's armed forces. An important component of this crusade was the campaign for admitting blacks to the Air Corps. One authority, historian Ulysses Lee, has characterized it as "the most widespread, persistent, and widely publicized of all the prewar public pressure campaigns affecting the Negro and the Army." The Air Corps participation campaign drew its strength from three important sources—black America's high regard for military service, a growing enthusiasm for aviation among the black public, and the emergence of civil rights as a national issue during the decade of the 1930s. These sources, which formed the foundation of the campaign for the admission of black Americans to the Air Corps, are the focus of chapters two, three, and four.

During the early months of 1939 the U.S. Congress enacted legislation to expand the Air Corps and to train thousands of college students in the rudiments of flying. Activists in the campaign for Air Corps participation lobbied for and obtained amendments in Public Law 18, which provided for a major expansion of the Air Corps, and the Civilian Pilot Training Act, which authorized the Civil Aeronautics Authority to establish the Civilian Pilot Training Program at a number of colleges and universities. Chapter five recounts the efforts of these black lobbyists and their congressional allies to amend the aviation legislation of 1939 and provide broader opportunities in aviation to black Americans.

Chapters six and seven describe the establishment of an aviation training program at Tuskegee Institute under the aegis of the Civilian Pilot Training Program. Under the sponsorship of this federally funded initiative, Tuskegee quickly developed its aviation curriculum, supplanting Chicago as the nation's center of black aviation. Although the instruction received under the Civilian Pilot Training Program was not military training, students at Tuskegee and several other black colleges eagerly participated, hoping that the racial barriers that excluded them from military aviation might soon fall.

In 1940 the campaign for Air Corps participation broadened and became a key component in black America's crusade for broader opportunities in the armed forces. Finally, in late 1940, the military establishment yielded to intense political pressure and directed the Air Corps to develop plans for a segregated air unit. In early January 1941 the secretary of war approved a plan to establish the Ninety-ninth Pursuit Squadron and base it at a new military airfield to be constructed near Tuskegee, Alabama. Chapters eight and nine examine these developments.

The War Department announced its intention to establish a segregated squadron in mid-January 1941. Although many black Americans welcomed the news, the decision met with criticism in some quarters. Chapter ten outlines the initial reaction to the announcement, particularly the opposition of the National Association for the Advancement of Colored People and the National Airmen's Association of America, a Chicago-based organization of black pilots.

The final chapters tell the story of the establishment of the military flight training program at Tuskegee through the graduation of the first class of pilots in March 1942. Tuskegee Institute's struggle to build its own airfield and inaugurate the first phases of military training with the Air Corps is addressed in chapter eleven. When the nation entered World War II in December 1941, construction of Tuskegee Army Air Field—which was to serve as the site of the final phases of flight training and the home station for the Ninety-ninth Pursuit Squadron—was incomplete. Moreover, no black pilots had completed their training and the War Department's promise to establish a black air unit remained unfulfilled. Despite frustrating delays and rumors that the program was to be abandoned, training continued and plans to establish a second black unit were announced. These final episodes in the establishment of the flight training program at Tuskegee are dealt with in chapter twelve.

This study could not have been written without the dedicated efforts of archivists and librarians who collected, cataloged, and preserved the records and publications upon which it is based. I gratefully acknowledge the patient assistance of the staff at the National Archives, the Manuscript Division of the Library of Congress, the Franklin D. Roosevelt Library, the United States Air Force Historical Research Center, and the Auburn University library. Thanks also to the NAACP for materials they graciously allowed me to use.

I owe a special debt to Dr. Dan Williams, archivist at Tuskegee University. The wealth of material he provided was only surpassed by the many courtesies he extended during the long hours of research in Tuskegee's archives.

I also acknowledge the assistance of the Office of Air Force History, Washington, D.C., and the Harry Merriwether Fellowship Fund, Auburn University, in the research and writing of this study. The generous financial support they provided, in the form of dissertation year fellowships, enabled me to conduct extensive research in

Washington, D.C., Cambridge, Massachusetts, and Hyde Park, New York.

Thanks are also due the history faculty at Auburn University. Their guidance and encouragement were crucial to the completion of this project. Professor Allen W. Jones deserves special mention and thanks; the effect of his prodding, encouragement, and commitment to excellence, on a sometimes discouraged graduate student, cannot be overstated.

I must also express my gratitude to the staff of The University of Alabama Press for their patience with a reluctant author who wreaked havoc on production schedules. Despite repeated delays and missed deadlines, Malcolm MacDonald and staff were ever gracious and supportive. Thank you.

Finally, I thank my wife and daughters for their constant patience and support throughout this project. They know that without their encouragement it would not have been completed.

The Divided Skies

Aviation and
Tuskegee Institute

The Early Years

On 22 May 1934 the first airplane to land on the grounds of Tuskegee Institute touched down in an oat field near the edge of the school farm. John C. Robinson, an aspiring Chicago aviator, had chosen the occasion of his ten-year class reunion to make a dramatic aerial return to his alma mater. Alvin J. Neely, registrar of the institute, welcomed the young pilot as he stepped from a small, single-seat monoplane before a cheering crowd gathered on the campus for spring commencement exercises.[1] An airplane landing on the grounds of the rural Alabama vocational school and college would have attracted a good deal of attention under any circumstances, but an airplane piloted by a former Tuskegee student was a newsworthy event indeed, especially when it coincided with spring commencement. Many black newspapers and at least one white paper took notice of the event and carried a photograph of Robinson seated in his cockpit shaking Neely's outstretched hand.[2]

Robinson's flight to Tuskegee was more than just a colorful adjunct to the commencement exercises of 1934; it marked the beginning of Tuskegee's first attempt to enter the air age. The young aviator used the occasion to urge the school's officials to build an airport that could serve as a site for an annual air show.[3] He met with G. L. Washington, director of the Department of Mechanical Industries, who confided that the inauguration of a course in aviation was under consideration.[4] During the next two years, Tuskegee exhibited a growing interest in aeronautics, and in 1936 the black and white press reported that Tuskegee planned to offer courses in aviation with Robinson, lately returned from duty in the Ethiopian air force, serving as instructor.[5]

Tuskegee Institute had much to recommend it as an aviation training center for blacks. Situated in the Deep South, it could offer excellent year-round flying weather, and its rural setting promised ample undeveloped land for an airfield. Moreover, aviation training, especially in aircraft mechanics, would complement the school's traditional emphasis on practical, task-oriented vocational education. Tuskegee also enjoyed substantial philanthropic support from northern industrialists and these connections might well prove useful in securing adequate funding for an aviation training center open to blacks. And most importantly, Tuskegee's reputation as a leading black educational institution could lend credibility to the idea of training blacks to enter a field that many whites believed them inherently unable to master.

Aviation was only a fantastic dream in 1881 when Booker T. Washington, a young graduate of Hampton Institute, arrived in Tuskegee, Alabama, to organize a normal school for the training of black teachers.[6] The sixty years between Tuskegee's founding and its emergence as the center of black aviation were years of growth and development.

Tuskegee was chartered by act of the Alabama legislature in early 1881, and three trustees—one black and two whites—were appointed to select a principal. They wrote Samuel C. Armstrong, a former Union general and the founder and principal of Hampton Institute, asking him to recommend a white candidate to organize and lead their new school. Armstrong told the trustees he had no qualified white candidate, but strongly endorsed Booker T. Washington as a capable and willing candidate for principal. The trustees had no other prospects, trusted the judgment of Armstrong, and selected Washington as their new principal.

Washington made his way to Alabama and quickly went to work. Despite extremely limited resources he managed to open the school on 4 July 1881. His ambition and resourcefulness became apparent as he steadily worked to place the school on a firm educational and financial footing. Although Tuskegee was founded as a normal school for training elementary school teachers, Washington soon added industrial training courses—similar to those at Hampton— such as carpentry, masonry, blacksmithing, and housekeeping. He envisioned Tuskegee as "a veritable cathedral of practical learning and black self-help, a Hampton run entirely by black people."[7]

The school was well established by 1895, the year Washington made his famous speech at the Atlanta Exposition, an address which catapulted him to national prominence and made him the leading black spokesman of his day, a role that had been filled by Frederick Douglass until his death earlier that year. Speaking before a racially mixed crowd, Washington urged blacks to stay in the South, to "cast down your bucket where you are," and to accept social segregation as a necessary condition for economic cooperation between the races.[8] Washington's speech, now known as the Atlanta Compromise, cast him in the role of a race leader for black America. In the years that followed, Washington's reputation and political influence grew; Tuskegee Institute shone in his reflected light and was acclaimed as a model for black education. Moreover, Tuskegee functioned as the headquarters for Washington's vast network of political influence and patronage, which came to be known as the "Tuskegee machine."

When Washington died in 1915 the trustees selected another Hampton graduate as principal, Robert Russa Moton. Moton, a close associate of Washington, had served for twenty-five years as commandant of cadets at Hampton. Although well qualified for the position, Moton was unable—and unwilling—to step into the broader leadership role that had been filled by Washington. Still, the position necessarily brought with it an implicit leadership status that extended beyond the Tuskegee campus, a reality Moton could not ignore. He brought to the job, however, a different set of priorities. While Washington had functioned primarily as a race leader and political boss and only secondarily as the head of Tuskegee, Moton saw himself first as principal, occasionally as a race leader, and only rarely as a political boss.[9]

When Moton took over in 1915, Tuskegee's reputation was well established as a vocational school that trained teachers, farmers, and tradesmen, while providing academic courses at the high school level. As Washington told his students in 1896, "we are not a college and if there are any of you here who expect to get a college training you will be disappointed."[10] By the mid-1920s Moton had introduced some college-level courses, but no degrees were conferred. Then in 1927 he organized a collegiate division, offering degrees in agriculture, home economics, and education.[11] When some of the school's trustees opposed the change, fearing that it signaled a shift away from vocational training in favor of a liberal arts course of study, Moton pointed to the increasing demand for college-trained teachers and argued that the institute was simply responding to that need, thereby improving job prospects for its graduates.[12]

Although Moton did not attempt to maintain the political aspects of the Tuskegee machine, he nevertheless sought to preserve Tuskegee's reputation and status. In 1925 he headed a joint fund-raising campaign with Hampton Institute and raised $10 million, tripling Tuskegee's endowment fund and permitting the construction of buildings for the new collegiate division. The closest Moton came to influence peddling was in the early 1920s when he played a key role in the selection of Tuskegee as the site for a veterans hospital for blacks. When he learned of plans to staff the hospital largely with white doctors and nurses, Moton once again used his position to see to it that the professional staff of the hospital was black instead of white.[13]

During Moton's tenure, Tuskegee's decline as the center of black political influence continued. Instead of trying to keep the Tuskegee machine running, Moton concentrated on solidifying and enhancing the institute's position as a leading black educational institution. He improved Tuskegee's financial situation, elevated the level of study in response to a changing job market, sought to meet the needs of southern blacks through farmers' conferences and agricultural extension, and broadened professional opportunities for black doctors and nurses. Thus when he retired in 1935, Moton had brought a number of changes to the school he inherited from Washington two decades earlier.[14] Although Tuskegee's political influence had declined, it was still highly regarded by both the black and white public for its efforts to foster black economic development while avoiding a direct challenge to segregation.

Tuskegee's third president, Frederick Douglass Patterson, differed markedly from Washington and Moton in both background and education.[15] Born in 1901 into a middle-class black family, he took office in his midthirties. Unlike his predecessors, who were products of Hampton Institute, Patterson brought to the position impeccable professional and academic credentials from recognized white institutions of higher learning. He graduated from Iowa State College in 1923 with a Doctorate of Veterinary Medicine and received his Master of Science from the same institution in 1927. In 1931 he entered graduate school at Cornell University and completed the requirements for a Ph.D. in bacteriology by the end of 1932. When not engaged in graduate studies, Patterson served as an instructor and academic administrator at Virginia State College and at Tuskegee Institute; prior to his election as president, he was serving as director of Tuskegee's School of Agriculture.[16]

Despite these differences, Patterson subscribed to the racial and educational philosophies of Washington and Moton. Echoing the ideas of Washington, he urged blacks to train for and seek technical jobs in areas where they traditionally found employment: agriculture, the laundry and cleaning industry, and the food service industry. He cautioned against continuing the "alarming neglect of trades education by Negroes themselves in earlier years, now ably reinforced by the exclusion policies of the trades unions" with the support of "directors of vocational education in the several southern states that administer federal vocational funds." Patterson believed the exclusion of blacks from technical training and employment deserved "far greater concern by Negroes than it is now receiving as one of the important stifling and thwarting influences to the realization of economic competence by the Negro people." Like Washington, he advised against a frontal attack on segregation, urging instead a policy of "segregated opportunity":

If we are to attain the objective of a large participation in technical fields, we shall need to be careful that our wise insistence upon the elimination of undemocratic practices does not lead us to a point of stupid insistence upon unsegregated opportunity in such a way as to defeat the chance for the employment of specialists through failure to attain the ultimate in one fell swoop. To counsel patience with the Negro is bringing coals to Newcastle but it is sometimes easy to overlook the fact that 95% of the professional competence or technical standing which Negroes have attained has come through segregated opportunity.[17]

Despite his ideological similarities to Washington and Moton, Patterson's election finalized a major change in the status and role of the president of Tuskegee. Washington had emerged from obscurity in 1895 to become a race leader and Tuskegee Institute served as his headquarters. Moton tried to strike a balance between race leadership and his role of chief executive of Tuskegee Institute. When Patterson came to office his age and background, together with the growth of organizations like the National Association for the Advancement of Colored People (NAACP), the National Urban League, and the Brotherhood of Sleeping Car Porters, made it inevitable that his primary role would be that of a college president and not of a race leader.

When Robinson introduced Tuskegee to aviation at the spring commencement in 1934, there was no hint that Patterson, then director of the School of Agriculture, would be the school's next president and ultimately play a key role in the development of an aviation program at Tuskegee. The hopeful report of talks about an airfield and annual air show there, which appeared in the *Chicago Defender*, had no sequel and nothing in the records of the institute suggests that the administration did anything more than toy with the idea of entering the field of aviation.

The reasons why Robinson's visit produced no immediate results are unclear. He had suffered an accident en route, and perhaps the news of this incident made the administration cautious. Maybe Robinson's approach, personality, or background failed to inspire confidence. In retrospect, however, the most important mitigating factor was likely poor timing. Robinson flew to his alma mater in the midst of the Great Depression, a time when virtually all segments of the nation, including black colleges, struggled to survive. Moreover, he came in the waning months of Moton's administration; by 1934 Moton, almost seventy, in poor health, and nearing retirement, conducted what amounted to a caretaker administration. In short, Robinson's return for his tenth class reunion came at an inauspicious time for the inauguration of expensive and risky ventures such as aviation programs.

Nevertheless, Robinson's flight highlighted dramatically for the Tuskegee community the growing black presence in aeronautics. This recognition of black interest and participation in aviation was perhaps crucial to Tuskegee's support of a proposal that came shortly after Robinson's flight, an initiative from an entirely different quarter and of an entirely different nature.

In September 1934 the Moton administration agreed to support two black aviators' ambitious plans for a Pan-American air tour by arranging for an airplane, recently purchased by one of the fliers, to be christened the *Booker T. Washington* in ceremonies on the campus. For the first time the name of Tuskegee Institute was linked publicly to a major aviation initiative. The event attracted the support of the campus community, who rallied behind the fliers and launched a successful fund-raising campaign. Moreover, institute secretary G. Lake Imes, a key member of Moton's staff, worked on the aviators' behalf in Washington and New York, and he subsequently proposed that Tuskegee Institute actively promote future international flights that the two pilots hoped to undertake.

The christening ceremony was part of a publicity and fund-raising campaign that was sponsored by a group in Atlantic City, New Jersey, known as the Interracial Goodwill Aviation Committee (IGAC). It was in anticipation of the third long-distance flight of Charles Alfred Anderson and Tuskegee alumnus Dr. Albert E. Forsythe. A year earlier Anderson and Forsythe earned a place in aviation history when they completed the first round-trip transcontinental flight by black aviators.[18] Several months later they made a round-trip flight from Atlantic City to Montreal, becoming the first American blacks to plan and execute a flight across international borders.[19] Plans for the third flight, touted as the Pan-American Goodwill Flight and the most ambitious of the series, called for a month-long, 12,000-mile circuit to more than twenty countries throughout South America, the Caribbean, and Central America.[20]

The flights were the brainchildren of Forsythe, who came from markedly different circumstances than his fellow alumnus and aviator, Robinson. Born in Nassau in 1897, the son of a civil engineer, Forsythe spent his boyhood in Jamaica and then came to the United States to attend Tuskegee. He continued his education at the University of Illinois and McGill University in Montreal, where he earned a medical degree. By 1932, after establishing his medical practice in Atlantic City, he became an aviation enthusiast, giving unselfishly of his time, energy, and money to the promotion of aviation among his race.[21] His flying partner, Anderson, described him years later as "a very, very aggressive and determined man and an ambitious person [who] wanted to advance aviation among the blacks."[22] A practical man, Forsythe did not fly merely for the sake of flying. Rather, he considered airplanes efficient and useful transportation devices, once declaring to Anderson that he liked "to go places in an airplane. That's what an airplane is for, to travel, not just fly around the home field."[23]

But Charles Alfred Anderson simply loved to fly. Ten years younger than Forsythe, Anderson spent his early years with his grandmother in the Shenandoah Valley near Staunton, Virginia, where he developed an intense fascination with airplanes and flying. His love of aviation remained strong after he returned to his parents' home in Bryn Mawr, Pennsylvania, near his birthplace. Unable to obtain flying lessons because of racial prejudice, he bought an airplane on borrowed money and taught himself to fly. In 1929 he became one of the first black pilots in the nation to earn a private license. Three years later he became the first black to qualify as a transport pilot; this was the highest rating then issued by the Department of Commerce, which authorized him to fly passengers for hire and teach others to fly. Forsythe, fifty miles away in Atlantic City, learned of Anderson's achievement from reports in the black press, contacted the young pilot, and asked for flying lessons.[24]

Thus began the Forsythe–Anderson partnership. Forsythe's medical practice gave him the financial resources to pursue his aviation interests. His association with Anderson afforded the nation's only black transport pilot flying experience far beyond what he could have accumulated on his own. By 1934 Anderson was still the only black holding a transport license; this, in combination with the wealth of experience gained from his fortuitous partnership with Forsythe, made him the most qualified black pilot in America.

Tuskegee's role in the publicity campaign for the Pan-American flight came almost as an afterthought. The IGAC had contacted Moton earlier in the year about serving as an advisor, along with other notable blacks such as Oscar De Priest, Emmett J. Scott, and Eugene Kinckle Jones, but Moton had not responded. Several months later, after plans for the flight were well under way, the

IGAC approached Moton again. On Sunday, 2 September 1934, the organization's chairman, Julia Goens, wrote to ask if Forsythe's new airplane could be flown to the Tuskegee campus at the end of the week for a christening ceremony.[25] Although Moton was away from Tuskegee when Goens's letter arrived, institute secretary G. Lake Imes realized the publicity value of the proposal; he wired Goens and asked if the ceremony could be held on the weekend of 15–16 September, shortly after Moton's return. Goens expressed reservations about the practicality of the proposal, asking whether arrangements could be made in Moton's absence. She hastened to add, however, that the IGAC was "most anxious that the plane should be christened at Tuskegee because Tuskegee represents true progress on the part of colored America, because the plane is to be named the Booker T. Washington and because one of the fliers, Dr. Forsythe, is a product of Tuskegee."[26]

Imes apparently convinced Goens that a proper ceremony could be arranged; on the eleventh she gave the go-ahead for Saturday, 15 September, with "program and features at the discretion of Principal [Moton] and Tuskegee officials." Moreover, she urged Imes to seek "immediately all possible nation wide publicity" and was pleasantly surprised when Imes wired back: "Trying to get Lindbergh for christening."[27]

To attract the most famous flier of all time would be a publicity coup worth great effort. Capitalizing on his Hampton connections, Moton wired Daniel W. Armstrong, son of Samuel C. Armstrong, asking him to invite the famous aviator to the ceremonies. Moton observed that the event was of "great importance to our people" and noted that Lindbergh's presence would "enhance [the] occasion and focus national attention on Tuskegee's work." Armstrong's response was disappointing. He told Moton the chances of Lindbergh's attending the ceremony were "remote," advising him instead to wire Lindbergh and request a statement on the value of the flight which could be publicized and "might well serve your purpose."[28] Subsequent efforts to secure a statement from Lindbergh through the services of James Edmund Boyack, a white public relations agent whose services the institute used on occasion, proved fruitless.[29] Nevertheless, the event received considerable press coverage. Boyack contacted six white aviators who had recently completed highly publicized long-distance flights, obtained statements supporting the flight and its objectives, and publicized the fliers' endorsements in both the black and white press.[30]

By the time Anderson and Forsythe arrived in Tuskegee on the evening of 14 September for the ceremonies the next afternoon, Imes had formed a program committee that included the future president of the institute, F. D. Patterson, and the head of the military department, Col. Benjamin O. Davis, Sr. A program booklet for the ceremony had been printed, containing brief biographical sketches of Anderson and Forsythe, a description of the aircraft, and the itinerary of the flight.[31]

The program opened with remarks by Moton, who served as master of ceremonies. Nettie H. Washington, granddaughter of the founder, welcomed the aviators on behalf of the students. She recounted the accomplishments of the aviators, noting that Anderson "is rated by the government as the best pilot in our race." Forsythe, she observed with pride, belonged to "the Tuskegee family." She reminded the audience that another former Tuskegee student, John C. Robinson, was a "licensed aviator connected with the Curtis [*sic*] Flying School in Chicago and that one of our young women [Doris Murphy], also a former student of Tuskegee, is now qualifying for her pilot's license at the same school." Following remarks by Anderson and Forsythe, Mrs. Moton christened the aircraft and an exhibition flight concluded the festivities.[32]

On the day of the christening, Tuskegee Institute went one step further in support of the Pan-American flight. When Forsythe indicated a need for navigational and weather flying instruments as well as emergency equipment such as parachutes and life rafts to enhance safety and reduce the likelihood of failure, President Moton immediately appointed a fund-raising committee and announced his action before the conclusion of the ceremony.[33] The committee set a goal of raising $200 toward the $3,000 needed for the equipment. By early October one of the committee members reported to Moton the $200 goal would be exceeded; ultimately $350 was sent to the IGAC.[34]

In addition to this fund-raising campaign, Imes put Forsythe in contact with sources of funds and technical advice. Shortly after the christening, Imes had occasion to travel to New York City, where he sought the offices of the Daniel Guggenheim Fund for the Promotion of Aeronautics. Although he discovered that the fund was no longer functioning, he managed to convince the head of the aeronautical engineering department at New York University, a Professor Karmin, to support Forsythe with technical advice.[35]

Imes also reported to Forsythe the results of his conversations with James Edmund Boyack, the publicity agent who had arranged for various aviators to send congratulatory telegrams to the christening. Boyack urged that the Pan-American tour start from Tuskegee with a group of prominent white aviators endorsing the project by their presence and technical advice. He introduced Imes to Robert G. Lyon, who had flown as copilot for J. Erroll Boyd on the first nonstop flight from New York to Haiti in 1933 and was affiliated with the Quiet Birdmen, a group that included Casey Jones and other prominent aviators. Imes asked Lyon for a written proposal outlining the public relations services he could provide in support of the Pan-American flight. Lyon obliged, proposing to introduce Forsythe and Anderson to aviators experienced in flying the air routes of South America. He also offered to fly to Tuskegee for the departure ceremonies with three other white aviators (J. Erroll Boyd, Roger Q. Williams, and Emile H. Burgin), and to solicit cooperation and congratulatory telegrams from other leading American aviators, all for a fee of $1,000.[36]

Boyack was optimistic about the public relations benefits that would accrue to the institute by beginning the flight at Tuskegee and securing the endorsement of the aviation community through the services of Lyon. Boyack believed that the "moral support and public sponsorship" of American aviators "for the first major Negro aviation venture will . . . prove immediately valuable in many ways." Moreover, he declared Tuskegee "the logical sponsor" of the flight and "the only place such an important development should begin. I know of no investment of time and money so pregnant with such magnificent possible results to both Tuskegee and the Negro race in America." He assured Imes that he would be happy to "co-operate in my professional capacity" to maximize coverage of the send-off from Tuskegee in both the North and South American press, newsreel, and radio media.[37] Imes, however, was skeptical about the feasibility of the program Boyack and Lyon proposed. Although he thought their ideas had merit and would certainly generate useful publicity, he doubted there was time or money to follow through.[38]

In early October 1934 Forsythe—encouraged by the success of the

Tuskegee Institute president R. R. Moton (far left) and his wife pose with C. Alfred Anderson (holding bouquet) and Dr. Albert E. Forsythe (far right) in front of the airplane Mrs. Moton christened *The Booker T. Washington*, 14 September 1934. (Courtesy Tuskegee University Archives, Anderson Papers)

This photograph appeared in the 1 October 1934 issue of the *Tuskegee Messenger*, which reported the christening of *The Booker T. Washington*. Col. Benjamin O. Davis, Sr., is on the left in the light uniform; F. D. Patterson is third from the right. (Courtesy Tuskegee University Archives, Anderson Papers)

christening, the formation of a Tuskegee fund-raising committee, Imes's efforts in New York, and Boyack's public relations ideas— suggested that Tuskegee Institute throw its full support behind publicity for the Pan-American flight. He wrote Moton and explained that problems in obtaining foreign landing permits had forced him to postpone the departure. This unavoidable delay, Forsythe went on, offered Tuskegee an opportunity to assume sponsorship of the flight and arrange for an official start from the campus. Most of the money and effort expended so far had of necessity been on the flight itself; little had been spent on "publicity that would multiply many fold the value of the achievement." With the school's reputation and connections, Tuskegee Institute could "secure with little outlay . . . effective, efficient and intelligent publicity before, during and after the Flight." Adequate publicity, Forsythe told Moton, would cost roughly $1,000; in return the institute could expect "tremendous returns in goodwill and prestige to the school and to the Race that would warrant any sacrifice or effort that might be called forth." He concluded by expressing his confidence in Boyack's ability to publicize the event if Tuskegee assumed sponsorship of the program.[39]

Forsythe wrote Imes the next day to advise him of the suggestions he offered to Moton. He also told Imes of his opposition to Lyon's plan to bring prominent aviators to Tuskegee "to the tune of $1000. I think there would be a lot of good ways of spending the thousand if we had it and that would not be one. Besides the public would quickly see through such a scheme." He thought the money would be better spent by hiring Boyack "to put over a high pressure publicity campaign flooding North and South America with news stories, mats, radio announcements etc. etc."[40]

Forsythe, of course, recognized that Tuskegee Institute might not be willing or able to invest $1,000 in publicity for the flight; he therefore submitted to Imes a program that Tuskegee could undertake independently of Boyack. The first portion of the program dealt with predeparture publicity. He suggested that endorsements for the flight be obtained from the Tuskegee Executive Council and the National Negro Business League. He also asked that Moton write public officials and other prominent individuals in the countries to be visited and urge that all courtesies be extended to the aviators. Finally, should the flight begin from Tuskegee, he urged efforts to secure flyovers by Air Corps aviators, newsreel and wire service coverage, and above all, a telegram of support from Pres. Franklin D. Roosevelt.[41]

The remainder of Forsythe's recommendations to Imes concerned postflight publicity. He suggested Moton use his influence to arrange for a meeting with President Roosevelt when the fliers returned to the United States. More important, he asked for help in arranging the homecoming tour:

> Machinery could be set in motion to arrange a tour to all principal cities of the U.S. and Canada with three aims in mind:
> a. The bringing to the attention of the largest group of whites possible in each city visited the fact that Negroes can achieve in any field and that we have aims and aspirations similar to those of other people.
> b. The bringing to the attention to people of our Race the fact that pride and confidence and the inspiration of our youth will follow in the wake of outstanding achievement.
> c. The raising of funds for the next flight of the series, the flight to Europe. (while enthusiasm is alive)[.][42]

Forsythe perceived correctly Tuskegee's reluctance to provide the funds for publicity. In late September Boyack wrote Imes again, recounting his efforts on behalf of the Pan-American flight. He tactfully informed Imes that he could do no more gratis work in support of the flight, having committed himself to the limit of his resources. Imes wired back: "Tuskegee cannot assume financial responsibility for the flight,"[43] and then wrote to explain Moton's attitude:

> We are, of course, in hearty accord with the purpose of the flight and will lend every assistance possible. This means that Tuskegee will be glad to be the scene of the take-off for this flight according to the plans suggested by you.
> Tuskegee Institute is not, however, in a position to advance the funds to close such a contract. Our present needs are really distressing and give us no opportunity to divert our funds to parties outside of those specified in our budget.[44]

Moton believed that the IGAC—not Tuskegee Institute—was the organization which should pay for Boyack's public relations proposals, and that Tuskegee's financial support for the flight should be limited to the efforts of the fund-raising committee—soliciting private, voluntary contributions from individuals associated with the institute.[45]

Imes did not answer Forsythe, however, for almost three weeks. In the interim the indefatigable Forsythe flooded Imes with more letters offering additional ideas for promotion of the flight and the homecoming. Finally, on 21 October 1934, frustrated and exhausted, Forsythe again wrote to Imes, saying he could not "account for the fact that I have had no word from you. . . . I wish you would oblige me by sending a day letter at my expense so that we can know just where Tuskegee will fit in."[46]

Forsythe was under pressure to learn of Tuskegee's intentions because of Atlantic City's plans to publicize the departure. The city had invited local and state officials to be on hand and had arranged for wide press coverage. A second departure ceremony at Tuskegee might be anticlimactic if not properly coordinated; in any event, an additional ceremony would require advanced planning and add an extra day of flying time within United States borders. If Tuskegee Institute was going to do nothing more than provide a departure site, there would be little advantage in flying to Miami via Tuskegee. He urged Imes to respond immediately, "so as to avoid giving the outside world the impression that even a group of supposedly intelligent Negroes cannot work harmoniously in an undertaking of this kind."[47]

Perhaps Forsythe regretted his peevish tone after receiving Imes's letter, written two days earlier on 19 October. The secretary apologized for failing to respond sooner, citing the pressures of travel and work at the institute. He assured Forsythe of his undiminished interest and enthusiasm. For the present, Imes continued, Tuskegee must limit its support to the fund-raising campaign already under way and the invitation to stop at Tuskegee on the flight south. The "various schemes for soliciting interest and support" should be directed toward the homecoming reception and subsequent flights. Imes assured Forsythe of Tuskegee's long-term commitment when he suggested that "the organization for the European flight definitely [be] centered at Tuskegee."[48]

Forsythe, reassured of the institute's support for the homecoming and the proposed European flight, decided that the inconvenient stop at Tuskegee on the flight from Atlantic City to Miami would not be necessary. By 1 November 1934 he had in hand all the necessary landing permits, had chosen a departure date of 8 November, and had invited Moton to attend the takeoff from Atlantic City. Unfortunately, the Tuskegee president's failing health kept him from making the trip and Forsythe settled for a congratulatory telegram.[49]

The *Booker T. Washington* left Atlantic City as planned, on Thursday morning, 8 November 1934.[50] Despite mechanical problems on the way to Miami—a broken gas line forced an unscheduled stop at Beaufort, South Carolina—the fliers managed to make Nassau just before nightfall on Friday. Automobile headlights lit the landing area for Anderson and Forsythe as they landed before a crowd of 5,000; the first leg of the tour was successful despite the earlier difficulties.

For the next nine days the flight proceeded according to schedule. On Saturday Forsythe and Anderson negotiated the 350-mile hop southwest from Nassau to Havana, the longest flight over open water of the entire trip, and perhaps the most hazardous part of the tour. From Havana they began their clockwise circuit around the Antilles, flying southeast across Cuba, down to Jamaica, and then east to the second largest island in the Caribbean, Hispaniola, shared by Haiti and the Dominican Republic. Once the duo departed Hispaniola, they planned to continue east to Puerto Rico, the last island of the Greater Antilles, and then work their way down the Lesser Antilles until they reached the South American continent. From there they would proceed west across the northern coast of South America and then head north to Mexico via Central America, reentering the United States at Brownsville, Texas.

The flight to Port-au-Prince, Haiti, on the western side of Hispaniola went smoothly, and on Sunday, 18 November the two fliers headed east across Hispaniola to Santo Domingo, the capital of the Dominican Republic. They had scarcely left Haiti when their engine failed, forcing them down in the mountains along the border. Unhurt, they managed to contact the authorities in Santo Domingo and settled down for a three-week wait while parts for repairs were shipped from the United States.

Forsythe used the delay to communicate once again with Imes, urging him to begin work on arrangements for the postflight tour of the United States. As usual, the physician-aviator inundated the institute secretary with suggestions and ideas. He requested information on final arrangements by the time that he and his flying partner reached Mexico City and began to make their way to Tuskegee, the "first real stop in the United States."[51]

By 8 December 1934 repairs to the *Booker T. Washington* had been completed; the goodwill pilots continued on to Santo Domingo and then proceeded east to Puerto Rico. After a gala reception hosted by the San Juan Tuskegee Club, complete with illustrated program, the pair flew to Saint Thomas in the Virgin Islands, where yet another enthusiastic crowd greeted them. While at Saint Thomas, they became the first American blacks to be received at the Government House since the United States purchased the islands from Denmark in 1917.[52]

After leaving Saint Thomas, Anderson and Forsythe spent a week island-hopping down the eastern periphery of the Caribbean. On 13 December the little airplane touched down on Trinidad, at the foot of the Lesser Antilles, just off the northern coast of South America. Port-of-Spain, Trinidad, marked the end of the Caribbean phase of the tour; from there the aviators planned to fly southeast to George-town, British Guiana, to begin the second phase of the tour—the continental phase—when they landed for the first time on the South American continent. Unfortunately, the Pan-American Goodwill Flight came to a premature end before the *Booker T. Washington* ever touched South American soil.

As Forsythe and Anderson had flown south along the string of islands that make up the Lesser Antilles, they had found no improved airfields and sought out any likely clear area for landings and takeoffs. After several very close calls, their luck finally ran out when they departed from an unimproved field at Port-of-Spain. Heavy with extra fuel for the long flight to Georgetown, the little Lambert monocoupe climbed so slowly after lift-off that the aircraft clipped a stand of bamboo and pieces became lodged in the landing gear. The trailing bamboo sections created extra drag, causing the overweight aircraft to stall and crash in a crowded residential area of the city. Miraculously, neither pilot was seriously injured, nor did the fuel ignite and pose a hazard to bystanders. But the *Booker T. Washington* was damaged beyond repair and the Pan-American Goodwill Flight came to an abrupt and untimely end.[53]

Embarrassed, tired, and disappointed, the Goodwill fliers quietly returned to the United States. The unfortunate conclusion of the Pan-American flight cut short their dreams of a triumphal home-

TUSKEGEE

"BOOKER T. WASHINGTON"

OFFICIAL PROGRAM.

BOOKER T. WASHINGTON

DR. FORSYTHE

ANDERSON

INTER-RACIAL GOOD WILL FLY

San Juan, P. R.

DICIEMBRE, 1934

Cover of the program published by the San Juan Tuskegee Club for the welcoming reception they hosted for Anderson and Forsythe on their arrival at San Juan, Puerto Rico, December 1934. (Courtesy Tuskegee University Archives, Anderson Papers) coming tour orchestrated by Tuskegee Institute. Anderson could rely on his steadily improving flying skills to sustain himself financially, but the loss of the *Booker T. Washington* ended Forsythe's active involvement in aviation, although he never lost interest.[54] The doctor, who had always borne the major financial burden of the flights, had to turn his attention to his medical practice, which his partner had patiently kept active while Forsythe pursued his aviation interests.[55]

When Forsythe withdrew, no individual or organization stepped in to provide leadership and financial backing for subsequent goodwill flights. The IGAC had never been the dynamic, enterprising organization it should have been, which was one reason Forsythe had turned to Tuskegee Institute after the school's administrators took an interest in the project and hosted the christening. Shortly before the Pan-American flight began, realizing that the IGAC would never be able to provide leadership of the caliber needed for the more ambitious flights he hoped would follow, Forsythe had proposed a new organization. The physician thought someone like "Mrs. Moton or Mrs. Mary Bethune could be National Chairman" with a "strong staff covering the entire country."[56]

An effective organization such as Forsythe envisioned might have emerged if the flight had succeeded and been followed by a homecoming tour that captured the interest of a broad spectrum of American blacks. But even if the flight had succeeded, obtaining the active support of Tuskegee along the lines proposed by Imes would have been difficult. The failure of the flight meant that once again circumstances had worked against Tuskegee's entrance into the air age. Certainly Moton had been willing to allow the limited, short-term participation required to arrange and host the christening and to solicit funds; but he would not be drawn in any further. Imes had held out the prospect of closer involvement following the flight, although he made no definite commitment. He knew that Moton's imminent retirement made it impossible to predict how far the institute might go in support of future goodwill flights. The crash in Trinidad made the question moot, of course, for Tuskegee was certainly unwilling and probably unable to resuscitate the goodwill flight series after the catastrophe in Trinidad and Forsythe's withdrawal.

In the months that followed the untimely end of the Pan-American flight, the Tuskegee Board of Trustees became preoccupied with an important decision with long-range implications—the selection of a new president. As the spring board meeting approached, the pace of speculation and editorializing quickened as the contenders maneuvered for position, overshadowing Tuskegee's brief flirtation with aviation the previous year. Shortly after the board surprised everyone by announcing the election of Patterson, who had not been considered a likely successor, Tuskegee alumnus John C. Robinson sailed for Ethiopia to begin an adventure that brought him fame and almost brought an aviation program to Tuskegee Institute.

When Robinson boarded a steamer bound for the horn of Africa, he reflected the concerns of many black Americans for Ethiopia, a country that would soon fall victim to the colonial appetite of the Italian dictator Benito Mussolini. Emperor Menelik II's decisive defeat of the Italians at Adowa in 1896 had preserved Ethiopia's independence for four decades, while most of Africa succumbed to European imperialism. The victory at Adowa profoundly affected black Americans: Ethiopia, which traced its roots back to the ancient kingdom of Abyssinia, had maintained its independence by force of arms, "the first time since Hannibal [that] an African people had successfully repulsed a major European army. And the significance of this battle was not lost upon Afro-Americans."[57]

The coronation of Haile Selassie in 1930, an impressive event which caught the attention of the international press, reawakened the interest of American blacks in the ancient African kingdom, especially those blacks of Robinson's generation who followed closely the new emperor's program of reform and modernization. After Italian and Ethiopian troops clashed along the Ethiopian–Italian Somaliland border in early December 1934, the horn of Africa became the center of world attention. The dispute was quickly referred to the League of Nations, but Mussolini had no interest in negotiated settlements—he was resolved to avenge the ignominious defeat at Adowa and add to his colonial possessions. By the spring of 1935, an Italian invasion appeared imminent and black Americans became increasingly concerned over the fate of Ethiopia. Most responded by raising funds and establishing organizations such as the International Council of Friends of Ethiopia.[58] Robinson, however, offered a different kind of aid; he volunteered to fly for Haile Selassie's air force.

Early in 1935, through the good offices of Claude Barnett, director of the Associated Negro Press (ANP), Robinson met Malaku E. Bayen, a member of the royal family attending medical school at Howard University in Washington, D.C.[59] When he told Bayen of his ambitions to serve the Selassie regime, the young medical student communicated Robinson's offer to the emperor. In April 1935, after Robinson provided satisfactory references and credentials, the Ethiopian monarch cabled him and offered the aviator a commission in his imperial army on the condition that he serve for at least a year; on 2 May Robinson departed for Africa, arriving in Addis Ababa on the twenty-ninth.[60] During the summer of 1935, while the League of Nations tried to negotiate a peaceful settlement, the young aviator from Chicago won the full confidence of the Ethiopian officials. By August he had been commissioned a colonel and given command of the minuscule Ethiopian air force.[61]

When the fighting broke out—the Italians invaded without a declaration of war on 3 October 1935—the circumstances in which Robinson found himself made it inevitable that the black press would portray him as a hero.[62] Except for the discredited Hubert F. Julian,[63] Robinson was the only American black in the Ethiopian armed forces; he held high commissioned rank and served in a new and elite branch, the air force. The black war correspondents quickly dubbed him the "Brown Condor," hoping he would deliver another embarrassing blow to the Italians, as Joe Louis had done in June when he took the heavyweight championship from Primo Carnera with a knockout in the sixth round and earned his well-known sobriquet, the "Brown Bomber."[64]

Throughout the war Robinson provided African American correspondents with colorful copy, and the articles on him frequently noted that Tuskegee Institute was his alma mater. The school, of course, took tremendous pride in his accomplishments, and shortly after he took command of the tiny Ethiopian air force, the student newspaper published an article on the new colonel. Under headlines that announced "Tuskegee Graduate Heads Ethiopian Air Force" was an emphatic endorsement of Robinson and his service in Africa: "It is unnecessary to state the degree of pride with which the whole Tuskegee family regards Mr. Robinson. He is more than a race pioneer in the field of aviation, he is a link in that chain which binds us to Africa."[65] Once the fighting broke out more articles appeared proclaiming pride in the Tuskegee alumnus and "the fine qualities that make him great in the eyes of the world" and declaring him "An Aviation Hero."[66]

Like most of the Ethiopian armed forces, the air force under the Brown Condor's command was no match for the Italians. At most, it consisted of nineteen obsolescent aircraft, some fifty Ethiopian pilots, and a handful of foreigner aviators. Such a force was power-less against the Italian fighters, who easily controlled the skies.[67] Consequently, Robinson flew mostly courier missions between the front and the capital and also served as the emperor's personal pilot.[68] By the spring of 1936, the superior Italian forces, using poison gas, had gained the upper hand; on 5 May Il Duce's forces took Addis Ababa and four days later Italy annexed Ethiopia. Fortunately for the Brown Condor, he was not in Ethiopia when the Italians took control; according to press reports he was en route to the United States on a fund-raising tour.[69]

The annexation of Ethiopia meant, of course, that Robinson's flying duties for the emperor had ended. When Claude Barnett learned of the Brown Condor's imminent return, he realized that the homecoming would generate tremendous excitement and interest and wanted to ensure that Tuskegee took advantage of the burgeon-ing enthusiasm for aviation which Robinson's activities in Ethiopia had begun to foster among black youth. Barnett had just orches-trated the homecoming of the *Pittsburgh Courier's* Ethiopian corre-spondent, Joel Rogers, an event which brought a great deal of publicity to the *Courier;* as a Tuskegee trustee and alumnus, Barnett hoped Robinson's return would produce similar public relations dividends for their alma mater.[70]

Before Robinson had left for Ethiopia in May 1935, he had renewed his efforts of the previous year to persuade the Tuskegee administration to add aviation to its curriculum. Despite the uncer-tainty due to Moton's retirement, G. L. Washington, head of the Department of Mechanical Industries, informally offered Robinson a position on the faculty. After he accepted Haile Selassie's invitation to fly for Ethiopia, Robinson apparently asked Washington to hold the position open until he returned. By November, convinced that Ethiopia would eventually fall to the superior Italian forces, Rob-inson advised Barnett that he would probably return to the United States at the expiration of his one-year commitment and asked him

G. L. Washington, head of the Department of Mechanical Industries at Tuskegee Institute. (Courtesy Tuskegee University Archives, G. L. Washington biographical file)

to find out if Washington's offer was still good. Barnett contacted both Washington and Patterson, determined that the position was still open, and assured Robinson that the job was his when he returned.[71]

The fall of Ethiopia shortly after Robinson departed for the United States made it obvious that the Brown Condor could not return to Ethiopia, and Barnett immediately took steps to ensure that the pilot would indeed have his chance to establish an aviation course at

Tuskegee. In early May he advised President Patterson that the aviator was en route to the United States and recommended that the plan to bring Robinson to Tuskegee be announced through a publicity campaign celebrating the flier's homecoming.[72] When Barnett learned that Robinson was due in New York harbor on 18 May, he pressed Patterson for a decision. Patterson advised Barnett that he was definitely holding open a position in the Department of Mechanical Industries for the aviator and approved Barnett's proposal to promote the homecoming and the establishment of a course in aviation. Barnett was gratified that Patterson had agreed to his proposal; he assured the Tuskegee president that there was "scarcely a single event in the offing which can be made to focus more favorable attention on the school" and directed his Washington correspondent to write a story on the exclusion of blacks from the United States Army Air Corps, which could be used "to lift the Tuskegee [aviation] effort into national prominence."[73]

After Barnett secured Patterson's approval, he contacted New York public relations representative James Edmund Boyack and instructed him to arrange the publicity for the Brown Condor's homecoming. Barnett had used Boyack's services for the Rogers homecoming and considered him an excellent public relations agent. Patterson was taken aback when he learned that Boyack's fee might be as high as $150, but Barnett insisted that Boyack's services were indispensable: "There is scarcely any way to duplicate what I want him to do; take the daily paper newspaper reporters down to the bay to meet the ship; prime them for stories, place photographs if possible, get news spots on the radio and a radio appearance if obtainable, also wire services. I know Boyack had the misfortune to make some sort of demand on Tuskegee, but he is by all odds the best man to do a job of this type, I have ever run across. A colored man without experience can not approach within a mile of him."[74]

Barnett traveled to New York several days before Robinson's arrival to work with Boyack on the publicity arrangements. Together they prepared advance press releases announcing Robinson's arrival on 18 May, "loading them up with names to attract attention," and taking care to point out that the returning pilot would be joining the faculty at Tuskegee to establish "the first aviation school under Negro auspices."[75] Barnett reported to Patterson that the white press and the radio networks accepted the story but complained that the

F. D. Patterson, president of Tuskegee Institute, 1935–
1953. (Courtesy Tuskegee University Archives, F. D.
Patterson biographical file)

New York Times shortened the release.[76] The *Times* did not, however,
fail to mention the projected aviation school at Tuskegee. In addition
to the advance press coverage, Barnett arranged for a reception
committee at the pier that included Dr. William Jay Schiefflin,
president of the Tuskegee Institute Board of Trustees, and Dr.
P. M. H. Savory, chairman of United Aid for Ethiopia.[77]

Robinson's ship arrived on schedule but a quarantine inspection
delayed docking for several hours. Undaunted, Barnett made his

way out to the anchored ship via Coast Guard cutter, found Robinson, and arranged for interviews and photographs on board. Barnett later told Patterson the aviator "gave a good account of himself" and boasted that "the Tuskegee angle was worked out to perfection," with all the news accounts of Robinson's arrival commenting on "the fact that the Colonel was going to Tuskegee to teach aviation."[78]

After he disembarked, Robinson attended a banquet in his honor that was hosted by United Aid for Ethiopia, but had to wait until Saturday for his first public appearance.[79] Barnett used the delay to seek funds and equipment for the new aviation school. Board president Schiefflin, "delighted with the result of the effort" to capitalize on the Brown Condor's sudden fame, launched a campaign to find a benefactor who would donate an airplane or fund the purchase of one. To this end he tried, unsuccessfully, to arrange for Barnett and Robinson to meet with Thomas J. Watson, president of IBM. Barnett used his own connections to solicit support from the Firestone Tire and Rubber Company, proposing that the Tuskegee Choir appear on the company's radio program with a portion of the proceeds earmarked for the aviation program.[80]

The climax of Robinson's stay in New York came on Saturday, 23 May 1936, at Harlem's Rockland Palace, where an enthusiastic crowd of 5,000 attending a public reception sponsored by United Aid for Ethiopia treated the colonel to a fifteen-minute ovation.[81] But the New York celebration was only a warm-up for the welcome Chicago gave its hometown hero the next day. On Sunday Barnett and Robinson flew to the Windy City, where a crowd of 3,000 greeted them as they stepped from a TWA airliner. The crowd swelled as Robinson led a procession of 500 automobiles from the airport to Chicago's Southside, where he addressed a cheering throng of 8,000 from the balcony of the Grand Hotel.[82] Caught up in the excitement, the *Chicago Defender* reporter covering the celebration wrote that never had "there been such a demonstration as was accorded the 31-year-old Chicago aviator who left the United States thirteen months ago and literally covered himself in glory, trying to preserve the independence of the last African empire."[83]

As a journalist, Barnett was attuned to black America's yearning for "race heroes" whose success might offer vicarious relief from the hopelessness of the depression and counter the virulent new strain

John C. Robinson arrives in Chicago, 24 May 1936, after his return from service in Ethiopia. (Courtesy Tuskegee University Archives, John C. Robinson biographical file)

of racism gaining strength in Nazi Germany. He had seen the racial pride fostered by the achievements of athletes such as Joe Louis and Jesse Owens, who excelled at the 1936 Olympic Games in Berlin. But the idol of white America after 1927 was Charles A. Lindbergh, who had flown alone across the Atlantic to become the "first real hero of the machine age."[84] When Robinson entered the service of Haile Selassie, no black aviator had achieved the right balance of flying expertise and public acclaim to become the nation's "black Lindy."[85] In the spring of 1936 Barnett's instincts as a newsman and observer of black America told him that Robinson might well emerge from his Ethiopian adventures as a bona fide air hero, and he had taken great pains to associate Tuskegee Institute with the young aviator. Indeed, the hero's welcome in New York and Chicago confirmed Barnett's judgment, but as Robinson settled into his new role, both Barnett and the Tuskegee administration realized that the Brown Condor had his own ideas about how to use his newfound fame.

For several weeks after the emotional homecoming celebrations in New York and Chicago, Robinson followed the scenario laid out for him by Barnett. Shortly after he arrived in New York, the John C. Robinson Aviation Fund was established to support the colonel's plans for teaching aviation and to purchase an airplane for his use. By early June 1936 he was scheduled for fund-raising appearances in Brooklyn, Washington, D.C., Pittsburgh, and Kansas City, and Barnett hoped to arrange additional speaking engagements to keep the flier in the public eye and stimulate further contributions. Equally important, he arranged for the aviator to visit Tuskegee and finalize the details of his employment and teaching responsibilities so that aviation courses could begin in the fall.[86]

Robinson visited Tuskegee at the end of June and President Patterson introduced him to the students as a Tuskegee graduate who was "blazing the trail for Negro youth in a new field of endeavor and proving to the world beyond doubt the Negro's capacity for accuracy, endurance, skill and courage."[87] The flier met with G. L. Washington and together they drew up a lengthy agreement that outlined the terms of Robinson's employment and the type of assistance Tuskegee would extend to establish an aviation school. Under the terms of the agreement, Tuskegee would provide land for an airfield, use of buildings, and administrative support. Although none of the school's resources were committed to operating expenses, it was anticipated that "through Tuskegee's influence will come finances which will support the enterprise . . . just as it has been since the founding of the Institute."[88]

In the weeks that led up to his visit to Tuskegee, Robinson had shown no inclination to chafe under Barnett's tutelage; instead he seemed an apt protégé, well disposed to the plans of his sponsor. Similarly, he left Tuskegee apparently satisfied with the terms of the agreement he had worked out with Washington. But shortly thereafter Colonel Robinson suddenly asserted his independence. He wrote Barnett and declared "I have not been satisfied . . . as to how things have been handled." He complained that Barnett and his associates were making decisions unilaterally, "instead of using my thoughts that are seasoned with much experience in aviation." He objected to Barnett's refusal to release funds from donations collected for the purchase of an airplane, believing he should be responsible for decisions regarding the use of money collected in his name. Robinson concluded by declaring that henceforth he would handle his own affairs, and hoped that he and Barnett could nevertheless maintain an amicable relationship.[89]

Robinson's relationship with Tuskegee also became strained. In early July 1936 Washington was shocked when Robinson returned their agreement unsigned.[90] After he left Tuskegee, Robinson had become dissatisfied with the agreement, believing its terms indicated a lack of confidence in his abilities and a reluctance on the part of Tuskegee to support fully an aviation training program. He not only objected to the terms of the agreement, but he also demanded the establishment of an autonomous aviation school—the J. C. Robinson School of Aviation—with all contributions and income under his personal control.[91]

In the opening sentence of his reply to Robinson, Washington scarcely hid his exasperation at the pilot's sudden turnabout: "I received your letter of July the first enclosing the Agreement which you did not see fit to sign." He scolded the colonel for not voicing his objections while they were negotiating the agreement and reminded him that on page four of the agreement "it specifically stated, 'It is understood that this agreement marks entrance into a new field, and that fixed agreement in every detail of the program and the employment of Col. Robinson is not possible or desirable at this time. The program and plan will be in the process of evolution for some time to come.'"[92]

His anger vented, Washington adopted an avuncular tone and explained the realities of the situation at Tuskegee. He admonished the headstrong pilot for even suggesting an independent aviation school, declaring that "any activity at Tuskegee Institute is a part of Tuskegee Institute, and I am certain that a separate school of aviation, named the Col. J. C. Robinson School of Aviation, would not be in line. I think you can understand the reason why." He told Robinson that Patterson and the entire Tuskegee administration were "very enthusiastic" about aviation, having already received numerous letters of inquiry from prospective students. And he apologized if the agreement "conveyed . . . that you would be circumscribed and handicapped," assuring the colonel that President Patterson allowed everyone "unlimited opportunity to demonstrate individual ability, and certainly you would be given every opportunity."[93] The administration favored the establishment of an aviation program at Tuskegee and would provide limited support, but its development and ultimate success depended on Robinson's willingness and ability to work harmoniously as part of the Tuskegee faculty and to cultivate financial support from sources outside the institute.

Perhaps without realizing it, Washington was advising Colonel Robinson that the approach outlined in their agreement reflected the typical pattern for establishing new programs at Tuskegee. The school, for example, had no effective agricultural program before 1896, when a Slater Fund grant financed the construction of an agriculture building and Booker T. Washington persuaded George Washington Carver to join the faculty. When he arrived, Carver found he was expected to build an agricultural program from scratch.[94] Similarly, the Collegiate Division had been organized slowly and deliberately, with a few college-level courses offered for several years before a formal course of study leading to the baccalaureate degree was authorized. Establishing an aviation program at Tuskegee would be no different: Robinson would be expected to "initiate and carry on the program."[95]

Washington concluded by inviting Robinson to return to Tuskegee so they could resolve their differences. "The whole matter," he told the flier, "hinges on whether or not you have confidence in

Tuskegee Institute and whether or not you think that we would do the fair thing throughout."[96] Apparently Robinson's confidence in the institute was shaken; instead of resolving his differences with Washington, he established his own training facility in Chicago—the John C. Robinson School of Aviation. He did not, however, totally abandon the idea of returning to his alma mater. In November 1936 he once again contacted Washington to propose that Tuskegee underwrite the transfer of his school and its equipment to the institute campus. Washington referred the proposal to Patterson without comment and the president took no further action on the matter, apparently unwilling to proceed without Washington's endorsement.[97]

The failure of the Robinson initiative marked the beginning of a three-year hiatus in Tuskegee's involvement in aviation. The events of 1934–36 had, however, heightened the institute administration's awareness of aviation and the growing enthusiasm of blacks for flying. Both Moton and Patterson had shown an interest in aviation; they were receptive to outside initiatives and were willing to support aviation activities so long as they did not detract from established programs or place undue demands on the slim fiscal resources of the institute. Of all the Tuskegee administrators, G. L. Washington exhibited the most sincere interest in aviation even though he lacked the technical expertise to establish and conduct an aviation program himself. He was, however, an able and energetic administrator who was well qualified to oversee an aviation program, if he had adequate staff and resources at his disposal. He had hoped Robinson would be willing and capable of establishing an aviation program, and when that fell through, he was forced to table the matter for three years. Not until the advent of federally sponsored initiatives— the Civilian Pilot Training Program in 1939 and the establishment of segregated flying units in 1941—was Tuskegee finally able to develop a viable aviation curriculum.

Black Americans
and the Military

2

During the 1930s civil rights and racial justice became issues of national importance.[1] By mid-decade black Americans were becoming increasingly vocal in their opposition to the endemic racism and discrimination that relegated them to second-class citizenship, and they began to challenge the institutional racism of American society on a broad scale. In the waning years of the 1930s, as conditions in Europe and Asia deteriorated and the nation began to rearm, civil rights advocates increasingly turned their attention to the issue of black participation in the armed forces. Although there were many areas of concern—blacks were allowed to participate in the nation's defense establishment only on a very limited and segregated basis—the exclusion of blacks from the United States Army Air Corps soon emerged as the issue of greatest importance to many black Americans. By 1939 public pressure had become so intense that two federal acts relating to the expansion of the nation's air arm were amended in an effort to broaden aviation training opportunities and open the Air Corps to blacks.[2] These amendments were the first fruits of the campaign for admitting blacks to the Air Corps, a movement that Ulysses Lee—author of a United States Army study on the role of black troops in World War II—has described as "the most widespread, persistent, and widely publicized of all the prewar pressure campaigns affecting the Negro and the Army."[3]

No attempt has been made by scholars to account for the preeminence of the campaign for black participation in the Air Corps.[4] The testimony of Edgar G. Brown—a black lobbyist, advisor to the Civilian Conservation Corps, and a member of President Roosevelt's unofficial Black Cabinet—before the Senate Military Affairs Committee in early 1939, however, highlights three major factors that formed the bases for the campaign. The committee was considering legislation that would facilitate expansion of the Air Corps and Brown appeared before it to urge that blacks be afforded the opportunity to serve as military pilots. The black lobbyist began his testimony by recounting the loyalty and bravery of black soldiers and patriots: he reminded the committee of men like Crispus Attucks, one of the first patriots to die at the hands of British soldiers in the Boston massacre of 1770, and Sgt. Henry Johnson, whose battlefield heroics during World War I earned him the *Croix de guerre* from the French government; he pointed to the long and distinguished service of the four black regiments of the United States Army; and he noted that several regiments of black volunteers fought in France during World War I. Then Brown related the growing interest of black Americans in aviation and, to prove that whites did not have a monopoly on flying aptitude, he submitted a bulletin issued by the Department of Commerce which listed the names of over one hundred black pilots. Finally, he called on Congress to "grant young Negro Americans . . . a new deal in the United States Army Air Corps." He urged that the Air Corps expansion bill include a stipulation that the nation's air arm be open to "all American youth, regardless of race, creed or color."[5] Thus Brown's testimony touched on the three crucial elements that gave rise to the campaign for black participation in the Air Corps—the long-standing military orientation of the black public, the growing air-mindedness of black Americans, and the increasing importance of civil rights issues to the national political agenda.

The military record of black Americans cited by Brown reaffirmed the high value they had traditionally placed on serving in the nation's armed forces. Collectively, blacks took pride in the fact that they had participated in virtually all the nation's wars, despite the sometimes overwhelming obstacles of slavery and racism, maintaining that their military record entitled them to enjoy the full benefits of citizenship. Individually, many black men found military life—either as regulars in the United States Army or as volunteers—an important source of economic security, upward mobility, and social prestige.

Brown could offer no more compelling evidence of burgeoning African American interest in aviation, the second element that underpinned the campaign for admitting blacks to the Air Corps, than the Department of Commerce data on black aviators that he presented to the committee. In the aftermath of Charles A. Lindbergh's historic transatlantic flight of 1927, blacks had taken to the sky in increasing numbers, despite the economic hardships of the Great Depression, and many more had become armchair aviation enthusiasts. The data Brown submitted showed 125 blacks holding pilot licenses, an increase of over 250 percent since 1935, when the Department of Commerce first began compiling statistics on black aviators.[6]

Finally, Brown's call for "a new deal in the United States Army Air Corps," reflected black America's growing reliance on political pressure and the power of the federal government in their struggle against racial discrimination, a product of what civil rights historian Harvard Sitkoff has called "the emergence of civil rights as a national issue."[7] During Roosevelt's first term, black Americans suddenly began to abandon their traditional allegiance to the Republican party and enthusiastically embraced the New Deal and the Democratic party. The presence of officials in the Roosevelt administration like Harold L. Ickes, Clark Foreman, and Will W. Alexander who cared about the civil rights and the welfare of blacks, the appointment of a record number of African Americans to important positions in the federal government, and above all, the commitment of First Lady Eleanor Roosevelt to the cause of racial justice established a new racial climate in the federal government which encouraged blacks to press for a wide variety of initiatives, all aimed at achieving racial progress. According to the eminent Swedish sociologist Gunnar Myrdal, who directed a comprehensive survey of blacks in America on the eve of the nation's entry into World War II, the New Deal "changed the whole configuration of the Negro problem. . . . For the first time in the history of the nation the state has done something in a substantial way without excluding the Negro."[8] For blacks interested in aviation, the emergence of civil rights as a national issue provided a catalyst for action; it galvanized the black community's traditional interest in military service and blacks' mushrooming interest in aviation into an effective pressure campaign, which ultimately resulted in the admission of blacks to the Air Corps, albeit on a limited and segregated basis.

Of the three elements that gave rise to the Air Corps participation campaign, the most venerable was "the military orientation of the Negro public," a phrase used by Ulysses Lee to describe black America's long-standing affinity for military service. Lee began his study on the role of the black soldier in World War II, *The Employment of Negro Troops*, with a commentary on the place military service held in the black community. A black American born early in the twentieth century and a scholar who held a doctorate in the history of culture from the University of Chicago, Lee observed that on the eve of World War II:

> The Army and military life had long occupied a position of relatively greater concern and importance to the Negro public than to Americans in general. Soldiering had been an honored career for the few Negroes who were able to enter upon it. In the restricted range of economic opportunities open to them, the military life ranked high. Thus the Army and its policies remained [after World War I] a significant center of interest to Negro organizations, to the press, and to the public as a whole. It was one of the few national endeavors in which Negroes had had a relatively secure position and which, at least in time of war, could lead to national recognition of their worth as citizens and their potential as partners in a common undertaking.[9]

Soldiering did not, however, emerge as an "honored career" among blacks, nor did they attain a "relatively secure position" in the army, until after the Civil War. During the era of slavery, blacks were generally excluded from military duties, except in time of crisis when all available human resources were needed to oppose a common enemy. The policy of exclusion in peace and service in war played an incalculable role in shaping the attitudes of antebellum blacks toward military service, attitudes inevitably passed down to succeeding generations and reinforced by developments after emancipation.

The pattern of exclusion was evident by 1639—only two decades after the first Africans arrived in the English colonies—when the Virginia General Assembly passed "the first discriminatory provision in American history," an act which required all residents of the colony, except blacks, to arm themselves.[10] Exempting blacks from the requirement to bear arms excluded them from membership in the colony's defense force, the militia. By the end of the seventeenth century the other English colonies had passed similar legislation against blacks bearing arms and serving in the militia.[11] The militia was an important institution in colonial society, the product of the white colonists' common English heritage which required all able-bodied male members of a community to defend the settlement against attack. To be excluded from the obligation for military service had serious implications, especially for free blacks, for it marked all those of African origin as a special group, not obliged to defend the community and therefore not entitled to the rights of citizenship.

The general pattern of exclusion continued until the Civil War. During the American Revolution, the official policy of the Continental Congress prohibited the enlistment of blacks in the forces of the central government, the Continental Army.[12] After the Revolution blacks were unofficially excluded from federal forces until 1820, when the army ordered that "No Negro or Mulatto will be received as a recruit of the Army" and then codified the policy with a revision to the General Regulations that restricted enlistments to "free white male persons."[13] The exclusion of blacks from state forces was perpetuated by the Militia Act of 1792. It mandated the enrollment of all able-bodied free white men but did not address the militia obligations of blacks; consequently, most states assumed that the act prohibited the enrollment of blacks and acted accordingly.[14]

Although the general policy before the Civil War was exclusion, the ban was sometimes lifted in time of crisis, when free blacks were allowed to volunteer and slaves were offered freedom in exchange for military service. A 1703 act of the colonial government of South Carolina, which promised freedom to any slave who killed or captured an invading enemy, provides the earliest example of rewarding combat valor with manumission.[15] Throughout the eighteenth century slaves fought and earned their freedom. In New England the practice was so prevalent that it caught the attention of the prominent clergyman and historian Jeremy Belknap, who attributed Massachusetts' shrinking slave population to black participation in King George's War and the French and Indian War.[16] Thus by 1775 it was widely understood that slaves could be attracted into military service by the prospect of freedom, and both the British and the Americans used the practice as a recruiting tool during the American Revolution.

Approximately 5,000 blacks, mostly slaves hoping to earn their freedom, served with American forces during the Revolution, despite the ban on black enlistments imposed by the Continental Congress. There were free blacks in the New England units that made up the original Continental Army established in June 1775, and these men were allowed to reenlist throughout the war. Some northern states occasionally met their recruiting quotas by enlisting slaves and promising them freedom. In 1778, for example, Rhode Island met its Continental Army quota by purchasing the freedom of a sufficient number of slaves to form a battalion; General Washington ignored the ban on blacks and accepted the unit, which fought until it was disbanded in 1783.[17]

The British also realized that slaves would fight to earn their freedom. In November 1775 the royal governor of Virginia, Lord Dunmore, offered freedom to all indentured servants and slaves who joined royalist forces. After several hundred slaves took advantage of the offer, Dunmore organized an Ethiopian Regiment, whose uniforms bore the inscription "Liberty to Slaves," but by the following summer disease and casualties decimated the unit and it was disbanded. British forces in Georgia also offered freedom to fleeing slaves, prompting South Carolina to pass legislation authorizing the death penalty for blacks who defected to British ranks.[18]

By the close of the American Revolution "military service had become a means by which a slave might earn his freedom or a free black enhance his standing in the community, discharge what he perceived as a duty owed his country, or simply satisfy a craving for adventure." The Revolution played a significant role in developing the Negro public's military orientation, for it "lent credence to the belief, which would later become almost an article of faith among blacks, that military service in wartime represented a path toward freedom and greater postwar opportunity."[19]

Events during the War of 1812 reinforced the connection between military service and freedom. Once again the British encouraged slaves to defect by promising them freedom. By 1814 the British commander of the Chesapeake region, Rear Adm. George Cockburn, had organized 200 escaped slaves into a unit of marines that saw action in Virginia and Maryland. Other slaves fled into British hands and served in labor battalions. Most estimates put the number of slaves who escaped to freedom during the War of 1812 at 5,000.[20] On the American side, opportunities for freedom did not materialize as they had during the Revolution. Although New York adopted legislation that authorized slaves to enlist and earn their freedom, the measure came so late in the war that the fighting was over before it took effect.[21] According to the black abolitionist and former slave J. Sella Martin, however, many southern slaves fought beside their masters against the British and some were rewarded with their freedom.[22]

In Louisiana several black militia companies defended American soil against British troops. Militia units of free blacks had been organized in New Orleans under Spanish rule and were maintained intact when France regained control of Louisiana in 1801. Shortly after the Louisiana Purchase of 1803, United States authorities disbanded the black militia companies at the behest of the white citizens of New Orleans. The units were revived in 1812, after war broke out with Britain, and approximately 600 free blacks fought against British regulars in the Battle of New Orleans. After the war the companies slowly deteriorated, and by 1834 they were legislated out of existence.[23]

Following the War of 1812, blacks found few opportunities for military service. Except for the Mexican War, the nation engaged no foreign enemy, and blacks participated in that conflict on a very limited basis.[24] Throughout the half-century between the War of 1812 and the Civil War, the United States Army continued to prohibit the enlistment of blacks and the states excluded them from the militia. Yet the memory of the black soldiers and sailors who fought against Britain was kept alive by the abolitionists, who hoped that a recounting of the military record of black Americans would foster their assimilation into American society.

In 1847 John Greenleaf Whittier, who had devoted himself to the abolition of slavery since 1833, extolled the heroism of black soldiers in an article he wrote for the abolitionist paper *National Era*.[25] He noted that despite "the services and sufferings of the colored soldiers of the Revolution, no attempt has been made to preserve a record. They have had no historian."[26] Whittier's brief account inspired William Cooper Nell, a black journalist and abolitionist, to write *Services of Colored Americans in the Wars of 1776 and 1812*, a pamphlet published in 1851, which he revised and expanded four years later under the title *The Colored Patriots of the American Revolution*.[27] Nell's enlarged work, described by historian Benjamin Quarles as "the first serious attempt by a Negro American to write scholarly history,"[28] reminded its readers of the opportunities for manumission that wartime service had provided black Americans and acknowledged the unique relationship between citizenship and military obligations that permeated American society.[29]

In 1857 the United States Supreme Court also addressed the issue of military service as an obligation of citizenship and inadvertently contributed to the growing military orientation of black Americans. In his majority opinion in the Dred Scott case, issued in 1857, Chief Justice Roger B. Taney asserted that blacks—freeborn, emancipated, and enslaved—were not, and had never been, citizens of the colonies and states. He cited a wide variety of colonial and state laws to support his judgment, among them a New Hampshire law specifying that only free white citizens could enroll in the militia. "Nothing," declared Taney, "could more strongly mark the entire repudiation of the African race." Why were naturalized white men enrolled in the militia and freeborn black natives "not permitted to share in one of the highest duties of the citizen?" The answer, Taney concluded, was obvious: the black man "is not, by the institutions of the State, numbered among its people. He forms no part of the sovereignty of the State, and is not therefore called on to uphold and defend it."[30]

Thus on the eve of the Civil War the issues were clearly drawn. The abolitionists pointed to the tradition of wartime service as evidence that African Americans were loyal to the nation, had fought and died for it, and therefore deserved to partake of its benefits. Those opposing the emancipation of slaves and citizenship for free blacks ignored their wartime service, focusing instead on the broader pattern of exclusion to justify the continued subjugation of blacks. As the sectional controversy deepened in the wake of the Dred Scott decision, Taney's words no doubt strengthened the resolve of many blacks to fight should war break out between the North and the South, in an effort to earn for themselves a "part of the sovereignty of the State" by volunteering to "uphold and defend it," just as their forebears had done in 1775.

Early in the war free blacks in the North eagerly offered their services. Typical were the sentiments expressed on 20 April 1861, only a week after the fall of Fort Sumter, by Alfred M. Green of Philadelphia, who spoke in support of attempts to raise two regiments of black Philadelphians. Green, perhaps influenced by the writings of Whittier and Nell, cited "the brave deeds of our fathers, sworn and subscribed to by the immortal Washington of the Revolution of 1776, and of Jackson and others, in the War of 1812," and admonished his fellow blacks to "let not the honor and glory achieved by our fathers be blasted or sullied by a want of true heroism among their sons."[31] Various abolitionists and opponents of slavery, such as William Lloyd Garrison and George H. Moore, cited the record of black participation in earlier wars as they urged Lincoln and Congress to free the slaves and enlist them in Union forces.[32]

No systematic plan for the enlistment of black troops was developed until after the Emancipation Proclamation of 1 January 1863. There were, however, early attempts in the spring and summer of 1862 to recruit black regiments by Generals David Hunter, James H. Lane, and John W. Phelps, all without the approval of Lincoln or the War Department. Then, in August, the War Department formally authorized recruitment of blacks in the South. A month later, on 27 September 1862 and only five days after Lincoln issued the Preliminary Emancipation Proclamation, Gen. Benjamin F. Butler, commander of the Union forces occupying New Orleans, mustered in the United States Army's first black regiment, the First Louisiana Native Guards. Black units were soon formed in Kansas, Massachusetts, and Rhode Island. After the Emancipation Proclamation was issued in January 1863, Lincoln called for four black regiments, and the following May a Bureau of Colored Troops was established to recruit and organize black units. The black regiments came to be known collectively as the United States Colored Troops.[33]

The first blacks to see combat were recruits from Kansas, who skirmished with Confederate forces at Island Mound, Missouri, in late October 1862. Seven months later, on 27 May 1863, two Louisiana regiments participated in the assault on the Confederacy's Mississippi River stronghold at Port Hudson, the first major engagement involving black troops. In the remaining two years of the war

blacks enlisted and fought on a scale that far exceeded their participation in previous wars. Estimates of the total number serving with
Union forces vary from 186,000 upwards to 300,000, with calcula-
tions of the total who lost their lives ranging from an improbable low
of 2,750 to the more widely accepted figure of 38,000.[34]

Until late in the war, Confederate policy limited black participation to labor in mines, fortifications, and other essential war-related
fatigue duty. After the reversals of 1863, when Confederate personnel shortages became acute, one Southern general, Patrick Cleburne,
tried to resurrect the old policy of offering freedom to slaves in
exchange for military service. Although Cleburne's proposal was
rejected, by 1865—with the Confederacy on the verge of collapse—
the idea resurfaced. Encouraged by Robert E. Lee's endorsement, the
Confederate Congress approved the recruiting of 300,000 slaves and
held out the prospect of manumission as an inducement to enlistment. The measure was adopted too late to have any military
effect—only a few companies of slaves had been recruited before Lee
surrendered at Appomattox—but it undoubtedly reminded black
Americans of the traditional linkage between military service and
manumission, one of the few legal means of escaping slavery open to
their forebears.[35]

The participation of thousands of black Americans in a war that led
to their emancipation and citizenship had a dramatic impact on the
attitudes of all African Americans toward military service. Many no
doubt agreed with Frederick Douglass, the leading black spokesman,
concerning the implications of bearing arms in defense of the nation:
"Let the black man get upon his person the brass letters U.S.; let him
get an eagle on his buttons and musket on his shoulder, and bullets
in his pocket, and there is no power on the earth which can deny
that he has earned the right to citizenship."[36] Just as white Americans honored their veterans, when black soldiers reentered civilian
life after the war, they generally became honored and respected
members of the black community, a constant reminder that black
men had fought and died for both the freedom of the race and the
cause of national unity.[37]

Equally important to the development of black America's military
orientation were the opportunities for military service that emerged
after the war. For the first time in the nation's history, postwar
demobilization did not prompt a return to the policy of excluding

blacks. As the nation abandoned its war footing after the Civil War, blacks were instead offered a place in the peacetime standing army, thereby attaining the "relatively secure position" referred to by Ulysses Lee. A permanent place in the regular army, together with a tradition of military service that inevitably associated soldiering with freedom and citizenship, made it inevitable that the army would become "a significant center of interest to Negro organizations, to the press, and to the public as a whole."[38]

Legislation guaranteeing black Americans a place in the regular army came before Congress in 1866 as part of a postwar army reorganization bill, which stipulated that ten percent of the sixty regiments authorized for the regular army be reserved for black enlistees. Under this plan, two cavalry regiments (the Ninth and Tenth) and four infantry regiments (the Thirty-eighth, Thirty-ninth, Fortieth, and Forty-first) were to be manned by black troopers and soldiers. In 1869, as part of a measure that reduced the overall size of the regular army by twenty infantry regiments, Congress cut the authorization for black infantry regiments in half. The army consolidated the personnel from the original four regiments into two and redesignated the units the Twenty-fourth and Twenty-fifth Infantry Regiments. The Ninth and Tenth Cavalry were unaffected by the change.[39] Following the 1869 legislation, the existence of the black regiments was codified in Sections 1104 and 1108 of the Revised Statutes of 1878 and "there was no express repeal of the Revised Statutes in any later legislation concerning the Regular Army."[40] Since the black regiments were created by Congress, the War Department could not unilaterally disband them; thus all four survived as separate black units in the regular army until the armed forces were desegregated following World War II. Throughout the nineteenth century and well into the twentieth, black Americans followed the activities of the black regiments, noting the treatment they received from the War Department and the white communities with which they had contact.

In the quarter-century after the Civil War the United States Army fought a series of engagements—mostly skirmishes—on the western frontier, known collectively as the Indian Wars. Black regulars saw service throughout this era of sporadic warfare and proved themselves capable and effective soldiers.[41] With only one exception— when a troop of cavalry was temporarily stationed at Fort Myer,

Virginia—the black regiments were kept west of the Mississippi River until the advent of the war with Spain in 1898.[42] Throughout the last three decades of the nineteenth century, the four black regiments of the regular army served with distinction on the western frontier, earning seventeen Congressional Medals of Honor and proving themselves dependable and trustworthy under extremely harsh conditions. The inevitable boredom and isolation of frontier life often destroyed the morale of white regiments, causing high desertion and alcoholism rates but such problems were conspicuously absent in the black units, whose soldiers were instead eager to reenlist.[43] Through their valor and steadfastness, the black soldiers gradually earned the respect of both their white officers and their Indian adversaries, who dubbed the black cavalrymen "buffalo soldiers," a sobriquet of respect because the Plains Indians considered the buffalo a sacred animal.[44]

When not fighting Indians, black soldiers performed other services essential to the settling of the West, including escorting stagecoaches and trains, protecting railroad work gangs and survey teams, stringing telegraph lines, and locating sources of water.[45] The honorable service of the black regiments on the frontier was perhaps best expressed by the reflections of a white officer who served many years with the Tenth Cavalry. He concluded that black soldiers "show a pride in being trusted to perform any special duties that may require courage and good judgment, and no troops can be more determined or daring, nor are any more deserving of the highest commendations. Any doubts as to their being efficient and trustworthy soldiers should be eliminated."[46]

Although their long years of service on the frontier earned black soldiers the respect of their officers, the Spanish-American War brought them acclaim and ensured the continued interest of the black public in military matters. The participation of black regulars and volunteers in the war with Spain and the ensuing Philippine Insurrection gave a new generation of black Americans a fresh set of heroes, as the Civil War had done for its parents. Once again hopes were raised that the loyalty and battlefield heroism of the black soldier would touch the conscience of white America and usher in a new era of race relations.

After the sinking of the United States battleship *Maine* in mid-February 1898, the War Department adopted a war footing

and began shifting troops south. The first units to receive movement orders were the Ninth and Tenth Cavalry Regiments and the
Twenty-fourth Infantry Regiment, a move that did not escape the
attention of the black public. One black clergyman, comparing the
situation to the early days of the Civil War, boasted that in 1861
black volunteers were turned away but in 1898 they were "the very
first" to be called upon in time of crisis.[47] He might have been less
enthusiastic had he known that the decision was based on a racial
stereotype—officials in the War Department believed "the Negro is
better able to withstand the Cuban climate than the white man."[48]
All four black regiments saw action in Cuba, and the Ninth and
Tenth Cavalry—brigaded with the First Volunteer Cavalry, better
known as the Rough Riders, in a division commanded by the former
Confederate general Joseph Wheeler—participated in what became
the most memorable episode of the war, the charge up San Juan
Hill.[49]

Roughly nine thousand black Americans entered federal service as
volunteers during the Spanish-American War, either in state regiments or in volunteer regiments recruited by the War Department.[50]
The black National Guard units organized after the Civil War
got their first opportunity for wartime federal duty when Pres.
William McKinley called on the state governors for a volunteer army
of 200,000. Three states—Ohio, Indiana, and Massachusetts—sent
fully manned black units directly into federal service. Five others—
Alabama, North Carolina, Kansas, Virginia, and Illinois—used their
black units as cadres to form the core of larger units of black
volunteers recruited for the emergency.[51] Blacks could also volunteer for federal service by enlisting in one of the four black "immune" regiments, congressionally authorized units composed of
men supposedly resistant to the tropical diseases of malaria and
yellow fever. None of these black volunteer units saw combat in
Cuba, although two state regiments (the Eighth Illinois and the
Twenty-third Kansas) and one immune regiment (the Ninth U.S.
Volunteer Infantry) arrived after the fighting was over and performed garrison duty.[52]

Black troops also helped suppress the Philippine Insurrection, a
nationalist movement that erupted in February 1899 after the
United States took possession of the islands from Spain and refused
to acknowledge Filipino claims of independence. By July black

regulars had landed, and as late as 1906 elements of the Twenty-fourth Infantry were still involved in skirmishes on the southern island of Leyte, even though the insurrection had officially ended in July 1902. Two regiments of black volunteers also saw duty in the Philippines, part of the volunteer army of 35,000 authorized by Congress in March 1899 to help suppress the insurrection.[53]

The Spanish-American War brought the black soldier into the public eye. The yellow press and jingoism that characterized the war with Spain infected the black press, and black soldiers no longer fought in obscurity on the Western frontier. Their service in Cuba and in the Philippines was widely covered in black newspapers, and lithographs of the Ninth and Tenth Cavalry charging up San Juan Hill were displayed in many homes.[54] In an era of deteriorating race relations, black soldiers became a symbol of hope for many Negroes. The late Rayford W. Logan, an eminent black historian trained at Harvard University, observed that "Negroes had little, at the turn of the century, to help sustain our faith in ourselves except the pride that we took in the Ninth and Tenth Cavalry, [and] the Twenty-fourth and Twenty-fifth Infantry. . . . They were our Ralph Bunche, Marian Anderson, Joe Louis, and Jackie Robinson."[55]

Between the Spanish-American War and World War I the status of the soldier remained high in the black community. Young men eagerly sought to enlist, but were often turned away: the traditionally high reenlistment rates in the black regiments meant that vacancies rarely occurred. An adage current among army recruiters before World War I gives some indication of the importance of the military to blacks in the early decades of the twentieth century: "To the white soldier the Army is a refuge; to the Negro a career."[56]

Black soldiers came to national attention in 1906 when Pres. Theodore Roosevelt dishonorably discharged an entire battalion of the Twenty-fifth Infantry. The episode began with the transfer of the battalion to Fort Brown, Texas, over the protests of whites in nearby Brownsville. Shortly after the black soldiers arrived a score of armed men swept through Brownsville under the cover of darkness, shooting wildly into buildings, killing one man and injuring several others. Town authorities immediately accused the black soldiers of the assault, even though the evidence was inconclusive, and army investigators quickly reached the same conclusion. After a grand jury found insufficient evidence to indict the suspects, the army investigating officer recommended summary discharge of the entire battalion unless those guilty confessed or were identified by their comrades. When the entire battalion denied knowledge of the incident, army authorities charged the black soldiers with a "conspiracy of silence" and urged dismissal. In late November 167 black enlisted men, including six Medal of Honor recipients and one veteran of twenty-seven years, were dishonorably discharged and barred from subsequent federal service in any capacity.[57]

News of the dismissals outraged the black public and many whites, including Sen. Joseph B. Foraker who questioned the propriety of such a punishment when no trial of any kind had been held. For three years Foraker fought on behalf of the dismissed soldiers and his efforts finally led to the establishment of a court of inquiry, which ruled that fourteen of the dismissed soldiers were eligible to reenlist. The episode, which had received national attention for three years, was soon forgotten by white Americans. For the black public, however, the Brownsville affair remained a bitter memory and showed the importance of closely monitoring the treatment of black soldiers.[58]

The approach of World War I, with its rhetoric of "making the world safe for democracy," raised the hopes of black Americans. Once again they rallied to defend the nation and drew strength from the long-held belief that loyal service in war was a prelude to freedom, the rights of citizenship, and greater economic benefits.[59] After the United States entered the war, W. E. B. Du Bois, a historian and officer in the NAACP, urged blacks to "Close Ranks" with whites and present a unified front during the national emergency. Writing in *The Crisis*, he editorialized "Let us, while this war lasts, forget our special grievances and close our ranks shoulder to shoulder with our own white fellow citizens and the allied nations that are fighting for democracy. We make no ordinary sacrifice, but we make it gladly and willingly with our eyes lifted to the hills."[60]

When critics challenged Du Bois, accusing him of abandoning the founding principles of the NAACP, he observed that closing ranks "is precisely what in practice the Negroes of America have already done during the war and have been advised to do by every responsible editor and leader."[61] He reiterated his call for wartime solidarity— "*first* your Country, *then* your Rights!"—and told those who complained "that [although] we have fought our country's battles for one hundred fifty years, we have *not* gained our rights,"[62] that loyalty in past wars had brought benefits to the race:

> No, we have not gained all our rights, but we have gained rights and gained them rapidly and effectively by our loyalty in time of trial. . . .
> God knows we have enough left to fight for, but any people who by loyalty and patriotism have gained what we have in four wars ought surely to have sense enough to give that same loyalty and patriotism a chance to win in the fifth.[63]

Although Du Bois obviously oversimplified history as he exhorted black Americans to their patriotic duty, his comments provide an important insight into the military orientation of the black public: he spoke the mind of black America on the importance of loyalty, military service, and patriotism to racial progress. Many lesser-known blacks were voicing the same sentiments, including a teacher from the South who described blacks as "soldiers of freedom" and promised that "when we have proved ourselves men, worthy to work and fight and die for our country, a grateful nation may gladly give us the recognition of real men, and the rights and privileges of true and loyal citizens of these United States."[64]

Unfortunately, such optimism was misplaced—the treatment black Americans received during the war and its aftermath showed the futility of relying exclusively on wartime loyalty to overcome racism and discrimination. Nevertheless, the sentiment Du Bois expressed so eloquently did not die after the war but was tempered by an attitude of caution. The experience of World War I warned against "assuming goodwill on the part of white authority" and instead fostered an attitude of trading "military service for measurable progress toward full citizenship, at times accepting promises, if reasonably confident the pledges would be honored, but continuing to press for civil rights, economic opportunity, and a useful role in the military."[65]

Although other black leaders, such as A. Philip Randolph, opposed the war, most black Americans overwhelmingly supported World War I, hoping that Wilson's call to "make the world safe for democracy" was indeed a portent of improved race relations. Du Bois's call to close ranks was scarcely necessary, for the "major concern [of black Americans] was that they might be excluded from the military and their hope for achieving greater democracy at home might be frustrated."[66] After the United States entered the war and the War Department authorized voluntary enlistments to bring all the regular army regiments up to full strength, the four black units were at their maximum authorized strength within a week. Over two million blacks registered for Selective Service with only one incident of draft resistance and that was initiated by whites. A Kansas editor observed that blacks "have responded more universally and cheerfully to the call of the Government than the white man. When called under the selective service draft they have rarely asked for exemption."[67]

When President Wilson delivered his war message in the spring of 1917, blacks constituted just under three percent of the nation's military personnel, with 10,000 on active duty in the four black regiments of the regular army and roughly the same number in the National Guard.[68] Blacks, constituting roughly ten percent of the nation's population, contributed more than their share of draftees to the war effort. Approximately 2.3 million blacks registered for the draft and 367,000 were called to duty, some thirteen percent of the nation's 2.8 million draftees.[69] In all, some 400,000 blacks served in the United States Army during World War I.[70] Although some saw combat, most served as labor troops in segregated units variously designated pioneer infantry regiments, stevedore regiments, labor battalions, and the like. As many as 160,000 worked as military laborers in France.[71]

Blacks were especially anxious that qualified Negroes have an opportunity for commissions. They had been excluded from the unofficial officer training camps begun in 1915, the so-called Plattsburg movement. After the United States entered the war, it became apparent that blacks would not be admitted to any of the army's officer training facilities. Joel Spingarn, a founder of the NAACP and its first president, campaigned vigorously for a segregated officer training camp—even though he philosophically favored integrated training—because he believed that the army would consent to training black officers only if the camps were segregated. Secretary of War Newton D. Baker agreed to Spingarn's proposal and a training camp for black officers was established at Fort Des Moines, Iowa, in the summer of 1917. Of the 1,250 candidates who entered training at Fort Des Moines, 1,000 were recruited from civilian life and 250 from the enlisted ranks of the four black regiments of the regular army; slightly more than half successfully completed the course and were commissioned.[72]

Blacks not only wanted to be included, but they also sought to play a meaningful role in the conflict beyond labor duty. They called for combat troops and officers, not just labor and service battalions. The obvious units for combat duty were the four regular regiments with their long tradition of combat service in the West, Cuba, the Philippines, and most recently, on the Mexican border. To the distress of the black public, however, these units did not see action. Instead, they were posted to stations along the southern border or in the island possessions for the duration of the war; their most experienced personnel were siphoned off to provide cadres for units of black draftees and candidates for the segregated officer training school.[73]

Some black Americans did, however, see combat in World War I; roughly 40,000 served in France in the Ninety-second Division and the Ninety-third Division (Provisional). The Ninety-third Division was designated a provisional unit because it was never brought up to full strength, consisting of only four infantry regiments. Shortly after the unit arrived in France it was transferred to French control and brigaded in French divisions, where it fought with distinction for the duration of the war. National Guardsmen made up three of the Ninety-third's regiments, and the fourth was manned with draftees from the South. The French gladly accepted the four regiments, treated the men as fellow soldiers, and were so impressed with their performance under fire that they awarded the *Croix de guerre* to three regiments, one company, and over five hundred individuals.[74]

The Ninety-second Division, composed of draftees and black junior officers trained at Fort Des Moines, remained with American forces. From the outset the unit was plagued by shortages of equipment, inadequate training, and poorly qualified personnel, all exacerbated by the racial prejudice of the division's senior white officers and the American high command. When elements of the Ninety-second failed to hold an important position in the early stages of the Meuse–Argonne offensive of September–October 1918, army authorities took the failure as proof that blacks were inherently unsuited for combat—especially under the leadership of black officers—despite the fact that white units also suffered setbacks.[75]

The poor showing of the Ninety-second Division in the Argonne was not the only incident that convinced the army leadership that blacks were not reliable soldiers. In August 1917, several months after the United States entered the war, racial violence erupted in Houston, Texas, between local whites and black soldiers of the Twenty-fourth Infantry Regiment, encamped on the outskirts of the city. At least a hundred armed black soldiers descended on the city after a Houston policeman administered beatings to two soldiers. For two hours the enraged soldiers shot up the town and by the time the violence ended, eighteen whites and four of the blacks were dead. In December 1917, after an investigation and court-martial, the leader of the mutineers and eleven others were hanged in secret, before the verdict was released. The following September six more of the rioters were convicted and executed.[76]

The effects of the Houston Riot and the disappointing performance of the Ninety-second Division in France were felt throughout the interwar years. After the war, the memory of these two episodes reinforced the army leadership's belief that black soldiers lacked the discipline and ability to function as effective combat troops. Perhaps the best evidence of this attitude on the part of the army leadership is a study completed in 1925 by faculty and students at the Army War College entitled "The Use of Negro Manpower in War." Reflecting the racist attitudes of the day, the study concluded that racial segregation was proper and inevitable because blacks were socially, morally, and mentally inferior to whites. The authors of the study believed that blacks lacked the qualities necessary for victory in combat—initiative, resourcefulness, and courage. Moreover, blacks were particularly unsuited for duty as officers, lacking the qualities of leadership and seeking only to advance the causes of the race.[77]

The views espoused in the war college study typified the racial attitudes of the army leadership between the wars.[78] They recognized, however, that in the event of another war, black personnel would have to be included in any general mobilization of the population, but they were reluctant to discuss the matter publicly. Moreover, throughout the interwar period, the strength of the four regular regiments was gradually reduced and elements of the units were scattered in a number of posts nationwide where they performed housekeeping duties. The army's failure to release information regarding the role of blacks in its mobilization plans, together with the strength reductions of four regular regiments, led to the general conclusion "both outside and inside the Army that no comprehensive plan for the employment of Negro troops in time of war existed."[79]

Despite limited opportunities and the fear of exclusion or relegation to labor battalions during a future war, military service continued to serve as a symbol of citizenship, and the military orientation of black Americans persisted between the wars. There were elements of black National Guard units in a number of northern cities, including New York, Chicago, and Washington, D.C. Senior Reserve Officer Training Corps (ROTC) units were located at Howard University in Washington and Wilberforce University in Ohio, and junior ROTC units could be found in black high schools and other postsecondary schools such as Tuskegee Institute, Hampton Institute, and Prairie View College in Texas.[80] The strength reductions in the regular regiments were closely monitored by organizations such as the NAACP and were widely reported in the pages of the black press.[81] Thus as the decade of the 1930s drew to a close and the prospect of a major expansion of the nation's armed forces loomed large, black Americans once again began to clamor for the opportunity to answer the call to arms and prove themselves as loyal and patriotic in the service of the nation as their white countrymen. In 1939 this resurging military orientation of the black public combined with a newer theme among black Americans—an emerging interest in aviation—to give birth to a pressure campaign to force the Air Corps to admit blacks to its ranks.

3

The Black Public
Becomes Air-Minded

A southern black newspaper observed in 1938 that a growing number of black Americans were becoming "airminded," a contemporary expression which "meant having enthusiasm for airplanes, believing in their potential to better human life, and supporting aviation development."[1] By the 1930s the airplane had come to symbolize the promise of the future, a portent of "a wondrous era of peace and harmony, of culture and prosperity," and many blacks were eager to participate in the new and exciting field of aviation.[2] Consequently, the air-mindedness of black Americans in the 1930s became a significant factor in the development of an effective pressure campaign for accepting blacks into the Army Air Corps.

Whites, for the most part, dismissed the notion that blacks should or could play a role in the air age. Many no doubt agreed with the preeminent air hero of the era, Charles A. Lindbergh, who heralded aviation as a "tool specially shaped for Western hands, a scientific art which others only copy in a mediocre fashion, another barrier between the teeming millions of Asia and the Grecian inheritance of Europe—one of those priceless possessions which permit the White race to live at all in a pressing sea of Yellow, Black, and Brown."[3] Another experienced white aviator, Kenneth Brown Collings, was more direct; he declared bluntly in the pages of the *American Mercury* that "Negroes cannot fly—even the bureau of Air Commerce admits that."[4] The black public, however, knew that Negroes could fly. Since the turn of the century Negroes had been reading in the black press of African American parachutists, inventors, barnstormers, long-distance fliers, and aerial soldiers of fortune; they were outraged at Collings's cavalier dismissal of their potential as aviators.[5]

Until the 1920s black aviation enthusiasts struggled in relative obscurity, with their exploits recorded primarily in black newspapers and magazines. In 1900, three years before the Wright brothers flew into history at Kitty Hawk, a black woman, Mary Doughtry, reportedly began a career as a parachutist when she leaped from a balloon before a crowd in New Orleans.[6] Six years later an obscure black laborer, identified only as Jackson until he assumed the dashing pseudonym Ajax Montmorency, thrilled crowds attending a Fourth of July celebration in Pittsburgh with a series of balloon ascensions and parachute jumps.[7]

In the decades before World War I, several black inventors designed various types of aircraft and some received patents for their efforts. One of the earliest was John F. Pickering, of Gonaives, Haiti, who submitted a design for a motorized, steerable balloon to the United States Patent Office in 1899.[8] After the Wrights demonstrated their flying machine at Fort Myer, Virginia, in late 1908 and the American public began to realize that the problem of heavier-than-air powered flight had been solved, a surge of interest in aviation swept the nation and at least five blacks designed new types of flying machines. In 1911 twenty-one-year-old Walter Swagerty of Los Angeles claimed to have invented a "heavier than air machine" that earned him the backing of a local millionaire.[9] Three blacks received patents for flying machines in 1912, although they scarcely looked airworthy judging from the design sketches submitted to the patent office: James E. Marshall of New York City;[10] Walter G. Madison of Ames, Iowa;[11] and James Smith of Oakland, California.[12] In 1914 a St. Louis black, whose name is variously reported as J. E. Whooter, H. E. Hooter, and John E. McWhorter, patented a strange, wingless contraption consisting of two huge rotating cylinders on either side of a central frame, with conventional horizontal and vertical stabilizers at the rear of the aircraft.[13] Like many of the flying machines designed and patented by white inventors, it is unlikely that any of these unwieldy devices ever flew, but they demonstrate that the dream of flight captured the imagination of black Americans from the earliest days of aviation.

Besides aeronauts, parachutists, and inventors, there is strong evidence which suggests that at least two blacks were flying airplanes before the outbreak of World War I. One was Lucian Arthur Hayden, a North Carolina native born in the early 1880s, who reportedly toured the South in 1912 giving aerial demonstrations and by 1913 held a French flying license. Hayden, also an inventor, reputedly developed and patented an aeronautical safety device that was accepted and used by the British in World War I.[14]

The other was Charles Wesley Peters of Pittsburgh, Pennsylvania, who has been described as the "first black to pilot a heavier-than-air craft and the first black designer and builder of an airplane."[15] Born in 1889, Peters developed an early interest in flying; by the time he was fourteen he had built a number of kites and gliders and in 1906 made his first successful glider flight. He subsequently designed and built a powered craft with a forty-foot wing span and an air-cooled automobile engine, making ten flights in the machine before it was destroyed by fire.[16] When news of Peters's achievement reached R. R. Wright, president of Georgia's State Industrial College at Savannah and organizer of the Georgia State Colored Fair held each fall in Macon, he added an aviation day to the schedule of events for the 1911 fair. Wright hoped to feature flights by Peters and ascensions by a black balloonist, and he took out advertisements that announced "For the First Time in the History of Fairs a Colored Man Goes Up in a Air Ship—Everybody Should See It," Thursday, 9 November was proclaimed "Airship Day. COLORED AVIATOR."[17] Unfortunately, a disagreement over money kept Peters from performing at the fair and Wright hastily secured the services of a white aviator so the crowds would not be disappointed.[18]

Although the United States armed forces trained thousands of military pilots during World War I, none was black. Early in the war, however, there was a glimmer of hope that qualified blacks would be accepted for training as aviators in the United States Army. In the summer of 1917, shortly after the country entered the war, at least one black newspaper reported that Pres. W. S. Scarborough of Wilberforce University, a black college in Ohio, had received a War Department telegram asking him to encourage his "best military students" to apply for army aviation training.[19] The wording of the telegram, however, suggests that similar requests were sent to all institutions with senior ROTC units (Wilberforce was then one of only two black colleges with a senior ROTC unit). That being the case, the telegram to President Scarborough was most likely sent inadvertently as there is no other evidence that the U.S. Army gave any consideration to training blacks as pilots during World War I. Instead, those who applied for duty with the Air Service had their applications returned without action and were told that no black squadrons had been established and none were being organized.[20]

The only blacks who even came close to serving in the air with American forces during World War I were black officers who entered training as airborne artillery spotters at the Aerial Observers School, Fort Sill, Oklahoma. One, whose light complexion did not immediately identify him as a Negro, remained in the school until two days before his scheduled graduation date. The others, obviously black, were immediately segregated and denied the normal military courtesies due officers until they withdrew in anger and frustration.[21]

One black American did fly as a military pilot during World War I, not with American forces but with the French. Eugene Jacques Bullard, born in Columbus, Georgia, in 1894, left the United States and joined the French Foreign Legion shortly after the war broke out in Europe and earned the *Croix de guerre* while serving as an infantryman on the Western front. In November 1916 he transferred to the French Flying Service, thus becoming a member of that unofficial brotherhood of American volunteers who flew with the French, the Lafayette Flying Corps.[22] Bullard completed his training as an enlisted pursuit pilot the following August and within a week was flying combat sorties with a French squadron at the front. By November he was credited with the destruction of one German aircraft and claimed a second, becoming the first black American to destroy an enemy aircraft in aerial combat. Bullard's career as a fighter pilot was cut short several months later after an altercation with an officer, and he finished the war as an infantryman.[23]

Bullard proved that blacks could fly and fly competently. Unlike Peters or Hayden, his flying skills were well documented; he had completed the French Air Service course of training and engaged the enemy in aerial combat. James Norman Hall's and Charles Bernard Nordhoff's *Lafayette Flying Corps*, published immediately after the war, gave a full and favorable account of his participation as an

Eugene Bullard, the first black American to participate in aerial combat, flew as a volunteer with France during World War I. In November 1916 he was credited with the destruction of a German aircraft. (Courtesy Tuskegee University Archives, Black Wings Collection)

American volunteer flier and, together with the accompanying photograph, left little doubt as to Bullard's complexion. He might have become black America's air hero, but instead he remained in Paris managing a nightclub until Hitler invaded France in 1940, his exploits virtually unknown to the American public, black or white.[24]

The aviator who in the early 1930s became the symbol of black America's aeronautical dreams was Bessie Coleman. In 1922 she made her debut as an exhibition flier in an air show at Chicago's Checkerboard Field. A native of Texas, Coleman joined the flood of blacks who migrated north during the World War I era. She settled in Chicago and decided to become an aviator. When several flying schools refused to admit her, she contacted Robert S. Abbott—owner of the *Chicago Defender*, one of the leading black papers of the period—who advised her to seek flying lessons in France. After two trips to Europe, she returned to America with a license from Federation Aeronautique Internationale, becoming the first black woman from the United States to hold a pilot's license. For four years "Brave Bessie" barnstormed, thrilling large crowds of blacks and whites curious to see whether a black woman really could fly, as she tried to raise enough money to establish a flying school that would be open to blacks. She died tragically in an air crash on 30 April 1926, while practicing for an air show in Florida, and became a martyr to the cause of air progress among blacks.[25]

While Bessie Coleman was learning to fly in Europe, a young black man from the West Indies migrated to the United States by way of Canada, claiming he learned to fly from the Canadian war ace Billy Bishop. Hubert Julian, the "Black Eagle of Harlem," was a controversial, flamboyant figure given to grandiose schemes and self-conferred titles. By the mid-1920s his audacious behavior and ambitious aviation projects kept him constantly in the news, a foil for white reporters and an embarrassment to the black public.[26]

Julian first came to the attention of American blacks in 1921, as "Dr. H. Julian, a Negro student at McGill University, Montreal, Canada," whose patented air safety device "brought an offer of $300,000 from the Curtis [*sic*] Aeroplane Company for patent rights and one for $150,000 from the Gerni Aeroplane Company of Montreal."[27] The following year he had settled in New York City and affiliated himself with Marcus Garvey, another controversial West Indian who founded the Universal Negro Improvement Association

(UNIA). Garvey had come to the United States in 1916 and captured the imagination of the black masses with his rhetoric of black pride and his grandiose schemes of resettling American blacks in Liberia.[28] In August 1922 the *Negro World* announced that Julian would head the aeronautical department of the UNIA, which came to be known as the "Black Eagle Flying Corps."[29]

Thus Julian launched the American phase of his long and checkered career as a flier, maintaining a ubiquitous presence in black aviation until the 1940s. In 1924 his fund-raising campaign for a solo flight from New York to Africa came under the scrutiny of postal authorities and the FBI, who suspected mail fraud. The investigators were apparently satisfied when, to much acclaim, Julian made a Fourth of July takeoff from Long Island, only to crash before the crowd of well-wishers and sightseers that had gathered to see him off.[30] In 1926 and again in 1928 Julian announced plans for a transatlantic flight, but on neither occasion did he even attempt a takeoff.[31]

Julian began the decade of the 1930s by traveling to Ethiopia, apparently winning the confidence of the emperor-elect and returning to the United States as Col. Hubert Fauntleroy Julian of the Ethiopian Air Force, accoutered in "white jodhpurs, blue tunic, tan pith helmet with royal crest, and high leather boots with spurs."[32] He ended it by challenging the head of Nazi Germany's Luftwaffe, Hermann Göring, to an air duel over the English Channel.[33] Throughout the decade he was constantly in the news, leaving Ethiopia in disgrace after crashing the emperor's prize airplane, claiming to be the personal pilot of Father Divine, performing in air circuses, brawling with John C. Robinson in an Addis Ababa hotel, and volunteering to fly for the Finnish Air Force after the Soviet invasion.[34]

As the only black aviator who came to the attention of most white Americans during the interwar years, Julian's posturing, swaggering, and blustering reinforced white America's preconceived notion that blacks were at best inept pilots, seriously undermining the credibility of legitimate black aviators striving to prove that whites did not have a monopoly on flying aptitude. By the mid-1930s he had become an embarrassment to serious-minded blacks and he was soundly condemned by black editors. The *Boston Chronicle* finally refused to advertise his public appearances, explaining that "we see

Harlem's 'Black Eagle' as a blatant jackdaw. We trust Boston will never be inflicted with him again. . . . Julian talks loudly about his being a 'black' man. We wish he were otherwise."[35]

Despite the handicap of Julian's antics, black fliers struggled on. After the news of Lindbergh's flight electrified the nation and the world, black Americans were attracted to aviation in increasing numbers. In June 1927, a month after Lindbergh crossed the Atlantic, an editorial in the *Pittsburgh Courier* asked rhetorically, "What Will the Negro Contribute to Aviation?" Those who complained that no black had gained fame in the air were reminded "that a Negro youth would have stronger prejudices to combat than a Lindbergh . . . [and] that no Negro youth has as yet become a serious part of any aviation service." Black Americans could nevertheless anticipate, the editorial continued, "that some youth of ours, inspired by this feat of Lindbergh, will begin a serious apprenticeship in aeronautics. And let us hope that he will show the same stamina as Lindbergh in the face of ridicule . . . and the same modesty in the face of success."[36]

Five years later James Herman Banning and Thomas Allen became the first blacks to complete a coast-to-coast flight across the United States, and many blacks proclaimed them their first air heroes. In the interim Bessie Coleman was immortalized as the aeronautical pioneer of the race, while African Americans took to the air in increasing numbers, establishing aero clubs, publishing a flying magazine, organizing traveling air circuses, and attempting long-distance publicity flights. White Americans remained, for the most part, quite oblivious to the burgeoning black interest in aviation during the post-Lindbergh era.

If whites thought about blacks in aviation at all, one of two unflattering stereotypes usually prevailed. Julian served as the archetype for one, the swaggering, boastful black who claimed to be an expert pilot but was actually quite incompetent, the aeronautical equivalent of Kingfish, the self-important black attorney of the popular radio show "Amos 'n Andy."

Most whites, however, assumed that blacks were simply overwhelmed by the technological complexity of airplanes and possessed an inherent fear of flying. Lindbergh reinforced this stereotype in his popular autobiography *"We"*, published shortly after his transatlantic flight, by devoting six pages to a condescending description of his experiences with rural Mississippi blacks while barnstorming through the South in the early 1920s. He reported that an elderly black woman asked "'Boss! How much you all charge foah take me up to Heaben and leave me deah?'" and he described an encounter with a young black man, who boldly agreed to take a flight when some conniving whites paid his fare, having conspired earlier with Lindbergh to "give this negro a stunt ride." As he climbed aboard, the unwitting black passenger reassured a group of black bystanders, telling them "he would wave his red bandanna handkerchief over the side of the cockpit during the entire flight to show them he was still unafraid." According to Lindbergh's account, the black youth panicked on the takeoff roll and "with the first deviation from straight flight my passenger had his head down on the floor of the cockpit but continued to wave the red handkerchief with one hand while he was holding on to everything available with the other, although he was held in securely with the safety belt." The handkerchief disappeared, Lindbergh reported, when he attempted a loop, and "it was not until we were almost touching the ground that the bandanna appeared again over the cowling."[37]

The portrayal of blacks as technological illiterates, who either feigned aeronautical competence or were overcome by irrational fears once they became airborne, was a serious obstacle to aspiring black pilots. They could expect little encouragement from the white aviation community, for "[t]hose who manufactured the early planes, or established aviation companies, or flew or serviced them, were a close-knit group into whose 'brotherhood' the black man could not be received as an equal."[38]

Despite these obstacles, by 1927 blacks were entering the field of aviation in increasing numbers. Several months before Lindbergh's Atlantic crossing, Joel "Ace" Foreman of Los Angeles attempted a transcontinental flight, the first bona fide attempt at a long-distance flight by a black. Foreman and his mechanic, Artis Ward, left Los Angeles for New York City in February 1927, flying a patched-up Curtiss JN-4 "Jenny," sponsored by a local black newspaper, the *California Eagle*, and the Los Angeles chapter of the NAACP. After an engine malfunction grounded the pair in Salt Lake City for several weeks, they eventually made Chicago where additional mechanical problems brought the flight to a premature end.[39]

After Lindbergh's flight, black America's hopes for an air hero continued to loom large. The readers of the National Urban League's monthly magazine, *Opportunity*, learned that a black man, Samuel V. B. Sauzereseteo, had flown from Moscow to Berlin, from Belgium to the African Congo, and from Paris to London.[40] When pineapple magnate James Dole offered a total of $35,000 for the first two nonstop flights from the United States mainland to the Hawaiian Islands, the *Pittsburgh Courier* announced that two black fliers from California intended to compete for the prize money, Walter E. Swagerty of San Francisco and Clarence E. Martin of Oakland. Swagerty, who had reportedly designed a flying machine in 1911, told reporters he learned to fly in 1914 and had appeared at county fairs throughout the Southwest. Martin claimed he had received his first flying lesson from stunt pilot Lincoln Beachey and then went on to become an air mail pilot.[41] Neither pilot was among the official entrants for the Dole prize; Martin subsequently announced that he would not compete and that there were no other black fliers in the race.[42] Another young black man, Jesse Boland, also made aviation news in 1927 when he reportedly built an airplane and made a demonstration flight of several hours duration over his hometown of Roanoke, Virginia.[43]

Toward the end of the 1920s the number of blacks interested in aviation had grown to the point that aviation clubs began appearing; by 1936 some thirty-seven black flying clubs had been organized and twenty-four were still active.[44] Although a black aviation club was reported in Los Angeles as early as 1921,[45] the movement did not really begin until late in the decade. Julian, who in 1922 had sought to organize an aviation arm in Garvey's UNIA, tried once again to attract a following in May 1929 when he announced the formation of the National Association for the Advancement of Aviation Amongst Colored Races.[46] This organization probably existed only in Julian's mind, but in the fall of the same year, the Universal Aviation Association held the "first national aviation meet of Negro flyers" at Checkerboard Field near Chicago, where Bessie Coleman had launched her aviation career in the United States six years earlier. At least six Chicago blacks demonstrated their flying skills at the meet, and Dr. A. Porter Davis, a physician from Missouri who had been flying for over a year, piloted his own airplane cross-country to attend the event.[47]

In May 1930 the Bessie Coleman Aero Clubs published the first number of a new monthly flying magazine, *Bessie Coleman Aero*

News.[48] The organization and its magazine were the brainchildren of William J. Powell, a Chicago black who became a thoroughgoing aviation enthusiast during the post-Lindbergh aviation boom. Rejected by the Air Corps and by civilian aviation schools in the Midwest, Powell moved to Los Angeles where he was accepted by a local flying school. He quickly formed a group of like-minded blacks—including James Herman Banning, who had already distinguished himself in 1927 by becoming the first black pilot licensed by the Bureau of Air Commerce—and began an ambitious campaign to establish a network of local black flying clubs organized nationally as the Bessie Coleman Aero Clubs, Inc.[49] The group received national attention in 1929 when black congressman Oscar De Priest visited Los Angeles: Powell arranged for an airplane owned by the club to be christened the *Oscar De Priest* and the congressman took a flight in the craft with Banning. De Priest took it all in stride and observed enthusiastically that the "field of aviation presents great opportunities to the Negro and he should enter it at once."[50]

Powell hoped that blacks could "get in on the ground floor" of the aviation industry. The first issue of *Bessie Coleman Aero News* contained his message to "The Negro Youth of America." He urged black youth to train for aviation careers, telling them with evangelistic fervor: "There is a better job and a better future for you in aviation than any other industry. The reason is this: Aviation is just beginning its period of growth. Aviation is going to be America's next gigantic industry, and if you can get into it now while it is still uncrowded, you can grow as aviation grows."[51] Powell believed that the demand for skilled fliers and mechanics would be so great that trained, competent blacks could overcome the racial barriers that had traditionally excluded them from responsible positions in other branches of the transportation industry.[52] Yet scarcely a year after Powell began proselytizing young blacks, the organizers of the Air Line Pilots Association included a clause in the union's bylaws that restricted membership to whites only.[53]

While Powell was organizing blacks in the West and attempting to establish a national flying organization, other air-minded blacks were also banding together to promote aviation. In Chicago John C. Robinson, who in 1931 had graduated from the local Curtiss-Wright School after he convinced its director to accept him, organized the Brown Eagle Aero Club,[54] subsequently reorganized as the Challenger Air Pilot Association.[55] In 1932 the International Colored Aeronautical Association sought to bridge national borders when it sponsored a flight by Leon Paris from New York City to his native Haiti.[56]

For five long years after Lindbergh emerged as the hero of the machine age, black America waited for the appearance of a "colored Lindy": Bullard had abandoned flying and was forgotten; Julian was an embarrassment; nothing more was heard of Sauzereseteo, who was apparently not an American; and Foreman, Martin, Swagerty, and Boland lapsed into obscurity. In the interim, the memory of Bessie Coleman inspired air-minded blacks; her untimely death, her flying credentials, and the fact that she was an attractive woman engaged in an extremely dangerous pursuit made it inevitable that she would become a virtual patron saint to blacks who aspired to fly. The most obvious tribute to her memory was, of course, Powell's use of her name for his club and magazine.[57] The Brown Eagle Aero Club and its successors also honored Bessie: they adopted the practice of flying over Chicago each Memorial Day and scattering flowers in her memory.[58] Black journalists who sketched the history of blacks in aviation invariably portrayed Bessie as the aeronautical pioneer of the race, a perceptive, petite young lady who saw her race's future in the air but died before her dreams could be realized.[59]

Perhaps the most eloquent testimony to both the black public's desire for an air hero and the role the memory of Bessie Coleman played until one appeared is Harry Levette's poem "Call of the Wings," published by the *Pittsburgh Courier* in 1931:

> Black men! Last to the call of wings,
>   As the myriads of ships course the skies!
> Each an Argonaut venturing brings,
>   Golden fleece from the land where it flies
> High over white peak, angry sea,
>   Man is fearlessly conquering the air,
> History making. The entry is free—
>   Black men! Say, why are you not there?
>
> Are you cowardly, spineless, and weak,
>   That your feet cling closely to the earth?
> Rise from your lethargy; this new field seek!
>   You've won others; in this, prove your worth.

A mere girl pioneered for the Race,
     But our men let her sacrifice fail.
Fly! Fly! With the nations keep pace!
     Let the sun glint your silver sail.[60]

A year-and-a-half later two young black men answered the "Call of the Wings." On 9 October 1932 James Herman Banning and his mechanic, Thomas Allen, landed at Roosevelt Field on Long Island. Eighteen days earlier, on 21 September, they had departed Los Angeles in an open-cockpit biplane hoping to become the first African Americans to cross the continent by air. Black America

In October 1932 James Herman Banning and Thomas Allen became the first black Americans to complete a transcontinental flight. Their exploits were serialized in the influential black newspaper, the *Pittsburgh Courier*. (Courtesy Tuskegee University Archives, Black Wings Collection)

rejoiced at their success, calling them "suntanned editions of the Lindy of yesteryear" and presenting them with commemorative medals to honor their achievement. For nine weeks the black public relived the trials and triumphs of their flight in a series of columns the pair contributed to the *Pittsburgh Courier*.[61]

The Banning–Allen flight ushered in a new era in black aviation, though many feared that Banning's untimely death in an air crash the following February might thwart further black achievement in the air.[62] Instead, the mounting interest and experience of air-minded blacks quickly brought more successes and attracted even more blacks to the ranks of the aviation enthusiasts. Less than six months after Banning's death, Anderson and Forsythe became the first black Americans to complete a round-trip transcontinental flight, and several months later they were in the news again with a goodwill flight to Montreal, Canada.[63] The year 1934 brought further publicity to black aviators with the Anderson–Forsythe flying team's widely publicized Pan-American Goodwill Flight in their newly christened aircraft, the *Booker T. Washington*.[64] John C. Robinson kept black America's interest in aviation alive in 1935 and 1936 with his widely publicized exploits as Haile Selassie's pilot during Ethiopia's conflict with Italy.[65] Thus by 1936, after five years of notable achievement during the most severe economic crisis in the nation's history and without any federal assistance, blacks came to believe that even greater aeronautical accomplishment could be theirs if only the opportunities routinely open to white fliers were opened to them. One result of this growing black interest in aviation in the wake of the exploits of Banning, Allen, Forsythe, Anderson, and Robinson was the rise of Chicago as the center of black aviation.

Three individuals were responsible for the emergence of Chicago as a center of black aviation in the late 1930s—Cornelius Coffey, Willa Brown, and Enoch P. Waters. Coffey's enthusiasm for aviation dated from 1931, when Robinson "converted [him] from an auto mechanic to an airplane and engine mechanic and interested him in becoming a flier." Until Robinson's departure from Chicago in 1935, they worked together to train "scores of pilots, mechanics, navigators and parachute jumpers who became the nucleus" of a Chicago-based national association of black aviation enthusiasts. With Robinson in Ethiopia, Coffey took over as the "top authority of the local group" of black aviators, but he lacked the Brown Condor's flair for publicity. Although "completely devoted to aviation," Coffey was a "quiet retiring man of few words . . . content being an unnoticed instructor because it allowed him to spend his days at the airport."[66] By early 1936, Willa Brown assumed the role of promoter of black aviation activities in Chicago and allied herself with the taciturn Coffey, who willingly deferred to his attractive and charismatic colleague. Shortly before Robinson's return from Africa, Brown approached Enoch Waters, city editor for the influential *Chicago Defender*, seeking publicity for an air show that Chicago's black pilots were planning. Waters recalled vividly the scene when she entered the newspaper's offices: "When Willa Brown, a shapely young brownskin woman, wearing white jodhpurs, a form fitting white jacket and white boots, strode into our newsroom, in 1936, she made such a stunning appearance that all the typewriters suddenly went silent."[67] When Waters learned that Brown represented some thirty black fliers, he proposed that the air show become an annual event sponsored by the newspaper. The owner of the *Defender*, Robert Abbott—the man who had encouraged Bessie Coleman to go to France for flying lessons—readily agreed.[68]

Although he never learned to fly, Waters became an avid supporter of black aviation. He urged Brown and Coffey to broaden their horizons and establish a national organization for black aviators and their supporters.[69] A nationwide organization, Waters maintained, would stimulate publicity in other black newspapers, would "give us better information about aviation activities elsewhere in the country and provide us with a vehicle to campaign for our goals." By 1937 Coffey, Brown, Waters, and nine other Chicago blacks had organized the National Airmen's Association of America (NAAA), chartered by the state of Illinois and headquartered at the offices of the *Chicago Defender*.[70] Waters and Brown handled the NAAA membership drive; Waters recalled later that they quickly established NAAA chapters "in several cities in the Midwest and East, with flying visits by Willa providing the impetus. These chapters weren't big, consisting of just six to a dozen fliers. Considering there were only about two hundred Negroes flying in the country, we were satisfied with the response."[71]

Chicago's role as a center of black aviation was reflected in the data on black pilots that the Division of Negro Affairs in the Department of Commerce began compiling at mid-decade. In June 1936 *Crisis* reported the results of the division's study, which showed that of the fifty-five blacks holding Bureau of Air Commerce licenses, all but a handful lived outside the South, with a high of fifteen concentrated in Illinois. Of primary concern to air-minded blacks were the conclusions of the study:

> Financial difficulties and limited training facilities have hampered the progress of the Negro in aviation. The majority of these men follow occupations which return only average incomes, but they have paid on an average of twelve dollars and fifty cents per hour for training. Many aviation schools are reluctant to receive Negroes, and many will not enroll them at all. Negro pilots have been reasonably resourceful: they have borrowed, stinted, bartered, and formed clubs to secure their training. . . . Their resourcefulness and the friendly interest of others have enabled them to obtain instruction.[72]

Despite the aeronautical successes of men like Banning, Anderson, Forsythe, Powell, and Robinson, such reports made it increasingly obvious to black Americans that their prospects for progress in the field of aviation were limited as long as opportunities open to whites—military aviation, flying the mail, and commercial flying—remained closed. At the beginning of the decade some black aviation enthusiasts had counseled racial solidarity and self-help. In 1931 Ed Sanders of New Orleans urged the *Pittsburgh Courier* to use its considerable influence to establish a National Aviation Fund, headed by Tuskegee president R. R. Moton, which would solicit one dollar annually from blacks across the nation to support aeronautical education among African Americans.[73] The National Aviation Fund never materialized, but in the waning years of the 1930s blacks were challenging the discriminatory practices that excluded them from aviation opportunities, especially those supported by the federal government. Black airmen were encouraged to press for broader opportunities in aviation after 1936 by the mounting emphasis on civil rights issues that emerged during the second term of the Roosevelt administration.

4

Civil Rights Emerges
as a National Issue

The third element that gave rise to the campaign for black participation in the Air Corps, the emergence of civil rights as a national issue, did not appear until after 1936. It was, however, a crucial element for it served as the catalyst that precipitated the campaign. Without it, the black public's long-standing military orientation and its burgeoning air-mindedness would not have been sufficient to launch "the most widespread, persistent, and widely publicized of all the prewar pressure campaigns affecting the Negro and the Army."[1] But when a new racial climate developed during Franklin D. Roosevelt's second term, conditions suddenly became favorable for the emergence of a viable crusade against the Air Corps' policy of racial exclusion.

One New Deal scholar has characterized Roosevelt's first term as "An Old Deal, A Raw Deal" for blacks, and his second term as "The Start of a New Deal."[2] Until 1936 the black press and black leaders found little of value in the New Deal, but their condemnations dwindled rapidly after Roosevelt was reelected. This sudden reversal occurred because of "significant developments . . . to reduce Negro powerlessness, to increase Afro-American expectations, and to alter white attitudes toward race relations."[3] Although the New Deal did not initiate these changes, it fostered their birth and growth "by substantively and symbolically assisting blacks to an unprecedented extent, by making explicit as never before the federal government's recognition of and responsibility for the plight of Afro-Americans, and creating a reform atmosphere that made possible a major campaign for civil rights."[4] Consequently, during Roosevelt's second term, the black public exhibited a new optimism, a new hope for the future, that heralded a "revolution in expectations" and a "revolution in confidence."[5] Thus the changes that occurred in the relationship of the federal government with the nation's largest and most economically distressed minority gave African Americans hope; they "believed the government now really intended to deal blacks into the game."[6]

When Roosevelt was elected in 1932, the status of black Americans was at a low ebb. In the decades since Reconstruction they had suffered the steady deterioration of their rights as American citizens, despite the provisions of the Fourteenth Amendment. Similarly, the provisions of the Fifteenth Amendment had initially enabled black voters and officeholders to wield a measure of political power during the Reconstruction era, but after 1877 their political influence dwindled; after 1901 no black representative sat in the halls of Congress for almost three decades, testimony to the political impotence of a tenth of the nation's population. Finally, the promise of economic achievement, touted by Booker T. Washington and his followers as a substitute for social and political equality, proved ephemeral as blacks remained the most economically deprived group in the country.

Although the Fourteenth Amendment and the Civil Rights Act of 1875 had given blacks reason to believe that the federal government intended to protect their rights as citizens, the willingness of the national authorities to support the cause of racial equality evaporated after the election of 1876. In 1883 the United States Supreme Court, in a decision that signaled the demise of Reconstruction legislation and amendments, ruled that the Civil Rights Act of 1875 applied only to states, not individuals. Thirteen years later, in the most significant decision affecting blacks since the Civil War, the Supreme Court held in *Plessy v. Ferguson* (1896) that segregation itself did not violate the Fourteenth Amendment, as long as the segregated facilities were equal. The Supreme Court's ruling that segregation was permissible under the Constitution stood for almost sixty years, finally being overturned in 1954 by *Brown v. Board of Education*.

Hard on the heels of *Plessy v. Ferguson*, the southern states adopted a series of measures that effectively disfranchised black voters. Even earlier, during the 1880s, as former Confederate Democrats regained control of the machinery of state governments, blacks had been excluded from the electoral process through violence, fraud, and intimidation. Then in the last decade of the nineteenth century and the first decade of the twentieth, the South institutionalized black disfranchisement by formally adopting a variety of devices such as the poll tax, literacy tests, and the white primary. By the turn of the century southern blacks were, for all practical purposes, politically impotent. Those in the North who voted allied themselves with the Republican party, even though the GOP showed little concern over issues important to African Americans.

Blacks in the post-Reconstruction South were not only deprived of their civil rights and the franchise; they also suffered economic deprivation at the hands of the agricultural labor arrangement that supplanted slavery, the sharecropping system. Plans for distributing confiscated lands to the freedmen never materialized and the former slaves found themselves at the mercy of an economic system that proved, in terms of material benefits, as oppressive as slavery. Blacks who abandoned agriculture and migrated to urban centers in the North found some relief from overt racism, but their hopes of economic advancement rarely materialized. Black workers were generally excluded from labor unions and thus were used by white capitalists as a source of cheap labor to undermine the unions. Lack of capital proved a severe obstacle to ambitious black businessmen who pursued the American dream of economic independence.

Despite the problems associated with migrating out of the rural South, some 200,000 blacks made their way north between 1890 and 1910.[7] They were followed by a half million during the second decade of the twentieth century and three-quarters of a million during the 1920s.[8] By 1930 forty percent of all blacks lived in cities, almost double the proportion in 1900, and of the five cities with the largest black populations, none was in the South.[9] African Americans came North to escape the oppressive and violent racial climate of the South, and despite the attendant economic hardships associated with finding employment and adjusting to urban life, they found certain advantages. Blacks in the North could vote, and by 1928 Chicago blacks elected Oscar De Priest to Congress, the first black to serve in the House of Representatives since 1901.[10] Blacks also found that educational opportunities were open to them in the North, often on an integrated basis. Finally, the concentration of large numbers of blacks in areas outside the South fostered a flowering of arts and letters for blacks during the 1920s known variously as the New Negro Movement, the Harlem Renaissance, and the Black Renaissance.[11]

During the 1920s, the optimism associated with these positive developments was offset by a resurgence of Ku Klux Klan activity and deteriorating economic conditions in the rural South, which had been in the throes of an agricultural depression since the end of World War I and had not benefited from the boom years of the 1920s. Northern blacks, although spared the severe economic hardships of black sharecroppers, had likewise benefited little from the prosperity after the war. In 1929, after the nation entered the greatest depression in its history, black Americans suffered immediately and severely. Northern blacks lost their tenuous foothold on the economic ladder as desperate whites competed for the menial jobs previously reserved for blacks. In the South the agricultural depression of the twenties deepened as cotton prices plummeted. The marginal gains of northern blacks—a limited political resurgence, improved educational opportunities, and an emerging intelligentsia—crumbled before the severity of the economic crisis that descended upon the nation. By 1931 conditions were so discouraging that the head of the National Urban League, T. Arnold Hill, declared that never "in the history of the Negro since slavery [has the] economic and social outlook seemed so discouraging."[12]

Despite the bleak conditions that black Americans faced in 1932, their traditional loyalty to the Republican party held firm. The Republican candidate, Herbert Hoover, won their support by default, for he and his party did little to garner the black vote; the percentage of black delegates to the 1932 Republican convention was the smallest in the twentieth century and the Republican platform contained nothing that addressed the problems of blacks in a meaningful way. African American voters had followed the lead of the black press, which warned that a victory by the Democratic candidate, Franklin D. Roosevelt, would mean an extension of segregation. The *Chicago Defender*'s pronouncements on the election typified the attitudes of the black press and the black public in 1932: "The future of the black man, so far as his civil rights are concerned, is at least safe in the hands of the Republican Party."[13]

There was strong historical precedent for black apprehensions that a Democratic victory would be inimical to their interests. The election of Woodrow Wilson, only the second Democratic president since the Civil War, had resulted in "the most Southern-dominated, anti-Negro, national administration since the 1850s."[14] As a northern Democrat, Roosevelt could hardly be expected to oppose the racial attitudes of a party that drew its strength from the Solid South, a party whose Alabama members emblazoned the phrase "White Supremacy" on their official party seal. Moreover, nothing in Roosevelt's record and nothing in his campaign rhetoric suggested that he would depart from the Democratic party's traditional pattern of northern deference to southern sensibilities on the race issue. Consequently, black voters supported Hoover in 1932 by an even greater margin than in 1928, despite the widespread unemployment and economic hardship that had been their lot under his administration.[15]

Thus the election of Roosevelt brought no ray of hope to the nation's black citizens, and no reason to believe that civil rights and racial justice would emerge as national issues by the end of the decade. The election of a northern Democrat who had no record of concern for blacks augured poorly for the future of African Americans, who were struggling just to obtain the bare necessities of life— food, clothing, and shelter. Most would have scoffed at the idea of embarking on a campaign to force the federal government to provide flight training to college-age blacks and open the Army Air Corps to blacks. Yet before Roosevelt's second term ended, such a campaign was launched, and it ultimately led to the establishment of the black air units that saw combat duty during World War II.

The first two years of the Roosevelt administration seemed to confirm the worst fears of the black public. Roosevelt's top priority was economic reconstruction, and he needed the cooperation of southern Democrats in Congress to ensure the passage of his bills and appropriations; he made it plain that he would not jeopardize economic recovery by backing civil rights measures. More important to the masses of blacks who suffered under the harsh realities of the Depression, African Americans received little assistance from the initial agricultural and industrial recovery programs, the Agricultural Adjustment Administration (AAA) and the National Recovery Administration (NRA).[16] Southern black tenant farmers, the most economically distressed group of farm workers in the nation, found the AAA of little benefit; indeed, it worsened their plight by encouraging landowners to reduce acreage and evict tenants.[17] Nor did black labor, the most distressed group of urban workers, receive any special consideration from the NRA. Many of the traditional "Negro jobs" were not covered by NRA wage codes; employers whose jobs were covered by the codes usually refused to follow them where black workers were concerned or, if forced to comply, fired blacks and hired whites.[18] Consequently, the black press bitterly asserted that, as far as blacks were concerned, the letters NRA might well stand for "Negro Run Around," "Negroes Ruined Again," "Negro Rarely Allowed," "Negro Removal Act," "Negro Robbed Again," or "No Roosevelt Again."[19]

By 1936, however, black criticism of Roosevelt's recovery program began to soften. Black condemnation of the New Deal, "so common between 1933 and 1936, soon stilled and turned to praise. Especially in the second term, new forces, both in and outside the Roosevelt administration, began to push the federal government in directions favorable to blacks."[20] The forces that brought this transformation came from four sources: pressure from racial improvement organizations such as the National Association for the Advancement of Colored People (NAACP) and the National Urban League (NUL); the development of black voting power; the support of nonblack groups such as intellectuals, the radical Left, labor unions, and southern liberals; and pressure from within the administration from the Black Cabinet and white liberals.

The most important factor in the transformation of New Deal attitudes toward blacks was the pressure applied by organizations like the NAACP and the NUL, the two largest and most influential groups working for the betterment of blacks during the 1930s. The NAACP, founded in 1909 by an interracial group of racial progressives, had focused its efforts on legal action, educational issues, and propaganda. The NUL, founded in 1910 with the support of Booker T. Washington, sought to assist southern blacks who migrated north in finding work and adjusting to urban life. During the 1920s both organizations had undergone changes that brought blacks into key leadership positions and strengthened their organizational structures. Thus on the eve of the Great Depression both the NAACP and the NUL had established themselves as important and influential institutions working for racial betterment. Nevertheless, both organizations felt the impact of the depression keenly, nearly going bankrupt after the 1929 crash.[21] They survived, however, and by 1935 had the common objective of forcing "the New Deal and its liberal supporters into extending the scope of its reform ventures and embracing more completely the black American perspective."[22]

After 1932 black voting power increased and became a factor in national politics. Although blacks were still excluded from the polls in the South, many obtained the franchise by migrating North— some 400,000 African Americans left the South between 1930 and 1940.[23] By 1936 there were enough blacks in key northern states "to make the black vote worth some attention."[24] For the first time it figured prominently in national politics, prompting *Time* magazine to observe that in "no national election since 1860 have politicians been so Negro-minded."[25]

The espousal of racial justice and equality by a wide array of groups and organizations after 1932 was a third factor that forced the administration to acknowledge the plight of black Americans.[26] During the 1930s the American Left added racial equality to their radical agenda, making the more conservative racial organizations like the Urban League and the NAACP seem moderate by comparison. At the same time it became apparent that a "new intellectual consensus" was emerging, promulgated by a new generation of social scientists, which challenged established theories of racial determinism and the inherent inferiority of blacks. The proponents of the new racial theories pointed to the literary and artistic efflorescence of the Harlem Renaissance as evidence that social and economic factors played a larger role in fostering creativity and intellectual endeavor than racial characteristics. Labor unions also began to address racial issues as Socialists and Communists gained national attention by organizing biracial unions in the South. By 1935 the American Federation of Labor's (AFL) long-standing apathy toward racial discrimination came under attack by black labor organizer A. Philip Randolph, and in 1936 his Brotherhood of Sleeping Car Porters received an international AFL charter. The Congress of Industrial Organizations proved even more receptive to blacks and, under the leadership of John L. Lewis, was soon recognized by black leaders and the black press as a champion of racial justice. Finally, southern liberals began to take a stronger stand on racial issues, a development which led to the establishment of regional interracial organizations such as the Southern Conference on Human Welfare.

The fourth and final factor that pushed the federal government in the direction of more racial equality was pressure from individuals within the Roosevelt administration itself. Shortly after his inauguration Roosevelt appointed several high officials who were outspoken advocates of racial justice and civil rights. One of the most prominent was Secretary of the Interior Harold Ickes, a former president of the Chicago branch of the NAACP. Others in the administration who were liberal on racial issues included Will Alexander, Clark Foreman, and Aubrey Williams. But the individual in Washington whom black Americans regarded as their most outspoken and influential advocate was Eleanor Roosevelt.[27] More than any other person associated with the Roosevelt administration, the first lady displayed a keen sensitivity to the problems of African Americans and she made civil rights one of her major concerns. Black Americans were immediately impressed with her many gestures of interest and genuine concern, and black leaders relied on her to intercede with the president and Cabinet officers. In addition to these white advocates within the administration, many of the New Deal agencies appointed black advisers to deal with racial issues.[28] By 1935 some forty-five black advisers were serving in federal agencies, known collectively as the Black Cabinet, and their presence became a powerful symbol of the New Deal's concern for black Americans.

The Roosevelt administration's response to these forces for change convinced many black Americans that something important had changed in the federal government's attitudes toward its black citizens. The elections of 1934 and 1936 demonstrated the shift in attitudes of blacks toward the federal government and the New Deal. The 1934 elections marked the first time a majority of the black electorate voted Democratic, and in the 1936 presidential election one poll reported that over seventy-five percent of northern blacks cast their ballots for Roosevelt.[29] Blacks recognized, of course, that the New Deal had not addressed all of their grievances, but many were convinced by 1936 that the federal government had finally acknowledged their problems and was making a sincere effort to solve them. The black electorate's overwhelming support for Roosevelt and the New Deal in 1936 demonstrates that a revolution in hopes and expectations had indeed occurred.

By mid-decade most blacks had reached virtually the same conclusions regarding the New Deal that Gunnar Myrdal would reach several years later in his monumental study *An American Dilemma*:

> The New Deal has actually changed the whole configuration of the Negro problem. Particularly when looked upon from the practical and political viewpoints, the contrast between the present situation and the one prior to the New Deal is striking.
>
> Until then the practical Negro problem involved civil rights, education, charity, and little more. Now it has widened, in pace with public policy in the new "welfare state," and involves housing, nutrition, medicine, education, relief and, lately, the armed forces and the war industries. The Negro's share may be meager in all this new state activity, but he has been given a share. He has been given a broader and more variegated front to defend and from which to push forward. This is the great import of the New Deal to the Negro. For almost the first time in the history of the nation the state has done something substantial in a social way without excluding the Negro.[30]

By 1936 the "practical Negro problem" referred to by Myrdal involved all the areas he cited, except the armed forces and the war industries. These, however, came to the forefront of the racial agenda in early 1938 and remained preeminent until the close of World War II.

The sudden emergence of civil rights as a national issue, so unexpected in 1932, and the federal government's unprecedented response, ushered in yet more demands and initiatives. By 1937 a virtual revolution of rising expectations was afoot among many urban, middle-class black Americans, as they became ever more strident in their demands for full participation in the myriad federal programs that had become dominant forces in the economic life of the nation. Inevitably, the issues of participation in the armed forces, a recurring theme in the history of black Americans, and support for aviation training became part of the civil rights agenda of the New Deal era. By 1939 they had become inextricably linked to both each other and the civil rights campaign.

Blacks had sought to gain admittance to the Air Corps since World

War I. In mid-1917 Charles S. Darden, a black from Los Angeles, made an unsuccessful attempt to enter the Air Service.[31] After the war, blacks urged that the Army Reorganization Bill of 1920 be amended to allow qualified blacks to enter the Air Service.[32] Scattered efforts like these had little effect on the resolve of the War Department to limit black participation in the army to the four black regiments authorized by Congress after the Civil War.

Then in 1931 black Americans were shocked to learn that the army planned to reduce drastically the strength of even these units to accommodate an expansion of the Air Corps. Legislation passed in 1926 had authorized a plan to expand the Air Corps in annual increments over the next five years, while keeping the overall strength of the army constant; thus increases in Air Corps personnel required corresponding reductions in the ground forces. For the first four years white ground units provided the spaces necessary for the enlargement of the all-white Air Corps. In the fifth year, however, the War Department decided to reduce the strength of the black regiments to complete the expansion of the Air Corps.[33]

When news of the plan became public in the summer of 1931, black leaders and the black press were unanimous in their condemnation of it and many questioned the policy of excluding blacks from the Air Corps. Walter White, executive secretary of the NAACP, challenged Army Chief of Staff Douglas MacArthur's claim that the army did not discriminate against blacks, asserting that if the army really practiced nonpreferential treatment, it "would result . . . in Negroes being enlisted in the Air Corps and every other service of the Army."[34] Maj. Gen. George Van Horn Moseley, a key member of MacArthur's staff, countered that the black ground units existed only because of congressional authorization, and no similar authorization had been made for black flying units. Moreover, Moseley doubted that enough qualified blacks could be found to staff a segregated squadron should congressional authorization become a reality.[35]

Despite black protests, the army proceeded with its plans to decimate the black regiments and steadfastly refused to accept blacks in the Air Corps.[36] The outcome of this episode highlights the importance of each of the three factors upon which the World War II–era Air Corps participation campaign was founded. The strident protests over Air Corps exclusion in 1931 had no effect on the War

Department because they were primarily the product of the military orientation of the black public. The NAACP's efforts to link the protests to broader civil rights issues failed in 1931 because civil rights had not yet emerged as an issue of national concern. Moreover, claims by the War Department that blacks lacked the requisite interest and aptitude for military aviation could not be refuted as effectively in 1931 as at the end of the decade because the third element, black air-mindedness, had not yet blossomed. The first major controversy over black participation in the Air Corps came, unfortunately, before the exploits of Banning, Allen, Forsythe, Anderson, and Robinson inspired the black public and imbued them with the confidence to challenge assertions that blacks lacked the attributes required to succeed as aviators.

As the civil rights issue began to emerge during Roosevelt's first term, blacks began to consider how the improving racial climate and the changing attitudes of the federal government might help to stimulate black participation in aviation. One of the earliest critics of the practices and policies of exclusion from federally supported aviation activities was the *Chicago Defender*. When the airmail system came under attack for corruption in 1934 and President Roosevelt canceled all airmail contracts, the *Defender* pointed out that black Americans also had complaints against the airmail system: "Why are there no black airmail pilots? What is there about this service that makes it all white. Dark men do just about everything else Americans do. Why can't they fly her mails? We are taxed to support this country, we help pay her bills, maintain her governments, contribute to her general welfare, but we can't take part in her aviation development. It's about time some person in the government authority explained this to us."[37] The *Defender* sounded a theme that was to be heard with increasing frequency as the decade wore on—why were black Americans not allowed to benefit from government programs that promoted aviation?

The first sign that the Roosevelt administration's new responsiveness to racial issues might be used to support black aviation came to public attention in 1936 with the news that William J. Powell—who had originally been forced to rely exclusively on racial self-help schemes—was offering aviation classes under the auspices of the Emergency Education Program, a recovery initiative begun during the first year of Roosevelt's administration.[38] In order to become eligible for the program, Powell had obtained a certificate to teach aeronautics from the California Board of Education and then applied for authorization to offer federally funded aviation courses at a local high school. By 1937 some 250 students, mostly Negroes, had received training under the program. Many continued their training under a private scholarship program sponsored by Craftsmen of Black Wings, a nonprofit, self-help organization established by Powell to promote aviation among blacks.[39]

The Black Cabinet provided another early link between the federal government and black aviation enthusiasts. In the early years of Roosevelt's administration the Department of Commerce appointed the executive secretary of the Urban League, Eugene Kinckle Jones, its adviser on Negro affairs and established a Negro affairs division. The Department of Commerce regulated the nation's aeronautical activities through its Bureau of Air Commerce, and as early as March 1934 Jones reported receiving inquiries regarding aviation. The following year the Division of Negro Affairs began compiling a list of licensed black aviators, and in August 1936 it issued the first of a series of bulletins entitled *Negro Aviators*, listing the names, addresses, license numbers, and expiration dates of all blacks holding Bureau of Air Commerce licenses.[40] The division also conducted a study on black pilots in 1936, which was "not available for general distribution but available to persons specifically interested in this phase of Negro development."[41] The fact that the Department of Commerce conducted a study of black fliers and maintained a list of black aviators was widely reported in the black press and gave the impression that the federal government was taking an active interest in the development of civil aviation among blacks.[42]

In May 1938 the airmail issue resurfaced when two black fliers, Dr. Theodore Cable and Grover Nash, received temporary commissions to fly the mail in celebration of National Air Mail Week. Nash flew an Illinois route from Chicago to Charleston via Mattoon.[43] Cable, an Indianapolis city councilman and Democratic candidate for the Indiana state legislature, also flew an intrastate route, from Indianapolis to Greencastle; the flight had been arranged at the behest of G.N.T. Gray and Percy L. Hines of the National Postal Alliance, an organization of black postal workers.[44] Like many of the New Deal's racial initiatives, the airmail flights were symbolic rather than real advances; from the perspective of the 1990s they are at best simply gestures of goodwill and at worst clumsy attempts to divert criticism while undertaking no substantial reforms. But in the context of racial relations in the 1930s, even symbolic gestures were welcomed as improvements over pre–New Deal practices. Such measures were widely seen as harbingers of racial progress, stimulating the black press and civil rights organizations to press even harder for equal opportunity and racial justice.

Thus by 1938 an increasingly air-minded black public began to hope that the federal government's new racial agenda might indeed bring African Americans a New Deal in the air. A growing interest in aviation and the emergence of civil rights as a national issue had, however, resulted in little more than limited support for aviation education, the collection of data about black pilots, and airmail publicity flights by private black pilots. Air-mindedness and civil rights agitation alone were not sufficient to galvanize black America and produce a public pressure campaign of sufficient intensity to break down the barriers of racial exclusivity which were keeping the Air Corps all white.

In 1938, however, another element of the Air Corps participation campaign, the military orientation of the black public, came to the forefront after the *Pittsburgh Courier* began a campaign for equal opportunity in the armed forces. The *Courier's* initiative was timed to take full advantage of the rising optimism and expectations of blacks during Roosevelt's second term. Almost immediately after it launched the campaign, in February 1938, blacks rallied to the cause in an unprecedented display of public concern. By the end of the year the black public's traditional interest in military participation had been reinvigorated, and because it drew strength from the emergence of civil rights as a national issue, it was stronger and more vital than ever before. Finally, all three components of the Air Corps participation campaign—the military orientation of the black public, a growing air-mindedness among blacks, and the emergence of civil rights as a national issue—were present in sufficient strength to crystallize public opinion and launch a public pressure campaign that would ultimately result in the establishment of separate black flying squadrons in the nation's air arm.

Prior to 1938 both the black press and civil rights organizations, particularly the NAACP, had complained about both the treatment of blacks in the military and the policies that restricted them to the black cavalry and infantry regiments. The NAACP had repeatedly called for an end to discriminatory employment practices in federal, state, and local governments. Resolutions and statements that attacked discrimination in government employment issued by the association generally criticized such abuses across the full spectrum of government service; they did not single out the armed forces for special criticism, but simply noted their conformance to the general pattern of discrimination. Typical was a resolution proposed at the NAACP's annual conference in 1936, which condemned the "systematic discrimination in employment because of race or color by the various federal and state governmental departments, including the army and navy and the National Guard."[45] In 1934 Charles H. Houston, legal counsel for the NAACP, visited the posts at which the black regulars were stationed, reported that the army had reduced the remnants of the black regiments "to the practical status of service battalions," and warned the War Department that blacks were becoming increasingly intolerant of such actions.[46]

Such abuses were also noted and widely reported by the black press. The *Pittsburgh Courier*, under the leadership of the noted black journalist and politician Robert L. Vann, had been particularly vigilant in monitoring status of the black regulars.[47] Vann, who by 1936 had built the *Courier* into the nation's leading black weekly, had a long-standing interest in the role of blacks in the armed forces. During the 1920s he had lobbied for the erection of a monument to the black soldiers of World War I. In 1934, at Vann's direction, one of the *Courier*'s reporters investigated the army's use of the black regulars and reported that men classified as combat troops were instead serving as orderlies and stable hands. By 1935 Vann had become influential in Pennsylvania politics and crusaded for the establishment of two black infantry divisions in the state's National Guard.

In February 1938 Vann realized that conditions were favorable for an all-out attack on the federal government's policies toward black participation in the armed forces. The previous October President Roosevelt had acknowledged in his "Quarantine the Aggressors" speech in Chicago that the situation in Europe and Asia was becoming critical—Germany was openly rearming, a civil war raged in Spain, and Japan had invaded China.[48] An expansion of the armed forces was becoming more and more likely, and judging from current army practices, Vann feared that blacks would be relegated to service in labor battalions on an even larger scale than in World War I. Moreover, his political acumen made him acutely aware of the new currency of civil rights issues in national politics. Thus when he announced his campaign in the 19 February 1938 issue of the *Courier*, he gave shape and direction to what had formerly been an inchoate movement. He seized the initiative and precipitated the coalescence of civil rights issues with the black public's traditional concern over military participation three weeks before Hitler annexed Austria and treated the world to an impressive display of military strength that announced Germany's return as a formidable military power. Roy Wilkins, assistant executive secretary of the NAACP, acknowledged Vann's superb timing in a letter to Walter White; Wilkins advised against challenging the *Courier*'s call for the formation of a separate Negro division: "Although we began the agitation years ago and have an excellent record on the question, the fact remains that the *Courier* at *the [opportune] psychological moment* whipped up the enthusiasm of the country into a campaign which has assumed such proportions that this Association could not issue any statement which would seem to be 'cold-watering' the *Courier's* crusade. I think you will realize how unpopular that would make us."[49]

Vann announced the campaign with bold headlines that proclaimed "COURIER CAMPAIGNS FOR ARMY, NAVY RECOGNITION" and with an open letter to President Roosevelt.[50] Citing the war record of black Americans, Vann declared that "the traditional loyalty of the American Negro remains unchanged. He wants to continue and to add to the service which has distinguished him in all our country's wars." Vann asked the president to use his "great influence to create greater opportunities for Negro citizens in America's fighting forces," noting that although blacks made up ten percent of the nation's population, "only one American fighting man in every 33 is black." Yet even this small number, Vann pointed out, "has been restricted in recent years, to a few combat troops and to quasi labor units. Even Negro combat troops have been made to feel that they are the 'domestic servants' of the army in peace time. This tends to stifle patriotism." Vann urged the president to institute a

MAY   1938   PRICE 15c

"Editor Vann: Please, Please Don't Stop *because* My Future Depends Upon Men Like You"

William J. Powell's Los Angeles organization, Craftsmen of Black Wings, published the *Craftsmen Aero-News* in the late 1930s. The May 1938 issue encouraged Robert L. Vann, editor of the *Pittsburgh Courier*, to continue his campaign for the participation of black Americans in the armed forces, including the Air Corps. (Courtesy Tuskegee University Archives, Tuskegee Institute Aviation Collection) four-point program to provide black Americans wider opportunities in the nation's armed forces that included the following recommendations:

1. That provision be made for increased enlistment of Negro citizens in the United States navy and that the color bar against Negro seamen be destroyed.

2. That openings be made for Negroes in the air corps of the Army and the Navy. The Negro is as daring a soldier as any man.

3. That steps be taken at this time for the formation of an entire division of Negro combat troops, composed of all the customary services.

4. That opportunities be provided for the training of Negro officers of such a division.[51]

Although Vann's first two points related to naval and air forces, the primary objective of his campaign throughout 1938 centered on the third and fourth proposals, the formation of an all-black army division and the training of officers to command and staff it. Thus the *Courier* campaign did not focus on the Air Corps, but did acknowledge and condemn the exclusion of blacks from it. In the months following the inauguration of the campaign, Vann kept the issue of black participation in the armed forces before the black public by publishing letters on the subject from a wide array of prominent blacks and whites—newspaper editors, civic and religious leaders, college presidents, congressmen, and state and local politicians. Their opinions were solicited in a survey letter in which Vann asked: "Do you believe that all branches of the army and naval service should be opened to Negroes (they are over 99 per cent native born)? Or do you think there should be an entire Negro division, including all arms of the service and officered, at least in the line, by educated colored men, in the army; and a squadron manned by Negroes in the Navy?"[52]

The responses varied widely, of course, depending on the background of the respondents, but public opinion generally favored a separate black division. Consequently, Vann and Emmett J. Scott, former secretary to Booker T. Washington and racial adviser to the secretary of war during World War I, sought the support of Cong. Hamilton Fish, a Republican from New York who had served as an officer with a black regiment in France. Together they drew up three bills, which Fish introduced on 5 April 1938, calling for an end to discrimination in the armed forces, the establishment of a black army division, and presidential appointment of two blacks to West Point annually.

For the remainder of the year Vann worked actively for the passage of the legislation, especially the bill calling for a black division. He had two conferences with the president on the matter, one in April and another in October. Although Roosevelt assured Vann of his support, he refused to publicly endorse the Fish–*Courier* bills because the passage of the Republican-sponsored legislation might encourage blacks to abandon the Democratic party and return to the Republican fold. Roy Wilkins considered the issue of an all-black division "so popular and is so recognized for what it is by the Democrats that they do not wish the Republicans, with Ham Fish's sponsorship, to get the credit for this move guaranteed to please great numbers of colored people."[53] The failure to secure presidential support for the legislation kept the bills buried in the Military Affairs Committee until December, when they died without action.

Thus Vann failed to achieve any tangible result during the first year of his campaign; it did, however, serve a vital function. As an initiative sponsored by the largest and most important black newspaper of the day—read by educated blacks across the nation—it focused the attention of an influential segment of the black public on the issue of participation in the armed forces. Although initially broad in scope, after April the campaign became preoccupied with the goal of securing an all-black army division; related issues, such as Air Corps participation, received only token attention. By December, however, as it became obvious that the Fish–*Courier* bills stood little chance of passage, there were signs that the intransigent Air Corps had not escaped the notice of an impatient black public, mobilized by the *Courier's* exhortations and anxious to see its youth afforded the opportunity to prepare itself for the much-anticipated air age.

Advocates for black participation in the armed forces began to shift their attention to the Air Corps in the fall of 1938, after President Roosevelt announced his support for a sevenfold increase in its strength when he called for a 10,000-plane air force.[54] Headlines soon appeared in the *Pittsburgh Courier* asking "Will A Negro Ever Fly Them?" over an article which asserted that twenty-five cents out of every tax dollar blacks paid in federal taxes would go toward the expansion of the Air Corps. "Since Negroes are helping to build and maintain these war machines," the article cynically concluded, "we are beginning to wonder just what we are getting out of it."[55] Mabe Kountze, Boston correspondent for the Associated Negro Press, noted that experienced blacks such as John C. Robinson and Willie "Suicide" Jones, who had recently completed a world record free-fall parachute jump, were denied opportunities in the Air Corps.[56] The NAACP once again challenged the Air Corps' exclusion of blacks and protested to the War Department after an otherwise qualified black applicant was rejected because of his race.[57]

By end of the 1930s a cadre of black Americans were poised to launch the campaign for Air Corps participation. Consequently, when bills came before Congress in early 1939 to expand the Air Corps and establish a federally funded program for training civilian pilots, black aviation enthusiasts made sure there were spokesmen on Capitol Hill to represent the interests of black Americans and influence the legislative process. Equally important, these black air pioneers had opened the eyes of a new generation of black youth to the possibilities in the air. Perhaps the most compelling testimony to their contribution was the ambition of a young black American who declared in the late 1930s: "I intend to fly some day at any cost. I'd like to . . . study under that fellow who flew for Haile Selassie in Ethiopia. That's the kind of life I like."[58]

The Aviation
Legislation of 1939

By the beginning of 1939 the United States began to emerge from the shell of isolationism that characterized the interwar years. Many Americans viewed with increasing alarm the course of world events after Italy annexed Ethiopia. Since the summer of 1936, civil war had raged in Spain, with Germany and Italy supporting the forces of Gen. Francisco Franco in their bid to oust the Soviet-supported Popular Front government, which had taken control in elections earlier in the year. Throughout the decade of the 1930s, tension between Japan and China was high; in July 1937 the Japanese launched a large-scale invasion into northern China, which by 1939 showed no signs of abating. By 1938 the European situation had deteriorated further. In March Hitler annexed Austria, and in the Czechoslovakian crisis the following September the British and French prime ministers capitulated to his demands that the Sudetenland be annexed to Germany.

One consequence of this rising tension in international relations was a growing public sentiment in the United States in favor of strengthening the nation's air defenses, viewed by many Americans as a favorable alternative to developing a credible land force. In November 1938—after British and French leaders ceded the Sudetenland to Hitler, largely from fear of German air attacks[1]—

opinion surveys of the American public showed that an overwhelming majority supported an expansion of the nation's air arm; a Gallup poll released in early 1939 reported that ninety percent of the respondents favored an immediate expansion of the air force.[2]

The American public's desire for a stronger air force was matched by President Roosevelt, who emphatically indicated he favored such measures. By mid-September 1938, in the aftermath of Hitler's inflammatory speech at the annual Nuremberg rally, one of President Roosevelt's closest advisers, Harry Hopkins, had concluded that "the President was sure then that we were going to get into war and he believed that air power would win it."[3] In October, after the president spoke publicly about the need to rearm and produce more military aircraft, the War Department began drafting plans for an expanded Air Corps. On 14 November Roosevelt met with a group of military and civilian advisers that included Secretary of War Harry H. Woodring, Maj. Gen. George C. Marshall, who would become army chief of staff in 1939, and Maj. Gen. Henry H. "Hap" Arnold, recently appointed chief of the Air Corps. The president declared that he wanted "airplanes—now—and lots of them." He initially insisted on 20,000 airplanes but finally scaled his demand down to 6,000 to accommodate a concurrent strengthening of the ground forces.[4]

In his State of the Union address of 4 January 1939, President Roosevelt emphasized the dangers to international peace posed by aggressors in Europe and Asia and urged all democracies to be prepared.[5] The following day he submitted his budget proposal for fiscal year 1940, which included $1.3 billion for defense, almost fifteen percent of the total budget request.[6] A week later, on 12 January, the president sent a message to Congress requesting a supplemental appropriation of $525 million for an emergency national defense program, with over half earmarked for the expansion of the Air Corps and for an aviation training program to be administered by the Civil Aeronautics Authority (CAA).[7]

President Roosevelt's call for greater federal support for aviation was welcome news to the nation's aviation community, but the prospect of expanded aviation training under federal sponsorship had a special significance to blacks interested in flying. Black aviators had long realized that their chances for progress in the field of aviation were limited as long as they were barred from military aviation, flying the mail, and commercial flying. Thus when new opportunities for government-funded flight training were proposed by the president, black aviation enthusiasts were determined to fight for the rights of black Americans to be included. Congress responded to President Roosevelt's appropriation request with two acts, one that authorized the expansion of the Air Corps and another that established the Civilian Pilot Training (CPT) Program, an initiative to provide federally funded flying training to tens of thousands of college-age youth.[8] When introduced in Congress, neither piece of legislation addressed the issue of pilot training for blacks; however, due to the efforts of black aviators and lobbyists and several supportive legislators, the bills were amended and the final versions of both acts contained provisos aimed at ensuring that black Americans were included in the new pilot training programs.

The civilian pilot training initiative was the work of Robert H. Hinckley, a New Dealer from Utah whom the president had appointed to the CAA in mid-1938, shortly after it was established as the successor to the Bureau of Air Commerce. Hinckley was concerned that the failure of the United States to actively promote aviation training among the nation's youth might have serious consequences, given the vigorous aeronautical training programs under way in Europe. He advocated the inauguration of federally funded aviation training, administered locally through colleges and universities, to foster an interest in aviation among American youth and enhance national defense by providing a pool of pilots who could be quickly trained for military duty in a national emergency. Hinckley's proposal called for the colleges and universities to select the students and conduct the ground training, and then contract with local, CAA-approved flying services to provide the aircraft and conduct the actual flight training. Thus the plan was at once a tool for stimulating interest in aviation, training pilots, and subsidizing local flying services scattered throughout the nation, known in the aviation community as fixed-base operators.[9]

The outlines of the CAA program had been released to the public in late 1938 at a presidential press conference. On 27 December 1938 President Roosevelt announced that the CAA had developed a "plan for the annual training of 20,000 civilian pilots." He described the initiative as a "vocational training project, utilizing existing educational institutions and present facilities of the aviation industry" to provide federally funded flying training to American youth.[10] The

CAA conducted a feasibility test of the Civilian Pilot Training Program during the early months of 1939 and the results were promising. Hundreds of students had applied for the 330 openings at the thirteen colleges and universities selected to participate in the test program, and 313 successfully completed the course.[11]

In mid-March 1939 legislation was introduced in the House of Representatives to authorize the establishment of the Civilian Pilot Training Program on a permanent basis.[12] At hearings before the House Committee on Interstate and Foreign Commerce on 20 and 27 March, CAA chairman Robert Hinckley explained the objectives of the program. He hoped that it might eventually serve as a substitute for the first phase of Air Corps flight training, arguing that the "training will be the equivalent of what is done in the primary courses at . . . the Army Training Center."[13] Hinckley subsequently explained, however, that the CAA was "not attempting to train military pilots. We do hope to create a reservoir from which there can be drawn either currently or in time of emergency young men who have the equivalent of primary training and start them out in the basic training, and from there [the Air Corps] can take them on to the advanced."[14]

Throughout the hearings, the relationship of the program to ROTC programs and the Air Corps was repeatedly discussed. Representatives of the Air Corps were careful to point out that they considered the CAA training valuable as a screening device for prospective aviation cadets but not as a suitable substitute for bona fide Air Corps flight training under their direct supervision.[15] On 27 June 1939 the CAA's pilot training program was authorized with the approval of the Civilian Pilot Training Act of 1939; funds for the program were appropriated in August.[16]

Less than a week after President Roosevelt's 12 January 1939 message, the Senate Military Affairs Committee opened hearings to consider the military aspects of his recommendations.[17] In early February a bill implementing the president's recommendations was introduced in the House of Representatives, and on 3 April 1939 it was approved as Public Law (P.L.) 18, "the primary legislative authorization for the Air Corps expansion program."[18] Public Law 18 authorized a maximum Air Corps strength of 6,000 airplanes, a significant increase considering that the total air strength came to only 1,401 in mid-1938.[19] The law also authorized the army to expand its pilot training program by permitting the use of facilities at civilian flying schools for portions of the Air Corps flight training curriculum.[20]

Gen. "Hap" Arnold, chief of the Air Corps, had begun to lay the groundwork for an expanded pilot training program in late 1938. Confident that both the public and Congress would endorse a stronger air force, Arnold had contacted civilian flying schools about the role he hoped they would play in the expansion. While the president thought in terms of "airplanes—now—and lots of them," Arnold knew that training crews to fly them would be a major undertaking which would require a change in the system of training military pilots that the Air Corps had relied on since World War I.[21]

During the interwar years, Air Corps aviation cadets had earned their wings at Kelly and Randolph fields, two Army flying fields at San Antonio, Texas, which comprised the Air Corps Training Center. The low staffing of the Air Corps between the wars had allowed high selectivity in recruitment and the emphasis of quality over quantity in the training of Air Corps pilots. The attitude of Air Corps training officers comes through clearly in the Air Corps' nickname for Randolph Field—"the West Point of the Air."[22] The system of flight training that had developed by 1939 consisted of four phases. In the initial phase, "primary training," the student learned the rudiments of flying in light aircraft, with little emphasis on military flying procedures. Primary students who showed an aptitude for flying continued to the second phase, "basic training," where they were introduced to military flying procedures in heavier, more powerful training aircraft. In the third phase, "advanced training," the aviation cadets perfected their military flying skills in high-performance trainers whose flying characteristics were similar to those of combat aircraft. On completion of advanced training the cadets received their wings and commissions, and then entered the fourth and final phase, "transition training," in which they qualified in the combat aircraft to which they were assigned for duty.[23]

When it became obvious in the fall of 1938 that the Air Corps' demand for pilots would soon exceed the capacity of the training center in San Antonio, Air Corps training officers and planners, anxious to preserve the quality of their training program, proposed no modifications in the training system. Instead, they urged that additional centers identical to the one at San Antonio be established to meet the pilot training requirements of the expanding Air Corps. General Arnold overruled his staff, however, concluding that it was not feasible or economical to establish and develop additional "West Points of the Air." He believed that the most expeditious and efficient way to expand the training capacity of the Air Corps was to use CAA-approved civilian flying schools for the first phase of training; this would allow military flying instructors to focus their efforts on subsequent phases of training where their expertise as military aviators was most needed. In October he discussed his ideas with representatives of the nation's top civilian flying schools, and together they developed the outlines of the primary training system that was used throughout World War II. While General Arnold warned the school representatives that he lacked formal authority and funding, he expressed confidence that congressional support would be forthcoming. He urged each to "go out and set up at his private school the facilities to house, feed, and train flying cadets for the Army Air Corps." By the end of November Arnold's staff completed surveys of the nation's civilian flying schools and concluded that by using the combined facilities of CAA-approved civilian flying schools and the Air Corps Training Center, over four thousand cadets could be trained in two years, more than doubling the normal pilot production rate of the San Antonio training center.[24]

As the details of the CAA's proposal for training civilian pilots and the Air Corps' plan to use CAA-approved civilian flying schools for its primary training unfolded in early 1939, it became increasingly difficult to differentiate between the two initiatives. The problem began with the president's 12 January 1939 message to Congress; he did not clearly indicate that the $10 million aviation training program that he advocated was indeed the CAA proposal he had announced on 27 December. He simply declared that "national defense calls for the training of additional air pilots," with the training "primarily directed to the essential qualifications of civilian flying." The funds requested, he noted, would provide "*primary* training to approximately 20,000 citizens."[25] Since the president's message focused exclusively on national defense, he failed to point out specifically that the $10 million he requested for pilot training would go to the CAA; thus he inevitably gave the impression that his call for expansion of the Air Corps and his proposal to train 20,000

pilots were somehow related. This impression was perhaps reinforced when he used the term *primary training*—which also applied
to the first phase of Air Corps flight training—to describe the CAA's
program. As the president's recommendations were debated in
Congress during the early months of 1939, the confusion that arose
from his vague explanation of the CAA's program became apparent.

Outside observers who tried to fathom the details of the two
programs often concluded, from the similarity in terms and the
sometimes imprecise language of the insiders, that the two programs
were one and the same or that there was a direct connection
between them. For most Americans outside the aviation community
and the War Department, confusion over the distinctions between
the programs of the CAA and the Air Corps was of little real
consequence. But for blacks seeking to dismantle the racial restrictions that barred them from the Air Corps and limited their opportunities in civil aviation, the difficulties in obtaining accurate and
precise information on the new aviation initiatives proved a significant problem. Consequently, the effectiveness of black spokesmen
and lobbyists who sought to influence pending legislation or urge
policy-makers to include blacks in the new programs was limited.
After the Civilian Pilot Training Act was passed in mid-1939, the
problems of distinguishing between CAA training and bona fide Air
Corps training persisted; on occasion those who opposed the idea of
admitting blacks to the Air Corps found it expedient to exploit the
situation. The confusion over the CPT Program and the Air Corps
remained a problem even after the air arm was opened to blacks,
primarily because of the close relationship between Tuskegee Institute, which conducted CPT courses and also trained black aviation
cadets for the army as a civilian primary contractor, and the nearby
Tuskegee Army Air Field built for the black Air Corps units.

Shortly after President Roosevelt's 27 December announcement of
the CAA's plans for a pilot training program, the White House began
to receive letters urging that blacks be included in the new initiative.
A southerner wrote a tactful letter to the White House suggesting
that black land grant colleges in the South be allowed to participate
in the program.[26] A northern correspondent was more indignant. On
4 January 1939 Helene Weissman of New York City wrote the
president to "protest the statement made by your Secretary of War to
the effect that there is no room for the Negro people in the new

aviation classes to be set up in American Universities." Comparing the plight of African Americans to that of the European Jews under Hitler and Mussolini, Weissman urged the president to use his influence to "put a stop to the discrimination planned against the Negro race."[27] The letter was referred to William J. Trent, Jr., a black serving as an adviser on Negro affairs in the Department of Interior. Following his inquiries to the CAA and the War Department, Trent curtly advised M. H. McIntyre, secretary to the president, that he found it "impossible to make adequate reply to the complaint of this person in view of the fact that the matter which she protests against seems to be admitted by the authorities concerned," prompting a member of the White House staff to ask McIntyre "What shall we do now?"[28] Apparently the matter was referred to the CAA for study, judging from subsequent replies to similar inquiries from the presidents of two leading black colleges, Wilberforce University and Hampton Institute.

In January 1939 Pres. D. Ormande Walker of Wilberforce University, wrote President Roosevelt to comment on the expansion of the Air Corps and the CAA's aviation training program. He asked the president to require the War Department to make "some provision for the training of Negro youths as pilots," noting, incorrectly, that the president's 12 January message to Congress called for the training of "20,000 young men for service in the *aviation corps*" in cooperation with the nation's colleges and universities. He concluded by suggesting that Wilberforce was a suitable site for the training of black pilots for the Air Corps.[29]

Shortly after President Roosevelt's 12 January message to Congress, Arthur Howe, president of Hampton Institute in Virginia, wrote to the White House and enclosed a letter from an organization at Hampton that expressed "what thousands of other young Negro students have been thinking about recently." Grady P. Anderson, writing on behalf of the members of the Hampton Aviation Club, endorsed the president's plan to train "20,000 pilots annually as part of our national defense program" and called for "equal opportunities for training and advancement in the aviation branch of the national defense." Taking their cue from the broader campaign for full participation in the armed forces that had been under way for almost a year, the Hampton Aviation Club asked that the "question of race be forgotten and the more important questions of rearmament and preparedness be given consideration and the 12,000,000 Negro Americans be included in this consideration." Grady concluded with a plea that an aviation training unit be established at Hampton. He cited the school's outstanding record in the field of black education, its close proximity to Langley Field, a major Air Corps installation, and the patriotism and loyalty of black Americans.[30]

When the letter from Hampton Institute arrived at the White House in late January, the president's staff inquired directly to the CAA about the matter. The CAA had dodged the issue when it responded to Trent's inquiries, sending its standard reply describing the testing period and advising that once the program was established on a permanent basis all interested educational institutions would be invited to apply for participation. This noncommittal reply had led Trent and at least one member of the White House staff to conclude that applications from black colleges would not be welcome. Trent's inquiry, however, had evidently prompted the CAA to consider further the question of black participation. Thus when Col. Edwin M. Watson, the president's military aide, asked Robert Hinckley to provide information for a reply to President Howe, Hinckley told him that "this is one of the matters that has already been called to our attention." He explained that the question of black participation was being studied for the CAA by Blake R. Van Leer, dean of Engineering at the University of North Carolina, and he was certain that qualified black colleges would be permitted to participate in the program.[31]

Hinckley's letter to Watson formed the basis of the reply to the presidents of both Hampton and Wilberforce. On 21 February 1939 McIntyre advised Howe that the training of black pilots was under study by the CAA.[32] He also replied to Walker on the same day, assuring him "that the question of allowing Negro youths to compete as pilots is receiving the most serious consideration. As you know, the President is very much interested in this development."[33] McIntyre did not disabuse Howe and Walker of their misconception that the CAA's goal of training 20,000 pilots was for the training of Air Corps pilots, probably because he himself did not fully understand the nature of the CAA's program. Nothing in McIntyre's response suggests that he intended to mislead the two college presidents. Nevertheless, his reply no doubt convinced Howe and Walker that the question of black participation in the Air Corps—the issue about which they had written—was being investigated. Thus the inherent confusion over the distinctions between CAA pilot training and Air Corps pilot training had begun to cloud the issue of black participation in the nation's air arm. By late 1940 the matter would come to a head and indignant blacks would accuse both the Air Corps and the Roosevelt administration of intentionally exploiting the situation in an effort to keep the Air Corps all white.

These initial responses to President Roosevelt's call for an expanded federal aviation program demonstrated that some black Americans had become air-minded and were anxious to participate in the expansion of the air arm, but they did little else. Of greater significance were the efforts of black lobbyist Edgar G. Brown described above. He appeared before the Senate Military Affairs Committee in late February to urge that the bill ultimately approved as P.L. 18 be amended to allow black Americans the opportunity to serve as pilots in the Air Corps. Brown, who had not previously shown an interest in aviation, was apparently influenced to appear before the committee by Emmett J. Scott, the conservative black leader who had collaborated with Robert L. Vann the previous year to seek congressional authorization for a black army division, and Dr. Theodore Cable, a black Democratic politician from Indiana and a private pilot who had flown an airmail publicity flight the previous year.[34]

Brown began his testimony by reminding the committee of black Americans' long record of military service, a record which entitled young blacks to "a new deal in the United States Army Air Corps." He asked the committee to add a proviso to allow "all American youth, regardless of race, creed, or color, to be employed in the building and flying of these ships [i.e., the new airplanes to be added to the inventory of the Air Corps], and to serve their country as Reserve officers and enlisted personnel, proportionately, in the 45,000 expansion for the United States Regular Army Air Corps." He recommended that the secretary of war be empowered to "designate and authorize aircraft training schools" at Howard University, Wilberforce University, Tuskegee Institute, and Hampton Institute, apparently convinced that the goal of the CAA's plan to train pilots at colleges and universities was to provide pilots for the Air Corps.[35] Despite strong Democratic representation, the committee took no action on Brown's recommendations and reported the bill to the floor of the Senate the following day, 22 February, without an amendment relating to black participation in the Air Corps.[36]

After the bill came out of committee, however, three amendments dealing with pilot training for blacks were considered, and the third was ultimately incorporated into P.L. 18. The first was introduced by Sen. H. Styles Bridges, an archconservative, anti–New Deal Republican from New Hampshire. The second, a weaker proposal, was offered by Sen. H. H. Schwartz—a New Deal Democrat from Wyoming—as an alternative to Senator Bridges's amendment. When the Senate voted on the two competing amendments, Schwartz's amendment carried. The third and last amendment was a weaker version of the Schwartz amendment, which was prepared by the House–Senate conference committee responsible for resolving the differences between the House and Senate versions of the bill. This revised version of the Schwartz amendment was approved by both houses of Congress and became part of P.L. 18. Unlike the original amendment proposed by Bridges, the final, watered-down proviso relating to the training of black pilots did not specifically mandate the admission of blacks to the air arm. Consequently, the Air Corps was able to maintain its policy of racial exclusion for two more years.

Senator Bridges introduced his amendment on 1 March. It mirrored the recommendations that Edgar Brown had presented to the Military Affairs Committee in February. Senator Bridges, a minority member of the Military Affairs Committee, proposed that the secretary of war be authorized to "establish at appropriate Negro colleges identical equipment, instruction, and facilities for training Negro air pilots, mechanics, and others for service in the United States Regular Army as is now available at the Air Corps Training Center."[37] His proposal also stipulated that there was to be no exclusion or discrimination on the basis of "race, creed, or color" in the training programs to be established under the bill.[38] Perhaps seeing an opportunity to embarrass the Democratic administration and attract black voters back to the Republican party, Senator Bridges's amendment was clearly intended to open the air arm to blacks. Although his proposal was apparently based on the erroneous assumption that the Air Corps intended to use colleges and universities to train aviation cadets, it nevertheless clearly stated that blacks should be permitted to serve in the Air Corps as part of the regular army.

Before the Senate debated Bridges's amendment, Senator

Schwartz—also a member of the Military Affairs Committee—introduced his proposal. He explained that he had been absent when Brown appeared before the committee and thus felt justified in offering an alternate amendment, as the matter had only recently been brought to his attention. Schwartz proposed that the section of the bill in which the Air Corps was authorized to use civilian aviation schools for primary training be amended to require the secretary of war to designate "at least one Negro school for the training of Negro air pilots." He suggested four black colleges—Howard, Wilberforce, Tuskegee, and Hampton—as likely sites for such training, which indicates that he too believed the CAA-sponsored training to be offered at colleges and universities was intended to prepare pilots for duty in the Air Corps. Schwartz's amendment differed from Bridges's in several respects. It did not address the issues of racial discrimination and exclusion in the training programs to be established under the bill's provisions. Perhaps the most significant difference between the two amendments, however, was that the Schwartz proposal failed to state specifically that the black pilots to be trained under its auspices would ultimately see duty in the Air Corps.[39]

Schwartz, a Democrat, had no doubt offered his substitute to the Republican-sponsored amendment in order to head off any appearance of Democratic intransigence on the issue of black participation in the Air Corps. To make his proposal more palatable to his party's Southern wing, he probably felt compelled to make its language weaker, although his subsequent statements on the matter show that he was indeed attempting to secure opportunities for blacks in the Air Corps.[40] Nevertheless, Senator Schwartz assured his Senate colleagues that his amendment did "not touch any of the controversial matters that have arisen from time to time in the Congress relating to the Negro race."[41]

Senator Bridges maintained that his amendment was more comprehensive and intended to accomplish the same end as the Schwartz proposal. Schwartz countered that his measure was "entirely satisfactory to Edgar G. Brown . . . and the leaders of the colored race, who desire to have this form of amendment. . . . All I am trying to do, without seeking to disturb social conditions here, there, or elsewhere, is to secure an affirmative, plain, simple declaration that at least one qualified colored college shall be given an opportunity to educate colored men as air pilots."[42] Brown may have realized that the Schwartz amendment was weaker, but probably thought that a milder, Democratic-sponsored proposal had a better chance of passage.

Following a round of parliamentary maneuvering, both amendments were put before the Senate for consideration, but not before another member of the Military Affairs Committee, Marvel Mills Logan, a New Dealer from Kentucky, commented on the issue.[43] Logan praised the loyalty and service of black Americans in the nation's wars, observed that a number of blacks had qualified as aviators, and asked that the Department of Commerce list of black pilots be printed in the *Congressional Record*.[44] When Bridges asked pointedly whether any of the aviators on the list were military pilots, Logan admitted that none "appear to be in the Regular Air Service of the United States."[45] Logan then outlined what he considered the appropriate procedure for providing training for blacks who applied for the Air Corps:

> If I am correctly advised, the Civil Aeronautics Authority designates certain schools, and the War Department accredits those schools. At present I am advised . . . that no institutions of learning for Negroes have been designated by the Civil Aeronautics Authority. . . .
>
> If either of the amendments should be adopted—and I do not think they differ materially, except that I think the Senator from New Hampshire has gone into the matter a little more fully—then such an institution as Howard University, or Wilberforce College, in Ohio, which has an R.O.T.C. unit, or Tuskegee Institute, in Alabama, or Hampton Institute, at Hampton, Va.—any of these could be designated by the Civil Aeronautics Authority and accredited by the War Department. Then, when Negroes asked for permission to enroll, they could be sent to one of those institutions. In my humble judgment, we ought to recognize this right, and I hope that one of these amendments will be agreed to.[46]

Like Schwartz and Bridges, Senator Logan had obviously failed to grasp the distinction between the CAA plan to train civilians using the facilities of the nation's colleges and universities and the Air Corps plan to train aviation cadets using the facilities of CAA-approved civilian flying schools. This was typical of the misunderstandings over the two initiatives, even though General Arnold and a CAA spokesman had taken great pains to explain the differences between the two programs.[47] Senators Schwartz, Bridges, and Logan all assumed that colleges were acceptable sites for training Air Corps pilots, even though General Arnold had made it clear in his statement before the committee that the Air Corps was not in favor of training its pilots at colleges, not even those with ROTC detachments.[48]

Senator Logan's remarks concluded the debate. The Senate voted on both amendments, and the weaker Schwartz amendment carried.[49] It had all the flaws of the Bridges amendment—it attempted to place training of black pilots for the Air Corps in colleges, in contrast to standard Air Corps practice—and none of its strengths: a definite statement that the purpose of the training was for service in the Air Corps.

Officials in the War Department, however, found the Schwartz amendment more to their liking. Immediately after Bridges introduced his amendment—but before Schwartz's alternative was proposed—a member of General Arnold's staff had cautioned against the measure "not only because it is superfluous but, also, because it attracts attention to the authorization and might result in political pressure being directed against the Secretary of War."[50] But when Secretary of War Harry H. Woodring sent a letter to the chairmen of the House and Senate military affairs committees commenting on the various amendments approved by the Senate on 7 March, he voiced no objections to the Schwartz amendment, perhaps satisfied that it had prevailed.[51]

The House and Senate each passed its own version of the bill, so a joint conference committee was appointed to resolve the differences and prepare a mutually acceptable bill.[52] By late March, the committee had completed its work. The committee's report contained the bill's third and final provision for the training of black pilots, the one which ultimately appeared in P.L. 18. Instead of requiring the Air Corps to select "at least one *Negro school* [i.e., a black college] for the training of Negro air pilots," the House–Senate conferees recommended that at least one of the *civilian primary schools* "be designated by the Civil Aeronautics Authority for the training of any Negro air pilot."[53] The ostensible reason for the change was, no doubt, to remove the Schwartz amendment's inappropriate reference to black schools, which would have required the Air Corps to train pilots at a college, a departure from normal flight training procedures. The conference committee's modification deleted the reference to black colleges and stipulated that one of the civilian primary schools be "designated" as a training site for blacks. Both the House and the Senate accepted the committee's recommendations, and the bill was signed into law by President Roosevelt as Public Law 18 on 3 April 1939.[54]

Although the conference committee's modification of the Schwartz amendment seems minor, it nevertheless provided a loophole that allowed the Air Corps to continue its policy of racial exclusion. Furthermore, there is evidence which suggests that this was not inadvertent. When the conference committee's report was submitted to the House at the end of March, it contained a statement which apparently indicated that the amendment was not designed to break the color line in the Air Corps. The revised Schwartz amendment, the statement explained, "*Merely* provides for the designation by the *Civil Aeronautics Authority* of one or more schools for the training of Negro air pilots."[55] In effect, the statement seems to be emphasizing the fact that the bill levied no requirement on the Air Corps to admit blacks, the intent of both the Bridges amendment and the Schwartz amendment. This conclusion seems all the more plausible in light of the fact that the statement was not included in the committee's report to the Senate, where it would have been subjected to the scrutiny of Bridges and Schwartz.

The Air Corps leadership understood the significance of the report's statement. The Air Staff had carefully monitored the progress of the bill while it was under consideration in Congress; when they learned that it might require the Air Corps to train black pilots, they prepared a contingency plan for the establishment and training of a segregated black squadron, in case it became necessary to accept blacks.[56] By the time the plan was completed, however, P.L. 18 had been approved, and the chief Air Staff planner noted that "further study of the act by several different individuals on the General Staff and in the C.A.A. has developed the belief that such steps will not be necessary."[57] The Air Staff, capitalizing on the loophole the conference committee had written into the act, concluded that the requirements of the law would be met when the CAA designated one of the civilian primary schools as a school for training black pilots: "The letter of the law would certainly be fulfilled, and it is believed that the spirit would also be fulfilled 100%. *There is absolutely nothing that directs us to enlist negro flying cadets. The original intent was to use the C.A.A. and the matter crept into this bill thru misunderstanding.* By being left in, it assures the Negro of training at a school of such high standards that 'personnel of the Military Establishment are pursuing a course' there."[58] The judge advocate general, chief legal counsel for the War Department, agreed that although the wording of P.L. 18 obliged the CAA to designate a school for the training of black pilots, "no duty . . . [was] imposed by such language on the War Department."[59] In short, the Air Corps believed that although the law required the CAA to designate one of the civilian primary schools as a site for training black pilots, there was no obligation on the Air Corps to accept blacks as aviation cadets.

By the end of May 1939 Air Corps officials communicated their interpretation of the law to Robert Hinckley, chairman of the CAA. He agreed to designate the North Suburban Flying Corporation at the Curtiss Airport at Glenview, Illinois (one of the civilian schools that conducted primary training for the Air Corps) as a school for "the training of any Negro air pilot" under the provisions of P.L. 18.[60] From the perspective of the War Department and the Air Corps, no further action regarding the training of black pilots was required. They hoped that the CAA's designation of the Glenview school would be enough to deflect the criticism of blacks seeking to open the nation's elite air arm to all qualified American citizens, regardless of color.

Lobbyist Edgar G. Brown had a different interpretation of P.L. 18. Acting on the assumption that it mandated the training of blacks as Air Corps pilots, he continued to lobby Congress. In mid-March 1939, shortly after the House–Senate conference committee had emasculated the Schwartz amendment, Brown appeared before the Senate Appropriations Committee. He urged that funds be set aside for the establishment of "a Negro Air Corps training center for the training of Negro pilots to serve in the United States Army as authorized by . . . an amendment to H.R. 3791 [and] agreed to by the conferees of the House and Senate."[61]

Two months later Brown was back on Capitol Hill to appear before the House Appropriations Committee.[62] He had testified alone in February and March, speaking only on behalf of the organization he headed, the United Government Employees. In May, however, he was accompanied by J. Finley Wilson, head of the Improved, Benevolent and Protective Order of Elks of the World, an influential black fraternal organization with over one-half million members. Wilson spoke to the congressmen about the military achievements of black Americans. He declared that "it is traditional with us to be loyal and patriotic." Brown told the committee that, in addition to the black Elks and the United Government Employees, those "interested especially in this air-training program of the United States Army Air Corps" included "the National Airmen's Association, several hundred weekly newspapers, including the Chicago Defender, its publisher, Mr. Robert S. Abbott, who recently sent two Negro Pilots from Chicago with a petition to the Congress; Mr. Robert L. Vann, publisher, Pittsburgh Courier; Dr. C. B. Powell, publisher, New York Amsterdam News; the Afro-American; Washington Tribune; and many others. As a matter of fact, it is really one matter they are all for, 100 percent."[63]

In his statement and in a subsequent letter Brown cited P.L. 18 and offered an interpretation which differed markedly from that of the War Department. He contended the law made it "mandatory on the part of the Secretary of War and the 'Civil Aeronautics Authority to designate one or more schools for the training of any Negro air pilots' *for service in the United States Army*."[64] Citing the Senate debates on the Bridges and Schwartz amendments and the action of the House–Senate conferees, Brown concluded that "the law provides that Negroes be given air-pilot training for service in the United States Air Corps." He told the committee that none of the civilian primary schools had been designated for the training of black pilots; he recommended that the flying school at Glenview, Illinois, "which has already graduated some Negro pilots," serve as the primary training school for black aviation cadets.[65] Recognizing that the Air Corps would be reluctant to send blacks to the training center in San Antonio for basic and advanced training, Brown urged that $10 million be appropriated to establish a separate training center in the South for black aviation cadets, where favorable flying weather prevailed year-round. Brown suggested Florida Agricultural and Mechanical College in Tallahassee, the state's black land-grant college, as a likely site, explaining that the Florida legislature would be willing to provide land for the project. He also suggested Tuskegee Institute as a suitable site, noting that a black West Point graduate—

B. O. Davis, Jr.—was assigned there to oversee the school's ROTC detachment and that a black veterans' hospital was located nearby.[66]

Senator Schwartz endorsed Brown's recommendations for a $10 million training center. He told his colleagues in the House that he had conferred with General Arnold and a key staff officer, Gen. Barton K. Yount, and they expressed reservations about the feasibility of training black pilots for the Air Corps. "Of course," Schwartz confided, "you understand the same as I do, whether we want to admit it or not, that back under this is a feeling in the Army and in the Navy that bringing [in] these Negro pilots and giving them this opportunity will result in some embarrassment one way or another on account of social or economic conditions." Arnold had told the senator that blacks could be given primary training at a civilian school outside the South with minimal difficulty because of the individual nature of the early training. Beyond primary training, however, the cadets flew in squadrons and "there would be a question of whether or not a white mechanic wanted to fly with a colored soldier, or whether a colored officer and a white officer could get along agreeably, or whether the social conditions would be embarrassing." Such problems could be avoided, Schwartz maintained, by following Brown's recommendation that money be appropriated for a separate training center for blacks. A special appropriation was necessary, the senator argued, because "the War Department needs a little urging." Like Brown, Schwartz considered the flying school at Glenview suitable for primary training, but only if separate quarters could be provided for the black cadets, "or else have them placed at one of the big schools such as Tuskegee Institute."[67]

The following month, June 1939, a coalition of New Deal Democrats and Republicans supported a scaled-down version of Brown's recommendation. On 21 June Hamilton Fish, a Republican representative from New York, urged that "some Member of Congress" offer an amendment to the supplemental defense appropriation bill to set aside $1 million to "establish and maintain an adequate training school for Negro pilots in our country . . . as an essential part of our national defense program." Fish's suggestion was supported by Republicans Albert J. Engel of Michigan and D. Lane Powers of New Jersey. Why Fish did not introduce such an amendment himself is unclear; perhaps he wanted to avoid a situation like that of the previous year when, at the behest of black newspaperman Robert Vann, he introduced three bills favorable to black participation in the military that were thwarted by the Democrats.[68]

Later in the day an Indianapolis Democrat, Louis Ludlow, offered an amendment along the lines suggested by Fish. Ludlow, who had not previously advocated black participation in the Air Corps, may have been influenced by Dr. Theodore Cable, the Indianapolis black who had flown an airmail publicity flight in 1938. Cable had served on the Indianapolis city council, and in 1938 he won a seat in the Indiana state legislature, campaigning as a New Deal Democrat.[69]

Ludlow's amendment specified that "Of the appropriations contained in this act for expanding military aviation, including appropriations for both personnel and material, $1,000,000 shall be available exclusively for training Negro pilots." In his remarks to the House, Ludlow noted that he found "nothing in the record of Negroes in war that serves as a reason for refusing to give Negroes their proportionate share of air-pilot training." Although Secretary of War Woodring claimed that the War Department was attempting to develop a program for training black pilots, Ludlow decried the delays in the implementation of the training. He told the House that "land and facilities for these Negro training units are being offered free of charge at Tallahassee, Fla., and Tuskegee, Ala." Finally, Congressman Ludlow acknowledged the cooperation and support of Edgar Brown and the National Airmen's Association in preparing his proposal and his remarks.[70]

For the most part, the ensuing debate on the Ludlow amendment followed sectional lines. Two northern Republicans—Everett M. Dirksen of Illinois and D. Lane Powers of New Jersey—and a northern Democrat—Raymond S. McKeough of Illinois—rose in support. Powers, recalling that the secretary of war had told him the day before that blacks would be trained as pilots, declared that he would support the "earmarking [of] certain funds for that purpose."[71] McKeough predicted that unless funds were specifically appropriated, the provisions of P.L. 18 would be circumvented by the War Department, and "Negroes will be denied the right to be trained as air pilots."[72]

Dirksen pointed out that Ludlow's proposal "is similar to an amendment I have pending on the Clerk's desk" and cited several reasons why he considered such a measure appropriate. First of all, Dirksen explained, it was fair, because blacks constituted roughly twelve percent of the population and the $1 million to be earmarked was about twelve percent of the total amount to be appropriated in the bill. Blacks had served honorably in all of the nation's wars and were therefore "entitled to have equitable consideration free from all discrimination." Moreover, Dirksen declared, "it can be conveniently done," noting that a special training camp could easily be established at Tuskegee Institute:

> At Tuskegee they have a Reserve Officers' Training Corps at the present time, which is officered by a Negro first lieutenant, who is a graduate of West Point, and whose father . . . is the colonel of the Fifteenth Infantry in New York, Col. B. J. [sic] Davis. In the second place, they have a mechanical school at Tuskegee. So this appropriation can be very conveniently adapted for the purpose of training civilian Negro pilots at Tuskegee in order to carry out and effectuate the purposes of section 4 of the act approved in April of 1939 [i.e., P.L. 18].[73]

Dirksen concluded by noting that blacks deserved the consideration provided in the Ludlow amendment because they paid taxes, had distinguished themselves as military officers, and had "shown aptitude for [flying] just like anybody else. There are today 350 Negro pilots in the country."[74]

The opposition to the Ludlow amendment came mostly from southern Democrats; two of the most outspoken were David D. Terry of Arkansas and R. Ewing Thomason of Texas. Terry reminded his colleagues that Secretary of War Woodring had specifically indicated that blacks would be trained as pilots. He pointed out that the bill under consideration funded the training of pilots, not "the training of white pilots or colored pilots or Indian pilots or any particular group, and I am wondering why we should come in here now [and amend] a bill which covers the citizens of this country generally."[75] Thomason, taking his cue from Terry, claimed that the amendment would invite the granting of special privileges to a host of ethnic groups:

> If we are going to give preferential treatment to certain classes and races of our people, I wonder if there would be any objection to including some very fine Mexican citizens in my part of the country

who have wonderful war records and who are fine American citizens? Where are you going to stop if you start this ridiculous policy? All American citizens should be treated alike. If you are going to include the Negroes, why not set aside so much for the Mexicans and the Jews and the Scotch and some of us Swedes? I suppose the Irish will not need help. This is no way to build up a great army, and you are doing a disservice to the Negroes, to whom I am friendly. This sounds like pure and unadulterated politics to me.[76]

Terry echoed his southern colleague's *reductio ad absurdum* attack on the Ludlow amendment with the facetious comment that because he was "partly Irish," he was "disappointed that we have not included a specific sum for the Irish."[77]

Despite the tactics of Terry and Thomason, the preliminary vote on the Ludlow amendment was favorable.[78] Later in the day, however, after Republican James W. Wadsworth of New York spoke against the measure, Representative Ludlow's proposal was defeated by a vote of 43 to 207.[79] The defeat of the Ludlow amendment marked the end of active efforts in Congress for a meaningful implementation of the provisions of P.L. 18 that related to the training of blacks for flying duty in the Air Corps. Those who supported black participation in the Air Corps were more successful, however, in amending the bill ultimately approved as the Civilian Pilot Training Act; their efforts guaranteed that blacks had the opportunity to participate in the CAA's pilot training program.

Although the Civilian Pilot Training Act of 1939 produced more tangible benefits for black Americans than P.L. 18 did, when legislation to establish a program for training civilian pilots was introduced in March 1939, it did not at first address the issue of training black pilots.[80] Similarly, the issue of black participation did not come up at the House committee hearings or at the 20 April hearings before the Senate Commerce Committee on a Senate version of the bill.[81] The day before the Senate hearings, however, Rep. Everett Dirksen raised the question of black participation in the CAA's pilot training program on the floor of the House. He offered an amendment which stipulated that "in the administration of [the CAA's program] none of the benefits of training or programs shall be denied on account of race, creed, or color."[82]

Dirksen's subsequent comments indicate that he too believed that there was a connection between the CAA's pilot training program and P.L. 18, which had been approved two weeks earlier. He asserted, incorrectly, that P.L. 18 prohibited discrimination and required that "all the benefits of a civilian training program should be made available to the Negroes as well." The Illinois congressman went on to extol the record of black Americans in the defense of the nation, from the American Revolution to the First World War. Noting that hundreds of blacks had qualified as pilots, he concluded that "the Negro . . . has an aptitude for flying the same as the members of any other race." Finally, he cited as precedent the various antidiscriminatory clauses included in other bills, such as the Works Progress Administration Act, and called for unanimous approval of his amendment.[83]

Unlike the racial amendments to P.L. 18 and the subsequent defense appropriations acts, Dirksen's amendment prompted little debate. Only the bill's sponsor, California Democrat Clarence F. Lea, spoke against it. Lea noted that the amendment proposed by Dirksen was unnecessary, because "the law authorizes no discrimination on the basis of race." Observing that there were "a few colored educational institutions in this country that may be able to qualify as training centers under this bill," Lea protested that "there is no contribution made to race equality to drag a thing like this forth that has no legitimate or necessary place in the legislation." Following Lea's remarks, Dirksen's amendment received preliminary approval by the House and went forward to the committee charged with resolving the differences between the House and Senate versions of the bill.[84]

What sparked Dirksen's interest in black aviation is unclear, but as an Illinois congressman, he may have been responding to the national organization of black pilots headquartered in Chicago, the National Airmen's Association of America (NAAA). Several weeks after Dirksen introduced his amendment, while the bill was awaiting the final vote of Congress, the NAAA sponsored a nationwide air tour to promote interest in the new federal aviation initiatives. As the two Chicago aviators selected to make the tour were preparing for the flight, one of the NAAA's founders, Enoch Waters of the *Chicago Defender*, suggested that they include a stop in the nation's capital on their itinerary. Waters put the two NAAA pilots, Chauncey Spencer and Dale White, in contact with Edgar Brown, who agreed to serve as the organization's Washington representative. After Spencer and White arrived in Washington, D.C., Brown introduced them to a number of officials, including Congressman Dirksen and Missouri senator Harry S. Truman.[85] According to Spencer, Truman was surprised to learn that blacks were excluded from the Air Corps and he helped to "put through legislation insuring that Negroes would be trained along with whites in the Civilian Pilot Training Program."[86]

The efforts of Dirksen, Truman, and the NAAA proved effective. The civilian pilot training bill, with the Dirksen amendment intact, was approved by the Senate on 15 June 1939 and by the House four days later.[87] When the final version of the Civilian Pilot Training Act of 1939 was approved on 27 June 1939, it contained, word for word, the proviso that Dirksen had proposed in April.[88] The ease with which the Dirksen amendment was incorporated into the final act contrasts sharply with the heated debate and legislative maneuvering that characterized the Bridges and Schwartz amendments, and subsequent attempts to earmark funds to ensure that blacks received training as Air Corps pilots. The Dirksen amendment succeeded where other legislative initiatives failed because it related only to civilian pilot training and did not challenge the Air Corps' policy of racial exclusion.

The inclusion of the revised Schwartz amendment in P.L. 18 and the Dirksen amendment in the Civilian Pilot Training Act were important victories in the campaign for black participation in the Air Corps. They focused the attention of black Americans on aviation and the racial policies of the Air Corps. Even though Air Corps leaders used a loophole in P.L. 18 that allowed them to perpetuate their policy of exclusion for several more years, the law served an important function. It provided black leaders and their supporters in Congress with an opportunity to challenge the War Department on the issue of admitting blacks. The CPT Program was equally important because it provided an opportunity, at government expense, for scores of black youth to learn to fly. The activities of the CPT units at black colleges like Tuskegee Institute, Hampton Institute, and Howard University were widely covered in the black press. The success of these black CPT students made the Air Corps' refusal to accept blacks all the more incongruous. By the fall of 1940 the campaign for admitting blacks to the Air Corps became so intense that the War Department directed the Air Corps to train blacks as military pilots, and in 1941 a segregated black squadron was established at a new Army flying field constructed near Tuskegee, Alabama.

6

The Civilian Pilot Training Program at Tuskegee

Throughout the congressional debates and hearings of 1939, black lobbyists and their congressional allies repeatedly suggested Tuskegee Institute as an ideal site for training black aviation cadets; institute officials, however, made no concerted or systematic efforts in this regard. Unlike the presidents of Hampton and Wilberforce, Tuskegee president F. D. Patterson did not inquire about the possibility of participating in the Civil Aeronautics Authority's pilot training program when President Roosevelt announced it in December 1938 and called for a $10 million appropriation in January 1939. Nevertheless, it was widely assumed that Tuskegee would participate. In late January one of the founders of the NAAA, Chicago pilot Cornelius R. Coffey, wrote Patterson to offer his services as a flight instructor, explaining that he had been "informed recently that your school has been named by the President of the United States as one that will be used by the government for the training of 20,000 youths each year."[1] Patterson advised Coffey that "no definite action has been taken by the United States government in regard to Tuskegee Institute as a training center in aviation at the present time." He told Coffey that "if and when such action is taken" the flier's application would be considered.[2]

The administration at Tuskegee did not take any action until

22 March 1939, the same day that Congress agreed to include the proviso relating to the training of black pilots in Public Law 18. In identical letters to Robert H. Hinckley of the Civil Aeronautics Authority and Harry H. Woodring, the secretary of war, Patterson offered "the facilities of the institution to the Federal Government for such use as it may deem wise in connection with an effort to train Negro pilots and mechanics in the field of aviation." The Tuskegee president claimed that the school's "long record of mechanical training" together with its "shops and landing areas" made it an ideal site for an aviation training program.[3]

The War Department notified Patterson that his letter had been referred to the CAA and made no comment regarding the issue of admitting blacks to the Air Corps. The CAA's response was routine. Hinckley advised Patterson that the criteria for selecting institutions to participate in the Civilian Pilot Training Program, tentatively scheduled to begin in the fall, would not be available until the results of the demonstration phase of the program, currently under way at selected colleges and universities, were analyzed. Hinckley concluded by noting that if the full-scale pilot training program was launched in the fall, institutions wishing to be considered for participation would be required to apply on a standard application form that would be made available.[4]

Several days after Patterson wrote the CAA and the War Department, John C. Robinson renewed his efforts to affiliate himself with Tuskegee Institute. In a lengthy letter to Patterson, he commented on the recent "publicity and agitation . . . carried on in Chicago" regarding black participation in the new civilian and military pilot training initiatives and noted that Tuskegee was frequently mentioned as a possible site for such programs. He enclosed a copy of the CAA's *Air Commerce Bulletin* of 15 March, which described the demonstration phase of the CPT Program, and pointed out that the actual flight instruction was in the hands of private operators, not the colleges and universities themselves. Robinson then recounted the results of his efforts over the last three years to establish his school, noting that he had "assembled over $19,000 of aviation equipment" and had "registered his school with the Civil Aeronautics Authority." Finally, he assured Patterson of his interest in his alma mater and offered to conduct the flying training if Tuskegee decided to participate in the CAA's program.[5]

Patterson's response was brief but encouraging. He explained that Tuskegee had inquired about participating in federally sponsored aviation training but that nothing was certain since "possibly only one unit will be given between the several Negro institutions." He assured Robinson, however, that if the institute was selected for aviation training, he would be "highly interested" in obtaining the flier's services.[6]

In the weeks that followed there was much speculation regarding Tuskegee's role in the training of black pilots for the Air Corps. The editor of the student newspaper announced that a decision was imminent:

> We have the information that any day it might be announced by the War Department that they are ready to establish a unit in some Negro school for the training of Negro college students to be flyers. Tuskegee has been mentioned as one of the schools where such a program will likely be established. The reason for the delay in deciding on which school shall have the unit has been the investigation which has been going on relative to the facilities which the prospective schools offer.
>
> Tuskegee's mechanical facilities, its admirable location in the heart of the deep South where the Negro population is concentrated, its nearness to Fort Benning, Georgia, and its present trades program make it the logical school for consideration.[7]

The paper also observed that the formation of a flying club was "one of the most significant steps taken by Tuskegee students in recent years," concluding optimistically that the organization's activities "may be one of the deciding factors in the determination of whether Tuskegee will be chosen for the important task of training the nation's youth for national defense."[8]

The Tuskegee student body's surge of interest in aviation had been fueled by veteran pilot Charles Alfred Anderson's most recent visit to the campus. In late April, Anderson and E. C. Wright, son of Major R. R. Wright, the founder of the Citizens and Southern Bank in Philadelphia, stopped at Tuskegee on the return leg of their flight to Cuba and Haiti. The flight, sponsored by Major Wright's Haitian Coffee and Products Trading Company, had been undertaken to foster commercial and cultural relations between American blacks, Haitians, and Cubans and to encourage their involvement in aviation. While on campus, Anderson had begun "an organized effort to obtain C.A.A. aid for those students who desire to study aeronautics" under the sponsorship of institute treasurer Lloyd Isaacs and G. L. Washington, director of Mechanical Industries.[9] Student enthusiasm for aviation remained at high pitch after Anderson left because of the efforts of Ralph W. Swaby, a white flier from nearby Columbus, Georgia, who flew to the institute on several occasions to discuss aviation with the students and to give flying lessons.[10]

In late May 1939 John C. Robinson once again contacted officials at Tuskegee Institute with information regarding the training of black pilots.[11] According to Robinson's information, Tuskegee's prospects for participating in federally sponsored aviation training were dwindling. He believed that three factors mitigated against the institute's participation. First, his sources indicated that "no Negro school will be designated in the South." Second, Tuskegee might be declared ineligible because only "schools that have done experimental work in the field of aviation in the past will be granted this training." Third, Robinson warned that Wilberforce University in Ohio was making a strong bid for the program, arguing that its close proximity to Air Corps engineering facilities at Wright Field near Dayton, its northern location, and its ROTC activities made it the ideal site for the training of black pilots. Robinson nevertheless had no doubts regarding the suitability of Tuskegee for such training. He assured the Tuskegee administration that "from my sixteen years [or] more of experience in the mechanical world of which twelve years have been devoted to the specialization of aviation and, from the experience I have gathered from my travels in over eighteen different countries and observing aeronautical schools and training programs that without any doubt, Tuskegee is the most logical place to be designated for Negro training."[12]

Robinson suggested that Tuskegee establish an aviation extension department using his facilities in Chicago. Students selected by the institute could complete their ground training in Alabama, travel to Chicago for flight training at the John C. Robinson National Air College, and then return to Tuskegee for graduation. Such a plan, Robinson asserted, would counter the arguments of the Wilberforce backers, since the training would be taking place in the North at one of the nation's centers of aviation development:

Chicago, being the hub of aviation, Chicago, having the busiest airport in the world, Chicago, having unlimited opportunities in all phases of aviation should be the most logical place for the student to have contact so that their ideas and views could be broadened and their imagination could be increased in the aeronautical world. Then too, they would have an opportunity to mingle with many outstanding people in the field of aviation—being able to go into the different factories—being able to meet many of the inspectors and being able to be exposed to every possible opportunity in the field of aviation to help carry them and help inspire them to do the bigger and better things in the aeronautical world.[13]

Robinson cautioned that the Wilberforce alumni were organizing a campaign to secure the government aviation program at their alma mater. Still, he observed, "the money for the Civil Aviation Training hasn't been appropriated but the money for the Military Aviation Training has . . . and I don't see why we couldn't handle the military as well as the civil program because the fellows in the civil program could probably graduate and be selected for the military program."[14]

Despite Robinson's enthusiastic proposals, the Tuskegee administration proceeded cautiously. In early June 1939, G. L. Washington, perhaps mindful of his negotiations with the Brown Condor three years earlier, wrote the flier that a connection between Robinson's school and Tuskegee Institute would be acceptable only if "adequate funds and teaching facilities for this work were available." He explained that he had to consult with the president's office, since "this whole matter—so far as promotion—rests with Dr. Patterson." Then, in a comment that accounts for Tuskegee's low-key approach to the whole idea of participating in flight training operations, Washington explained that he was "more concerned and interested in training for servicing airplanes. I feel that the Negro has better opportunity of immediate employment in that connection."[15]

Washington and Patterson had indeed been exploring the prospects of obtaining federal support for the training of aviation mechanics. Washington had discussed the matter with President Patterson in early January 1939, after T. M. Campbell, head of the institute's Agricultural Extension Service, had notified him of a new program to offer vocational training in aviation mechanics in federally supported vocational programs nationwide. Campbell suggested that Patterson secure the support of President Roosevelt, to ensure the cooperation of "certain of our white friends here in the South who could be depended upon to lend their support, but the matter, I believe, should be 'nailed down' in Washington first before any publicity is given to it."[16]

Apparently, Patterson could not obtain a definite commitment on the matter and no action was taken. In June, however, he did act on the advice of Mary McLeod Bethune, director of the National Youth Administration's Division of Negro Affairs. Bethune notified Patterson that the War and Navy Departments were currently transferring equipment to the National Youth Administration (NYA), which would be used to establish units for training aviation mechanics. It was crucial, Bethune advised, that Patterson contact the Alabama NYA director immediately; otherwise the opportunity might be lost: "Let us not sleep at the switch. Negroes cannot be prepared for these great advances in aviation unless they are up and on the job to see that our young people are well qualified and trained for openings."[17] Patterson immediately contacted NYA officials in Birmingham, explaining that a unit for training black pilots and mechanics "will be established somewhere in the nation, and I would appreciate your good offices in helping us to secure it for Tuskegee Institute." Apparently proceeding on Robinson's information that only those schools with an aviation component in their curriculum would be eligible for government-sponsored aviation training, Patterson argued that "the presence on our grounds of certain equipment, and at least a small effort in teaching this work, will increase our chances of getting the national unit."[18] Alabama officials asserted that they were unaware of any initiatives in the NYA relating to aviation mechanics training and suggested that Patterson contact the Civil Aeronautics Authority.[19]

Patterson took no further action on either pilot or ground crew training until September. In the interim the Civilian Pilot Training Act was passed (on 27 June 1939) and by August funds had been appropriated. Had it not been for the courtesy of Charles Alfred Anderson and G. L. Washington's brother-in-law, James C. Evans of the West Virginia State College for Negroes, Tuskegee might well have missed the application deadline for the first full-scale CPT class, which began in the fall of 1939. At the end of August, Anderson wrote Edmund H. Burke, institute comptroller and secretary to the Board of Trustees, with the news that Congress had authorized the

CPT Program and that application forms were available from Grove Webster of the CAA's Private Flying Development Division. He urged that the officials of the institute act promptly "as all plans have been made and the only steps necessary now are negotiations on the parts of the Colleges to get the program."[20] Two weeks later Evans wired his brother-in-law with similar information, noting that two black institutions, West Virginia State College and North Carolina Agricultural and Technical College, had already been approved for participation and that he had spoken to CAA officials on behalf of Tuskegee Institute when he learned that the school had not applied.[21] Washington immediately turned the wire over to Patterson, unaware that only the day before, 12 September 1939, the Tuskegee president had taken Anderson's advice and contacted the CAA.[22]

Despite the best efforts of Anderson and Evans, Tuskegee's application was not submitted until the end of September. Perhaps fearful that the application deadline had passed, Patterson acted as if his letters of inquiry to the Civil Aeronautics Authority and the War Department the previous March amounted to a formal application for participation in the CPT Program, and sent the following cryptic wire to the CAA on 12 September: "Hope Tuskegee's application for training unit will receive favorable consideration. Can offer outstanding facilities for such program."[23] The CAA official in charge of CPT Program applications, Grove Webster, responded that the authority had not received an application from Tuskegee Institute and advised Patterson that application forms were being forwarded.[24] More delays ensued when the forms did not arrive and a second set had to be requested, but finally, on 25 September 1939, the CAA acknowledged receipt of Tuskegee Institute's application to participate in the Civilian Pilot Training Program.[25]

Before Tuskegee could submit its application, however, a qualified flight operator had to be secured to conduct the flying training. Despite Robinson's earlier bid to serve as Tuskegee's flight operator, Patterson did not approach the Chicago flier in this regard. Instead he turned to Alabama Air Service, owned and operated by Joseph Wren Allen out of the municipal airport in Montgomery, forty miles west of Tuskegee.[26] Patterson's reasons for seeking out a local flight operator in lieu of Robinson are unclear. Perhaps his experience with Robinson in 1936 made him reluctant to stake Tuskegee's participation in the program on a man who had proven unreliable in the past. Perhaps the haste with which the application had to be prepared and submitted ruled out any consideration of establishing an "extension course" at Robinson's school in Chicago. Maybe Patterson believed that the CAA would not approve such an arrangement. At any rate, the Tuskegee president decided against approaching Robinson and sought out a local operator, who was necessarily white, as there were no licensed black aviators in Alabama.

Recognizing that a white flight operator might be reluctant to cooperate, Patterson contacted Allen personally to solicit his assistance. Allen readily agreed, recalling later that he did not find the Tuskegee president's request unusual and that he had no reservations about teaching blacks to fly.[27] Tuskegee was indeed fortunate that Allen had no qualms about cooperating because the pool of qualified flight operators in Alabama was extremely limited. In April 1939 the CAA reported only eighty-two certified aircraft in the state and a total of 168 licensed pilots, one with a transport license and fifty-nine with commercial licenses.[28] How Patterson came to contact Allen is unclear. Washington speculated that Ralph Swaby, the aviator from Columbus who had worked with Tuskegee students in the summer, might have recommended Alabama Air Service. Allen, however, discounts this possibility, noting that he and Swaby were competitors. He suggests that Patterson may have learned of him through the administration at Alabama Polytechnic Institute (API) twenty miles away in Auburn, as Allen negotiated contracts with both API and Tuskegee at about the same time.[29]

In the midst of Tuskegee's last-minute effort to submit an application, Tuskegee trustee and alumnus Claude A. Barnett learned that the school had not applied for the CPT Program. Concerned that Patterson might be overlooking an opportunity, he wrote his protégé and friend a concise and perceptive assessment of the new opportunities in the field of aviation which had recently opened up for blacks. Newsman that he was, Barnett's information regarding the distinctions between CAA and army flying training was extremely accurate. He was pessimistic about the Air Corps training program, noting that blacks were still not being accepted as aviation cadets for army training. Moreover, he had no illusions that the CAA program might be an avenue for admitting blacks into the Air Corps; he characterized the CPT Program as "more or less a sop, its chief purpose being the stimulating of interest in civil participation in aviation. . . . Insofar as the preparations for war flying are concerned, it does not seem important." Unaware of Patterson's earlier efforts, he suggested the NYA-sponsored aviation mechanics training program as an alternative that Patterson might consider.[30]

Patterson's response to Barnett reveals clearly the reasons for the Tuskegee president's complacent attitude regarding the CAA program. He persisted in characterizing his approach to the CAA the previous March as an application, declaring to Barnett that the "truth of the situation is that we made application so long ago . . . that the Civil Aeronautics Authority apparently failed to consider it in the recent bill." Reassuring Barnett that Tuskegee's application was again on file, Patterson went on to explain the administration's attitude, which closely paralleled Washington's earlier statement to Robinson: "We are not enthusiastic about this particular program because it does not go sufficiently far and it is my plan to follow up the matter to the extent of trying to get the mechanics program here as that is where the real opportunity seems to lie rather than in [a program that] merely [provides] an appreciation for flying. I hope that it will be possible for me to go to Washington soon on this matter."[31] Thus Patterson agreed with Barnett that the CPT Program was indeed a "sop" and had little bearing on the military aviation program; this attitude, together with his interest in training aviation mechanics, accounts for his failure to monitor subsequent developments closely, obtain application forms, and submit an early application in the months after his initial inquiry to the CAA in March 1939.

Patterson made the trip to Washington in late September or early October and conferred with Secretary of War Woodring on several matters, including the equipment that he had made little headway in obtaining through Alabama NYA channels.[32] While in Washington, D.C., he tried to schedule an appointment with Robert Hinckley of the CAA, but the meeting could not be arranged in time to coincide with Patterson's itinerary.[33] Consequently, Patterson sought the assistance of Edgar G. Brown and asked him to monitor the status of Tuskegee's application.

On 6 October 1939 Brown advised Patterson that Grove Webster was hesitant about accepting Tuskegee into the program because there was no airport nearby. The plan to use Alabama Air Service in Montgomery as the flight contractor and conduct the training at the city's municipal airport was not looked on favorably because of the inconvenience and expense of transporting the students. Brown, however, speculated that one of Webster's assistants, Leslie A. Walker, might influence his chief to grant an exception for Tuskegee: "He [Walker] is a native of Alabama and is well and very favorably acquainted with your great institution. He seemed particularly hopeful, but not conclusive, in his genuine interest in being able to overcome the situation resulting from the forty miles separating Tuskegee and the airport."[34]

On 7 October Brown managed to contact Webster, who had been away from Washington. When he learned that Webster planned to disapprove Tuskegee's application in the upcoming week, Brown immediately telephoned Patterson to forewarn him of the impending rejection. Patterson conferred with G. L. Washington and agreed to let his director of Mechanical Industries drive to CAA headquarters in the nation's capital and try to reverse the unfavorable decision. When Washington arrived at the CAA office, he met with Leslie Walker, the Alabamian whom Edgar Brown had suggested might prove helpful. He was pleased to learn that the CAA official was from Notasulga, a community just ten miles north of Tuskegee. Washington later recounted that he carefully observed the patterns of racial etiquette that had proven so effective in securing the cooperation of local whites since the days of Booker T. Washington: "The justifications I gave for making an exception [in] Tuskegee's case were, as I thought, [valid] in themselves. However, I wrapped them up in the type of approach I would make to a southern white man for a favor, though this may not have been necessary."[35]

By the end of his meeting, Washington felt certain that Tuskegee's application would be approved. Walker asked for assurances from city officials in Montgomery that the institute would be granted access to the city's municipal airport and from the fixed base operator, J. W. Allen, that he would conduct the flying training. Washington wired back to Tuskegee and asked Patterson to have Montgomery officials and Allen send telegrams to the CAA indicating they supported Tuskegee's application. Patterson complied and the next day, 10 October, W. A. Gayle, Montgomery's commissioner of public works, and Allen sent the requested telegrams. Once Washington ascertained that the messages had been received at CAA headquarters and all was in order, he returned to Alabama.[36]

Washington's confidence was well founded. By the end of the

week Robert Hinckley formally approved Tuskegee for participation in the Civilian Pilot Training Program for the academic year 1939–40, with a quota of twenty students.[37] Once final approval was received, a multitude of administrative details had to be handled before the training could begin: a budget had to be drawn up, transportation to Montgomery arranged, students selected, physical examinations administered, ground school courses scheduled, and funds raised for scholarships to assist qualified students who lacked the financial resources to participate in the program. Responsibility for these and many other details fell to G. L. Washington. Under his guiding hand Tuskegee's CPT Program quickly became a reality and the first of many CPT classes began training.

Immediately after Tuskegee received notification of acceptance, Washington issued the first of a series of announcements providing prospective students with information about the program.[38] They learned that only United States citizens, eighteen to twenty-five years of age, who had completed at least one year of college, were eligible. Since the program was for beginning pilots, applicants with previous solo flight experience would not be accepted. In addition to these basic eligibility requirements, applicants would have to pass a physical examination. Parental permission was also required, in the form of a statement releasing the institute, the flight operator, and the federal government from liability in the event of an accident. The expenses of ground and flight instruction would be borne by the federal government and amounted to roughly $300 per student. The only cost to the students was a forty-dollar fee to cover miscellaneous expenses such as medical examinations, insurance, and transportation.

The training program itself was extracurricular and included seventy-two hours of classroom instruction—ground school—to be taught at evening sessions on the campus during the fall and winter academic quarters. The bulk of the ground school training was in four areas—Theory of Flight and Aircraft, Civil Air Regulations, Practical Air Navigation, and Meteorology—with additional study in subjects such as Aircraft Power Plants and Instruments and the History of Aviation. Flight instruction at Montgomery's municipal airport, scheduled to begin approximately three weeks after the start of ground school, involved thirty-five to fifty hours of training in three thirty-minute lessons per week for the first eight weeks and two one-hour lessons per week thereafter.[39]

Finally, the announcement told prospective aviation students that enrollment was limited to twenty students, with qualified applicants to be selected in the order of application.[40] In the issue of the student newspaper that announced Tuskegee's selection as a CPT Program participant, the editors congratulated "those who worked to get this program at Tuskegee" and admonished their classmates to show their appreciation by participating in the program: "Students! Now we have it; let's support it."[41] Such an exhortation was hardly necessary; by the time it appeared in late October, some sixty students had applied, including three women.[42] Two factors mitigated against an even greater student response—financial considerations and parental consent.

G. L. Washington confided to President Patterson that obtaining parental consent was "the most outstanding problem" in recruiting students into the program. Some parents were concerned about the safety of the program. Others, alarmed by the outbreak of war in Europe after the German invasion of Poland on 1 September, feared that their sons would incur a military obligation if they participated in a government-funded flight training program. Washington sought to allay both concerns by writing each parent "a letter explaining very briefly the program, pointing out that it has no connection or obligation with the Army or Navy Air Corps, [and] emphasizing the steps that the government has taken to make the program safe."[43]

To meet the financial needs of students unable to pay the forty-dollar fee, a considerable sum in a day when a week's wages averaged just over twenty dollars, Washington established a scholarship fund and set the maximum amount of each scholarship at twenty-five dollars.[44] Washington, Patterson, and A. J. Neely, dean of men and executive secretary of the General Alumni Association, each contributed twenty-five dollars to inaugurate the scholarship fund drive. In a letter soliciting contributions, Washington appealed to the racial pride of potential donors: "Considering the few Negro colleges selected to date and with Tuskegee Institute being the only college in the Lower South approved to date, you can readily understand that we cannot afford to lose the unit or suffer it to be curtailed."[45] The response was gratifying. By 28 November 1939

$515 had been pledged, enough to provide scholarships for the entire quota of twenty.[46]

Obtaining physical examinations for prospective students was a minor problem and the way in which Washington and Patterson sought to handle it is an early indication of a general pattern that was to emerge regarding Tuskegee's approach to the CPT Program and, later, to the military unit. The pattern was in the tradition of Booker T. Washington's vision of Tuskegee as a "veritable cathedral of practical learning and black self-help . . . run entirely by black people."[47] Patterson had not, of course, followed this precept when he selected a flight operator, but the use of Alabama Air Service was from the outset viewed as a temporary arrangement that would be terminated when Tuskegee developed the resources to assume full responsibility for the program and run it without white assistance.[48] When Tuskegee received approval to participate in the CPT Program, the nearest CAA medical examiner was in Montgomery. Consequently, the CAA asked Patterson to recommend a local physician who could be designated an official CAA medical examiner, and he suggested Dr. George C. Branche, one of the black doctors at the Negro Veterans Hospital adjacent to the institute. The CAA, however, rejected Dr. Branche because of his status as a full-time government employee and contacted instead John W. Chenault, a physician at Tuskegee Institute's hospital.[49]

Another major task Washington faced was arranging for the ground school. Initially he had planned to use institute faculty almost exclusively, calling on the assistance of outside experts only for courses in meteorology and aircraft instruments.[50] As the Tuskegee administrator began to organize the ground school and as scholarship money began to accumulate, ensuring that the students made a good showing on the standardized CAA ground school tests became his central concern. Even though the CAA had approved two Tuskegee faculty members as ground school instructors, Washington became apprehensive about the quality of instruction they could provide, because they had no experience or training in aviation. By mid-November he abandoned his original plan to use Tuskegee faculty for the ground school, at least for the first class, and sought experienced aviation instructors, once again departing from Tuskegee's traditional policy of relying on black expertise and self-help. Washington later recalled that he "had the conviction that

Negro students, individually of course, would do an outstanding performance in flying; at the same time, however, I frankly was not too sure they would do as well when CAA inspectors examined them on ground school instruction."[51] Aware that API, only twenty miles away at Auburn, offered courses in aeronautical engineering, he contacted the two professors responsible for instruction in that field and asked them if they would be willing to conduct the ground school until Tuskegee could develop its own faculty.[52] As it turned out, both men at Auburn were highly qualified. Robert G. Pitts, an Alabama native, had recently earned a master's degree in aeronautical engineering at California Institute of Technology, one of the leading institutions of the day in the relatively new field of aeronautical engineering. He had also completed primary training as a Marine Corps aviator and was a CAA-certified aviation mechanic.[53] Bloomfield M. Cornell was a graduate of the Naval Academy and had served as a naval aviator until he was medically retired in the mid-1930s.[54] By the end of November Pitts and Cornell agreed to teach the four principle units of instruction, amounting to sixty of the seventy-two hours, at a rate of five dollars per hour, the same rate that the Tuskegee instructors were to be paid.[55]

Washington also had to solve the difficult problem of arranging transportation to the municipal airport at Montgomery. One of the CAA's conditions for permitting Tuskegee to participate in the CPT Program was that the institute would bear the financial burden of transporting the students to Montgomery.[56] Funds for transportation were limited; in his initial budget estimate, Washington calculated that only $300 would be available.[57] Allen did what he could to help, agreeing to conduct all the flight instruction in the afternoons so that the students could attend their morning classes before making the one-hour drive to Montgomery.[58] In the final analysis, the problem of commuting to Montgomery for flight instruction proved to be the most vexing that Washington faced in the establishment of Tuskegee's CPT program, and showed the validity of the CAA's policy of requiring participating institutions to have access to a local airport. Nevertheless, transportation was arranged so that the students could commence their flight instruction at Montgomery.

Flight training could not begin, however, until Alabama Air Service received CAA approval. By early December 1939, when Tuskegee's twenty CPT students began ground school, Allen still had not finalized his contract.[59] At last, on 22 December 1939, the CAA approved Allen's contract to train twenty Tuskegee CPT students, "provided that you have complied with the terms of the contract pertaining to insurance compliance, rerated instructors, and certified aircraft."[60] Insurance and certified aircraft presented no particular problems, but obtaining "rerated" instructors—that is instructors certified by the CAA for CPT—was another matter. Like many of the small flight operators who became involved in the CPT Program, Allen had to expand quickly what had been essentially a one-man operation so that he could fulfill his commitments under the program. As the only flight operator in the central Alabama area, he had agreed to provide flight instruction not only to Tuskegee but also to API in Auburn. With each school allotted a quota of twenty students, Allen had to find additional flight instructors, since the CAA required a student-instructor ratio of ten to one. Although he had wide contacts in the aviation community, Allen initially was able to secure the services of only one instructor pilot, Don S. Porter, who had been working out of Key Field in Meridian, Mississippi.[61] In late December or early January, Tuskegee's students began to commute to Montgomery for flight training; because of Allen's difficulties in finding instructors, only ten students began training rather than the full quota of twenty authorized by the CAA.[62]

Once flight instruction began, it became apparent that the CAA's reservations about conducting the flying phase of the program at an airfield forty miles from the campus were well founded. The students had been commuting to Montgomery only a few weeks when Washington frankly admitted to Leslie A. Walker, the helpful CAA official who had facilitated the approval of Tuskegee's application, that the plan to conduct flight training in Montgomery was plainly unworkable: "We are running into the difficulties you clearly pointed out in October. Each student must make 70 to 80 trips to Montgomery with accompanying road hazards. To obtain one half-hour flight period he must leave the campus here at noon and return with the group at 6:00 afternoon. Though he may have one or more free periods in the morning, the morning cannot be used."[63]

By late January the problems associated with commuting to Montgomery had become so severe that Washington feared that it would be impossible for all twenty students to complete their flight training by the end of the academic year.[64] Consequently, he began to investigate the feasibility of transferring flight operations from Montgomery to Kennedy Field, a private flying field just south of Tuskegee. The field had been constructed by three local white fliers, Forrest Shelton, Stanley Kennedy, and Joe Wright Wilkerson, for their personal use.[65] Shifting the flight training to a local field would certainly alleviate the strain and expense of commuting to Montgomery, but Washington had several problems to solve before the move could take place. Kennedy Field, strictly a private facility, was not a CAA-approved flying field, and federally sponsored flight training could not be conducted there unless CAA inspectors gave their approval. Equally important was the matter of obtaining permission to use the field, which was owned by John Connor and leased by Stanley Kennedy's father.[66]

At the end of January 1940, Washington invited the regional CAA director, Edward C. Nilson, and his assistant, Dale E. Altman, the general inspector of flying fields, to visit Tuskegee and inspect Kennedy Field. After several test flights Altman and Nilson indicated that they would permit the field to be used for CPT operations if certain improvements were made, and CAA headquarters in Washington concurred. Washington estimated that between "$1200 and $1500 would be spent in transportation to Montgomery this year and that the improvements on [the] field locally would not take that much. Efficient use of [the] students' time and elimination of road hazards in transportation would be [a] great gain [even] if no savings were made financially." Kennedy agreed to turn over his lease for the field to Tuskegee on the condition that he be allowed to keep his airplane there and continue to have access to the field for his private flying activities.[67]

The day after the meeting, Washington wrote Leslie Walker to solicit his support in securing approval for the use of the Kennedy Field. He explained that regional CAA officials as well as members of the Alabama Aviation Commission all agreed that the field could serve as a satisfactory interim facility until Tuskegee could complete construction of its own field. Washington concluded by outlining his vision for the future of aviation training at Tuskegee: "Permission from the Authority to use the local field [i.e., Kennedy Field] will help us over the one obstacle in permanently establishing the promotion of aviation here. Being a school nationally [prominent] rather than locally, the Institute bids fair to become an 'Air Center' for Negroes and should not like to lose out because of the present handicap in going to Montgomery in face of suitable flying field facilities locally."[68]

Perhaps more important to the final approval of Kennedy Field for CPT operations was Regional Director Nilson's enthusiastic report to Grove Webster, the CAA official in charge of the CPT Program, regarding his impressions of the work being done at Tuskegee. Nilson admitted that before he visited the Alabama school he was skeptical about the idea of an aviation training program for blacks. Once he inspected the operations at Tuskegee and learned of Washington's plans, he confessed to "a very serious and outstanding interest in the possibilities observed [regarding] this higher institution of learning and training for the negro race," and he urged Webster to approve the use of Kennedy Field immediately.[69]

The CAA headquarters in Washington approved Kennedy Field for CPT operations in late February 1940, after the necessary improvements were completed. In the tradition of Tuskegee Institute, the aviation students themselves had volunteered their labor to bring the facility up to CAA standards, working under the supervision of Royal B. Dunham, a Tuskegee instructor. They improved the grass landing area by cutting trees, filling holes, grading high spots, and erecting runway markers. A wooden hangar that could accommodate three airplanes was erected, along with a lavatory, fuel depot, and postflight briefing shack. Once the improvements were complete and CAA headquarters approved the field, Allen relocated the flight training for Tuskegee students from Montgomery to Kennedy Field.[70]

Washington's vision of Tuskegee as an "Air Center for Negroes" was no doubt bolstered by the performance of the first class in their CAA ground school examinations. On 25 March 1940, CAA inspector George A. Wiggs arrived to administer the standard written examination required of all CPT students. Washington knew that the performance of the Tuskegee trainees could well determine the future of the CPT Program at Tuskegee. Abnormally low scores would certainly prove embarrassing to the institute and might be the basis for limiting or denying black participation in future CPT classes. Washington later recalled his anxiety while the tests were under way: "While the students were taking the examination I was in and out of the room, and in my office most of the time, wondering about the results and if my care in selecting ground school instructors was going to pay off."[71] The CAA inspector agreed to grade the examinations on the spot, apparently as eager as Washington to learn how the black students' performance compared to their white counterparts. The proceedings took on an aura of high drama as Inspector Wiggs scored each test:

> Everyone gathered around the inspector as he graded the exams. The excitement heightened as . . . he finished paper by paper and put down the grade. It was a moment beyond description when at the very end of scoring every student had passed every subject!
> Up to that time, in the . . . seven southern states, no college had a record of 100% passing on the first examination (reexaminations were given those failing in a subject the first time).[72]

The Tuskegee CPT students not only passed, but they did so by a wide margin. The average score was 88 percent, with one-third scoring above 90 percent. The lowest score was a respectable 78 percent and competing for top honors were Charles R. Foxx, who averaged 97 percent, Alexander S. Anderson with 96 percent, and Elvatus C. Morris with 95 percent.[73] Washington believed that without the assistance of the instructors from Auburn, the Tuskegee CPT students could not have achieved such an outstanding record: "These ground school instructors worked hard with the students. . . . I will always feel the cooperation of Auburn's faculty made the difference, as I reasoned it would when engaging the services of Pitts and Cornell."[74]

The outstanding performance of the Tuskegee students was widely recognized. A southern white daily, the *Chattanooga News–Free Press*, ran a feature article on Tuskegee's aviation program the following month and gave full play to the "scholastic excellence of the students in their ground work." The author emphasized that the uniformly high scores and Tuskegee's unique status as the only southern school with a 100 percent pass rate had been achieved in competition "against such technological schools as Georgia Tech, Auburn, and North Carolina."[75] When John P. Davis, secretary of the National Negro Congress, learned of the performance of the Tuskegee students on the CAA examinations, he asked President Patterson to invite Charles Foxx, the student with the highest overall average, to speak at the organization's upcoming meeting. Davis suggested that

Foxx speak on "Negro Youth in Aviation," which he believed would "heighten the interest of hundreds of Negro organizations in the Civilian Aviation Training program, and, as a consequence, we could expect the rapid expansion of that program at Negro colleges."[76] Patterson agreed and wired Davis that Foxx would attend the meeting.[77]

When the time came for flight evaluations, the Tuskegee CPT students almost equaled the 100 percent pass rate they had achieved in the written examinations. Allen believed the performance of the Tuskegee students in the air compared favorably to others he had taught to fly, especially in light of the conditions at Kennedy Field.[78] Although the CAA had approved the field for CPT operations it was barely adequate for teaching beginning pilots to land. The runway was fairly short and, as the writer of a feature newspaper story on Tuskegee's first CPT class observed, "because of trees, [the] slope of [the] field, and other obstacles, every landing for these colored boys is a spot landing."[79] Allen gave considerable credit to the first CPT class at Tuskegee, because without their solid performance in both the ground and flying phases of the program, subsequent developments in civil and military aviation at the institute might not have been forthcoming. Allen discounted his role in training the black pilots, claiming that he did nothing more than provide the Tuskegee students with a solid foundation in the fundamentals of flying: "But the basic thing that I did, if I did anything up there, was to be sure that they got a good start . . . and then they could go on from there. . . . I did the best that I could and I hope I did a good job."[80] Allen and his instructors apparently did an excellent job. By the end of May the class finished the flying phase of the course and all but one of the trainees passed the flight examination administered by CAA inspectors and received their private pilot license.[81]

Perhaps the highest recognition accorded a member of the first CPT class was the selection of Charles Foxx as one of seven students in the southeastern region to compete for the Shell Intercollegiate Aviation Scholarship. Selected by the Institute of Aeronautical Sciences for the regional competition at Birmingham, he was one of only forty-nine students in the nation chosen to vie for the $1,500 scholarship, underwritten by the Shell Petroleum Company.[82] Foxx had been selected not only for his near-perfect score on the written examinations, but also on the basis of his superior flying skill, which was apparent to his instructors and fellow aviators.[83] Veteran pilot Charles Alfred Anderson had high regard for Foxx's abilities in the air, recalling that in his fifty years of flying he had "never seen a person as slick a pilot as Charlie Fox[x]. He was just a natural born pilot."[84] Also a talented musician, Foxx sang in the Tuskegee Institute Quintette, which was scheduled for a fund-raising tour on the same date as the scholarship competition. Forced to choose between vying for the aviation scholarship and raising money for the institute, the pilot-singer unselfishly declined to compete and fulfilled his obligations to his school.[85]

By spring of 1940, as Charles Foxx and his fellow CPT students completed their training and received their private pilot licenses, Tuskegee Institute was fully committed to the development of its aviation program. The administration's complacency toward the CAA's pilot training program, so apparent throughout most of 1939, had vanished. Tuskegee almost paid a dear price for its nonchalant attitude toward the CPT Program, perhaps lulled into a false sense of security by the frequent public references to Tuskegee as an ideal site for black pilot training. Ironically, the Tuskegee administration exhibited no real sense of urgency toward participating in the CPT Program until it was almost too late to be included and its application was on the verge of being rejected. Only the last-minute, personal efforts of G. L. Washington prevented Tuskegee from being excluded from first CPT class in the fall of 1939. Once Tuskegee was accepted, however, Washington became enthusiastic about the program and it was largely through his efforts that it succeeded. Still, without the cooperation of local whites like Walker, Allen, Pitts, Cornell, and Kennedy, Tuskegee would have found it difficult, if not impossible, to have participated in the CPT Program from its inception in academic year 1939–40. But all concerned, black and white, expected that Tuskegee Institute would assume full control of the program as soon as possible.

7

Tuskegee Emerges as the Center of Black Aviation

By the spring of 1940, thanks to the Civilian Pilot Training Program, Tuskegee had the beginnings of an aviation program. After a very precarious start, G. L. Washington's efficient and enthusiastic work as Tuskegee's CPT coordinator, together with the creditable performance of his students, had won the confidence of CAA officials in Atlanta and Washington. Consequently, there was every indication that Tuskegee would continue to participate in the CPT Program and might ultimately make aviation an integral part of its curriculum. But no one anticipated the startling growth and expansion of Tuskegee's aviation program that was to take place in the coming months. By the end of the year the institute owned a small fleet of airplanes, had hired a cadre of flight and ground school instructors, and offered a wide variety of flight training courses. In short, during the summer and fall of 1940 Tuskegee Institute emerged as the center of black aviation, a development that could scarcely have been foreseen the year before, when the school was almost rejected as a participant in the CPT Program.

The blossoming of aviation at Tuskegee was due largely to the leadership and ideas of G. L. Washington. By March 1940, Washington had proven himself an able director of Tuskegee's CPT Program. He persuaded CAA officials to accept the institute into the program, despite the lack of a suitable local airport, and managed to overcome the myriad problems associated with seeing the first class through its training. By April, after its success was assured, he turned his full attention to the larger, more difficult problem of establishing a permanent flight training program at Tuskegee, one that did not rely on subleased airfields, flight training contracts with private operators, and borrowed ground school instructors.

Throughout the remainder of 1940, Washington focused his energies on securing an airport, assuming responsibility for the flight training, and expanding the CPT Program. Unrestricted access to a suitable airfield near the campus was crucial, since the sublease on Kennedy Field, several miles to the south, ran only through 1940, and there was no guarantee of renewal. Nor could Tuskegee count on the continued cooperation of Joseph Wren Allen to provide the flight instruction. Although Allen had been willing to help Tuskegee establish its CPT Program, he had found it difficult to recruit instructors, and moreover, he had the added responsibility of the CPT Program at API. Finally, Washington had learned that the CAA planned to expand its pilot training program in the coming year. Anxious to avoid a repetition of the events of the previous fall, when the institute's failure to stay abreast of the CAA's plans almost kept the school out of the CPT Program, he was resolved to see that Tuskegee was included in any new CAA initiative. Washington began his efforts to develop an airfield in late January 1940, at the same time that he was making plans to shift flight training from Montgomery to Kennedy Field. He had contacted the Alabama Aviation Commission's director of airfield development, Asa Rountree, Jr., and asked him to visit Tuskegee and confer with institute officials and representatives of the city regarding the establishment of an airport.[1] Washington speculated that the "City of Tuskegee might be interested in joining with Tuskegee Institute in the development of a municipal airport at a suitable location in the town of Tuskegee." He told Rountree he was "certain that Tuskegee Institute would place at the disposal of the project any land that it has available" and urged him to allow time on his visit to examine Tuskegee's land as well as possible locations in the town of Tuskegee.[2]

Washington advised President Patterson, however, that "the land owned by Tuskegee Institute is not available in sufficient area, without a tremendous amount of grading to be done."[3] He explained that his real objective in seeking a meeting with Rountree and city officials was to "interest the town of Tuskegee in purchasing a desirable piece of land—which has already been spotted." Washington considered the participation of the town crucial, not only for the purchase of the land but to ensure eligibility for federal funding, which Tuskegee, as a private entity, would have difficulty obtaining. Washington warned, however, that Tuskegee Institute should "become a co-sponsor with the town of Tuskegee in order to assure itself of the use of the airport in its future program."[4]

Rountree responded to Washington's request immediately, agreeing to visit Tuskegee on 31 January 1940 and to stay as long as necessary. But he warned the Tuskegee administrator that he was not optimistic about the prospects of the town officials participating, since "one of the obstacles to the establishment of a municipal airport in Tuskegee has been the unwillingness of the town to purchase a site."[5] Commissioner Rountree's reservations were well founded. When the meeting took place at the end of January, town officials—who were all whites—were conspicuously absent from the proceedings and were apparently uninterested in developing a municipal airport. Thus what had begun as a meeting aimed at convincing the town of Tuskegee to develop a municipal airport became instead a meeting to find a suitable site for Tuskegee Institute's own airfield.[6]

Rountree and his engineer, Owen Draper, examined the institute's property and located a site on the south side of the Franklin Road that seemed to offer excellent prospects for development. If further study showed it to be suitable, then the Aviation Commission would provide cost estimates on grading and construction. In order to qualify for federal funding under the Works Progress Administration (WPA) or the NYA, Rountree suggested that the land be deeded to the Aviation Commission "for fear of future possible difficulties if deeded to the city of Tuskegee." Construction of the flying field would take roughly two years and in the interim Tuskegee would use Kennedy Field.[7]

By early April 1940 the analysis of the proposed site was complete. Draper, the state airport engineer, advised Washington on 3 April that an airfield consisting of two perpendicular grass landing strips with a hangar to accommodate eight airplanes could be constructed on the site for approximately $22,900; this amount was based on cost estimates for similar projects across the state.[8] On receipt of the cost estimates from the state, Washington, in collaboration with institute registrar Alvin J. Neely, immediately drafted a funding proposal.

Washington and Neely suggested that the Board of Trustees finance one-third of the airfield construction costs, that the Alumni Association provide an additional third, with the remaining third coming from contributions by institute alumni and supporters. They emphasized that the construction of an institute airport was crucial to the future of Tuskegee's aviation program. "Once this need is met," they asserted, "aeronautical training will grow by leaps and bounds at our Institute and without further net expense to Tuskegee Institute."[9] Finally, the establishment of an airfield under the complete control of the institute would avert any threat of local white interference: "Negro youth flying in Alabama is something new. While no definite disturbance has been allowed to develop in Montgomery, we have deemed it expedient to appreciably expend funds that our students might train on the local Kennedy Airfield. A threat of the same problem locally coupled with uncertainty of use of this field through the 1940–41 government program in flight training, contributes to the immediate necessity of establishing a flying field on our own property fully meeting government requirements."[10]

Washington and Neely urged the trustees to appropriate their portion immediately so that grading and excavation could begin at once, with additional work initiated as Alumni Association funds and private donations became available. The board, however, did not commit itself to underwriting the construction of the airfield and the matter was temporarily stalled while Washington and Patterson considered alternative sources of funding.[11] Both, nevertheless, remained firmly committed to obtaining a flying field for the institute, especially Washington. With a proper airfield at his disposal, the energetic CPT coordinator believed that he could expand the CPT Program, develop a cadre of Negro flight instructors, and ultimately establish an "'Air Center' in aeronautics for Negroes both in flight and mechanics."[12]

Washington's comment clearly shows that, despite the distractions associated with establishing and expanding the CPT Program and developing an airfield, he had not lost sight of his original goal of aircraft mechanics training, the vocation which both he and Patterson believed would offer blacks the best opportunity for immediate employment in aviation. Patterson had discussed the matter when he met with the secretary of war the previous fall, hoping to learn more details about the arrangement whereby surplus War Department equipment was transferred to the NYA for use in training aviation mechanics.[13] In November 1939, after he learned that such a program had been inaugurated at the West Virginia State College for Negroes, he renewed his efforts to obtain a similar program through the Alabama NYA office, but received little encouragement.[14]

Washington's goal of training black aviation mechanics had impressed regional CAA director E. C. Nilson when he visited Tuskegee at the end of January. After Washington explained that he planned to survey "manufacturers and commercial transport lines with respect to the employment of Negroes in maintenance service," Nilson agreed to provide him with a list of companies for that purpose.[15] Indeed, the regional director was so taken with Washington's plans in that regard that he felt compelled to inform his superiors at CAA headquarters of the endeavor: "I . . . want to particular[ly] commend highly Professor Washington's plan for the training of 'EXPERT AND PROFESSIONAL' precision aircraft and engine mechanics, not only professional in respect to their mechanics, but trained as this institution does, in personality, tact, diplomacy, and courtesy. It is my opinion that this phase of Professor Washington's original idea has enormous potentials, and it is recommended and urged that you and others give this very serious thought, analysis, and consideration."[16]

Washington's plans for training aircraft mechanics continued, however, to be overshadowed by the CPT Program and efforts to build an airfield. Several factors account for this. First, both he and Patterson had come to realize that the effectiveness of such training was in large part contingent upon the availability of a local flying operation, which could provide students with an opportunity to gain vital practical experience. Second, there was no federally funded program in aviation mechanics equivalent to the CAA's Civilian Pilot Training Program. So even though both Patterson and Washington believed that the real opportunities for blacks lay not in flying but in ground support work, they were determined to take full advantage of the CPT Program and expand their flight training operation to its full capacity.

If Tuskegee Institute was ultimately to develop a practical curriculum for training aviation mechanics, the school would not only have to build an airfield, but it would also have to assume responsibility for flight training. From the outset, Washington had considered Alabama Air Service's role as flight operator a temporary expedient. Allen, who continued to find it difficult to recruit and retain sufficient instructor pilots to meet his obligations at both Tuskegee and API, was anxious to relinquish responsibility for the Tuskegee CPT Program as well. When he discussed the matter with Washington in early April, Allen counseled the Tuskegee administrator to encourage the top students from the first CPT class to continue their flight training and prepare themselves to qualify as CAA-certified flight instructors, in light of the nationwide shortage of instructor pilots. Moreover, he suggested that such measures would certainly be "in line with the Institute's policy of providing Tuskegee students with the opportunity to become leaders in their chosen [field]." Allen advised Washington that a course of instruction leading to a CAA flight instructor's rating would require 200 hours flying time and 145 hours of ground school, at a cost of $1,500 per student.[17]

Washington scarcely needed convincing, although he knew that the steep cost of the training would be prohibitive unless he could secure outside funding. In mid-April he briefly outlined for President Patterson the difficulties in securing instructor pilots and suggested that steps be taken toward "introducing Negro flight instructors to the program."[18] To accomplish this end he suggested to Patterson:

(1) That opportunity be made for four (4) students to do the additional 200 hours of flight and ground study to qualify for [a] commercial license and meet government regulation[s] for primary course flight instructors. If necessary arrangements are completed without delay and in order that such instruction may begin by June 1, 1940, these students will be ready as instructors in the Fall and there will be no question of employment at a salary of approximately $150.00 per month.

(2) That the cost of $1500 be born[e] jointly by the student and scholarship aid, and that suitable arrangements with a bank be made to enable [each] student to borrow his half with [a] definite schedule of repayment after employment in the Fall. Four students at $750 scholar-

ship each would total $3,000 . . . which . . . might be secured from a Foundation if the problem is effectively placed.[19]

Patterson fully agreed with the thrust of Washington's proposals. In May 1940 he asked for a $2,000 grant from the General Education Board, a nationally prominent educational foundation established in 1902 by John D. Rockefeller with a history of supporting training programs for black teachers. Departing slightly from Washington's original proposal, Patterson explained that the funds he requested would provide $1,000 each for two students and that Tuskegee Institute would extend a loan of $500 to cover the balance of the training expenses.[20] The foundation rejected Patterson's request, calling the training of black pilots a project which is "of very timely importance but which falls outside of the program of the Board." Support for such an initiative, they observed, "would seem to come within the Government's program."[21] While the General Education Board's refusal to fund flight instructor training was another temporary setback, this early attempt to cultivate a cadre of black instructor pilots demonstrates that by the spring of 1940 both Washington and Patterson were fully committed to the development of an aviation program completely under the auspices of the institute.

Although Washington's initiatives to build an airfield, to establish a program for training aviation mechanics, and to train black instructor pilots produced no immediate results, his efforts to expand Tuskegee's pilot training program were a resounding success and ultimately allowed Tuskegee Institute to establish itself as the preeminent black institution in the field of aviation education. Indeed, the expansion of Tuskegee's CPT Program came much sooner than Washington had anticipated, and once again the resourceful Tuskegee administrator found a way to fashion a successful training program on short notice with limited resources.

From the outset of the CPT Program, Washington had sought to enlarge the quota of trainees allotted to Tuskegee Institute. In late October 1939, shortly after Tuskegee received approval to participate in the CPT Program, Washington hoped that enough students would apply to "justify [an] appeal now for [a] considerable increase in quota."[22] By the beginning of November 1939 Tuskegee's ambitious CPT coordinator advised CAA headquarters in Washington that the number of qualified applicants far exceeded the assigned quota of twenty and asked for an increase. In light of the serious difficulties encountered in completing the flight training for the first class, it was probably fortunate that the request for an immediate increase in the quota for the first class was denied.[23]

Students unable to participate in the first CPT class had another opportunity in the summer of 1940, when Tuskegee was allotted a quota of fifteen for a concentrated summer session, which ran from 15 June to 15 September. The forty-dollar fee was eliminated, as the CAA had raised the rate of payment for each student from twenty dollars to fifty dollars. As before, Tuskegee Institute administered the ground school with the assistance of Pitts and Cornell from API, and Alabama Air Service provided the flight instruction at Kennedy Field.[24]

In mid-April Washington renewed his campaign for more elementary students, asking for an increased quota for the fall 1940 CPT class. When he presented his case personally at CAA headquarters, Leslie Walker told him most of the increases would go to the noncollege CPT units, and "that only in a very few and exceptional cases would the college programs be increased." Washington countered that the institute's program was indeed "exceptional inasmuch as Tuskegee was the only Negro college in the whole lower south having this program and that twenty would not be a fair representation." He went away from his conference with Walker with the "feeling that when we officially apply for [an] increased primary [i.e., elementary] quota it will be given consideration."[25] A formal request for an enlarged CPT elementary course was submitted in mid-May and in September Washington was notified that Tuskegee's quota for the 1940–41 class had been raised from twenty to thirty students.[26]

Although the elementary summer session and the increase for the 1940–41 class were certainly important, they were soon overshadowed by the inauguration of advanced CPT courses, the single most important factor in Tuskegee's headlong rise to leadership in black aviation circles. Washington had begun to consider advanced training as early as mid-April 1940, when he declared to President Patterson that Tuskegee had a "real possibility of being approved for an advanced unit."[27] Initially he saw advanced training as simply another phase of the CPT Program. After he conferred with CAA officials, however, he realized that the black college which could offer advanced CPT courses would likely become the nation's center of black aviation.

Advanced CPT was actually a series of courses designed to provide promising elementary students with an opportunity to go beyond the introductory training received in the initial elementary CPT course. The first advanced training open to elementary graduates was the secondary course, consisting of a 146-hour ground school and fifty hours of flight training, designed to familiarize the student with acrobatic flight maneuvers. Secondary graduates could continue their aviation education with additional advanced training courses, such as the cross-country course, the instructor course, and the commercial course.[28] Although the CAA originally intended to offer advanced training as part of the first full-scale CPT Program in academic year 1939–40, these plans had to be temporarily abandoned because of funding cuts.[29] Nevertheless, ninety students received secondary training under an experimental program conducted during the first half of 1940, which was the basis for developing the full-scale advanced training program for the academic year 1940–41.[30] By April 1940 the CAA's plans for advanced training included a secondary course, with students taking the ground school at their own colleges during the 1940–41 school year and completing the flight training during the summer at selected institutions in their area authorized to conduct the secondary flight curriculum.[31] The CAA believed that students who had completed CPT elementary and secondary courses would have received training equivalent to Air Corps primary, the first phase of the army's flight training program.[32]

Although Washington had begun to explore the possibilities of offering advanced CPT courses at Tuskegee in early April 1940, it was not until the end of the month, after he met with CAA officials in Washington, D.C., that he began to realize the full significance of advanced training to the future of Tuskegee's aviation program.[33] Word of the Tuskegee students' outstanding performance on the written examinations had preceded Washington to CAA headquarters. Dean Brimhall, assistant to CAA chairman Robert Hinckley, met with the Tuskegee CPT coordinator and indicated that he was impressed with the work in aviation being done at the Alabama school. While at CAA headquarters, Washington also spoke with Leslie Walker, the cooperative CAA official from Notasulga. From his discussions with Walker, Washington got a clear picture of the plans for CPT advanced courses.

Walker explained that experimental advanced courses were currently under way, and once the results were analyzed, permanent advanced courses would be established. The advanced training would not be offered at every college participating in the CPT Program, however, but only at certain specially selected institutions. Washington immediately understood the implications of the CAA's plans for Tuskegee: "It was suddenly clear to me that there would be only one point where Negroes would be sent for advanced flight training. I proceeded to show why Tuskegee was the logical place." Walker fully agreed, confiding to Washington that he was "certain that an advanced program would be carried out quite satisfactorily" at Tuskegee, especially since the school planned to build its own airport. Washington was confident he had sufficient time to make the necessary arrangements for advanced courses; he understood they were not scheduled to begin until the summer of 1941. Walker advised Washington to submit a proposal for advanced training which emphasized the outstanding record of the first class, the school's plans to build an airfield on institute property by 1941 and establish a program for training aviation mechanics, and Tuskegee's reputation as a national center for black educational activities.[34]

In early May Washington prepared a proposal for advanced training and sent it to President Patterson for his approval and signature. The Tuskegee president was in full accord with the plan and forwarded it to CAA headquarters virtually unchanged. Washington was careful to point out that he based the institute's request on the understanding that "advanced ground instruction will be offered to those [students] having completed primary instruction in the five Negro colleges during the academic year, 1940–41, and that . . . a number will be selected to pursue advanced flight training during the summer of 1941 at a single location." The arguments Washington advanced to support Tuskegee's bid to become the advanced CPT center for blacks closely followed Walker's recommendations. He contended that "in the minds of the Negro populace Tuskegee Institute is centrally located and is the logical point for any regional or national movement affecting Negroes." He cited the record of Tuskegee's first CPT class, especially their outstanding performance on the written examinations. He noted that the institute "proposes construction of an air field on its own property within one mile from the center of its campus." Finally, Washington explained that Tuskegee's interest in aviation went beyond flight training and that the administration was "equally interested in the development of instruction in aviation mechanics. The development of flying and construction of an airfield is considered the first step in the direction of training in aviation mechanics."[35]

The CAA acknowledged Tuskegee's proposal for advanced training several days later and advised Patterson that detailed information on policies and procedures would be forthcoming.[36] Nothing more was heard from the CAA on the matter until the end of June. In the interim Washington traveled to Chicago and Detroit to meet with alumni clubs in both cities and discuss recent activities in the School of Mechanical Industries. The trip gave him an opportunity to relate the recent progress of the institute in aviation, and he attempted to interest the alumni in sponsoring a fund-raising campaign for the construction of the proposed airfield.[37]

While in Chicago, Washington visited the Coffey School of Aeronautics, the private flight school at Harlem Airport on the south side of Chicago that was owned and operated by black pilot and aircraft mechanic Cornelius Coffey and his partner, Willa Brown. The Coffey school ran two flying programs, a "demonstration unit" under the supervision of the Chicago School of Aeronautics at Glenview, just north of Chicago, and a noncollege CPT program. Washington later recalled that it was "an inspiration to see Negroes conducting the flight training, and to be flown over the city of Chicago by Miss Brown, the highest ranking colored aviatrix in the United States. She impressed me as a well educated and most progressive and aggressive advocate of the Negro in aviation."[38]

Washington returned to Tuskegee in time to oversee the start of the elementary summer session in mid-June. Two weeks later he received a call from Edward C. Nilson, the CAA's regional director in Atlanta, regarding Tuskegee's application for advanced training. Nilson told him that "an institution in Chicago" was also attempting to secure advanced training, that only one black flying school would be authorized to offer advanced CPT courses, and that the program was to begin in just a few weeks, rather than in the summer of 1941 as originally planned. The regional director said he would support Tuskegee's bid for advanced training if the school was still interested, but only if there were plans for an improved airfield at Tuskegee and the institute intended to conduct the flight training itself, rather than contracting it out to a private operator. When Washington replied that the institute was most interested in securing advanced CPT training and assured Nilson that the school intended to develop a flying field and take over responsibility for the flight training, "employing Negro personnel to the maximum extent possible," the regional director agreed to back Tuskegee.[39]

Nilson's arguments to CAA headquarters regarding the suitability of Tuskegee as the advanced training center for blacks were apparently persuasive; by the beginning of July the school was granted a quota of ten secondary CPT students, authorized to begin training on 15 July. The selection of Tuskegee marked the beginning of a rivalry that pitted the black aviation community in Chicago against the supporters of Tuskegee, a conflict that became increasingly bitter as Tuskegee rapidly came to be recognized as the center of black aviation. When Nilson called, Washington assumed that the "institution in Chicago" competing for the advanced training was the Coffey School of Aeronautics, even though he carefully avoided mentioning the name. Having visited the facility only a few weeks earlier, he knew that Coffey and Brown had a number of advantages over Tuskegee. But he was confident that Tuskegee's reputation would give the institute the edge:

> When Mr. Nilson mentioned a Chicago school bidding for the [advanced training] Center I immediately thought of Miss Willa Brown's organization, though I made no comment to Mr. Nilson. Chicago might logically have been the place for the Center. Miss Brown had a going flight operation, and what she lacked might easily have been supplied by her sponsor,—the Chicago School of Aeronautics. . . .
>
> Tuskegee had a good ground school staff, but no such supporting equipment as Miss Brown's. It had no flight equipment or personnel, nor [did] it have a suitable airport for advanced training.
>
> What Tuskegee did have was a reputation for cooperating with state and Federal Governments; an excellent tradition and setting for training; a name known throughout the world; widespread national publicity at the time for achievement of Civilian Pilot Training students on CAA examinations and in flying; the Goodwill of people of the south and north in key positions . . . whose opinions could be determinative; a geographical setting favorable to flying training; a favorable impression with CAA-Washington for overcoming obstacles; and possibly

other attributes which caused decisions in [its] favor. *Further, it could, as a partner in a segregated project, be quite relieving and comforting.*[40]

After the CAA approved Tuskegee for secondary training in early July, Washington worked feverishly for two hectic weeks, coping with the multitude of details that had to be handled in order for the training to begin by the middle of the month. By 11 July, satisfied that he had managed to take care of most of the major problems, the harried CPT coordinator jokingly told President Patterson:

> I had to ask my staff in the office what day this is. Had lost track of time. There were a thousand details on this advanced training and I have stuck with it day and night. I believe things are well in hand and I have been getting 100% cooperation from everyone here on the matter of preparations. . . .
> You say I am a detail man, and were there details on this thing in order that it might start without a hitch and have the impression of being well organized and taken care of.[41]

When the CAA selected Tuskegee as the black institution for advanced training, Washington once again had to organize a training program on short notice, just as he had done the previous fall when the institute first began CPT operations. This time, however, his problems were multiplied. He had to find and hire qualified black flight instructors, purchase an airplane, find a flying field that the CAA would approve for advanced training, and organize the ground school—all in less than three weeks.

One of Washington's first concerns was flight instructors. Since the CAA required a five-to-one student to instructor ratio for secondary training, two were needed, and Washington wanted to use black pilots if at all possible. He immediately thought of the two fliers who had sought to interest Tuskegee in aviation since the middle 1930s, Charles Alfred Anderson and John C. Robinson. Anderson was teaching CPT elementary for Howard University in Washington, D.C. President Patterson was scheduled for a trip to Washington, D.C., and he agreed to contact the veteran pilot about coming to Tuskegee to serve as a flight instructor. Anderson, enthusiastic about the prospects of teaching advanced CPT courses at Tuskegee, readily agreed. Before he could begin, however, he had to complete a refresher course in acrobatic flying that the CAA required of all new secondary instructors. Washington arranged for Anderson to take the training at the Chicago School of Aeronautics at Glenview, and Anderson subsequently began the training in early July.[42]

Washington contacted Robinson in Chicago, offered the Tuskegee alumnus a position as a flight instructor, and urged him to begin the secondary instructor course at the Chicago School of Aeronautics immediately. But once again Robinson and Tuskegee were unable to come to terms, even though Washington had found the pilot "enthusiastic" about coming to Tuskegee when he spoke with him by telephone. Washington subsequently learned from Grove Webster that Robinson had encountered difficulties renewing his expired commercial license and speculated that this was the real reason for his failure to accept employment with the institute.[43] When negotiations with Robinson began to bog down, Washington contacted another black pilot with a commercial license, Lewis A. Jackson, who was teaching elementary CPT at the Coffey School of Aeronautics in Chicago, and who had expressed a keen interest in coming to Tuskegee.[44] Procuring a second instructor was important but not crucial, because the secondary flight training could begin with only one instructor; a second instructor would, however, be required two or three weeks after the flight training began.[45]

Tuskegee also had to obtain an airplane suitable for secondary training. It had to be a heavier, more powerful craft than the Piper J–3 "Cubs" that Alabama Air Service used for elementary training, and one suitable for acrobatic flying. Nilson, the CAA official in Atlanta, advised Washington to purchase a Waco UPF–7, manufactured by the Waco Aircraft Company of Troy, Ohio. As the UPF–7 was in short supply and could only be sold to CPT flight operators, Tuskegee could not place its order until the CAA sent a letter of authorization to the Waco Company. Nilson again proved helpful and quickly provided the necessary letter. On 3 July Tuskegee placed an order for a Waco trainer and made a down payment.[46] The CAA set the delivery priorities for the aircraft and Webster advised Washington that the plane would be ready on 17 July.[47]

Washington encountered little difficulty in arranging for the ground school. Pitts and Cornell, the two aeronautical engineering professors from API, were already engaged for the elementary ground school and they agreed to take on responsibility for the secondary classes.[48] They were assisted by William Curtis, a Tuskegee faculty member who taught in the School of Mechanical Industries.[49] A more difficult problem was obtaining access to a flying field that could accommodate the larger Waco trainer, and Washington's contact with Cornell proved extremely helpful in this regard.

When Nilson had originally contacted Washington about advanced training, he suggested that Kennedy Field be improved to make it suitable for larger aircraft. Washington believed that J. W. Allen, the flight contractor conducting the summer session elementary training, would not be willing to allow advanced training at Kennedy Field while he was teaching beginning students to fly; so Washington explored another option.[50] He knew that the airfield at Auburn, which had recently been taken over by API, had been improved to meet CAA specifications for secondary training.[51] He asked B. M. Cornell to approach API president L. N. Duncan about allowing Tuskegee to conduct secondary training at the field on a temporary basis. Cornell reported that at first President Duncan was against the proposal, but he relented after his secretary, who overheard the discussion, admonished the president that: "We have been going to Tuskegee Institute as guests to various cultural and other affairs all these years and now we can't let its students fly on our airfield!" Duncan's approval, however, was contingent on the willingness of API's CPT students to share the field with their Tuskegee counterparts. Cornell polled the white students and the vote was unanimous in favor of cooperating with the institute.[52] When he learned of the arrangement, President Patterson assured Duncan that Tuskegee would not abuse the privilege of using the Auburn field and pledged to "work as rapidly as possible in order to get our own facilities ready in a short time."[53]

Thus Tuskegee students once again had to commute to another airfield for flying training. Although not ideal, the arrangement was far more workable than the previous attempt to conduct flight training at a remote airfield, the municipal airport in Montgomery. Auburn was twenty miles to the east, half the distance to Montgomery. Moreover, there were only ten students to transport rather than the twenty that made up the first elementary class. Because the secondary students were not taking college courses during the summer, they could devote full time to their CPT work and take advantage of the morning hours for flying. Washington planned to use some special funds in the School of Mechanical Industries to purchase a station wagon to transport the students between Tuskegee and Auburn.[54] When he approached CAA officials about conducting the secondary flight training from the Auburn field, they readily approved the arrangement.[55]

Washington did not have to contend with the problem of recruiting students. The CAA paid all the students' expenses except for travel to and from Tuskegee Institute. Grove Webster, at CAA headquarters in Washington, D.C., selected ten students from the five black colleges that had conducted elementary flight training during the 1939–40 academic year, and on 22 July 1940 they reported to Tuskegee to begin secondary training. Howard University and Tuskegee Institute each contributed three students, Hampton sent two, and West Virginia State College and North Carolina A. and T. each sent one.[56]

When the secondary students arrived, Tuskegee still did not have a flight instructor or an airplane. Anderson was scheduled to finish his training at the Chicago School of Aeronautics near the end of July, but Washington feared that the Waco Company, which was swamped with orders, might delay delivery of the secondary trainer. After several inquiries, Washington learned that the Waco would be ready on 22 July, and he had it delivered to Chicago so that Anderson could have an airplane for his exclusive use, hoping that such convenience would perhaps hasten the completion of his training.[57] The decision to have the Waco sent to Chicago proved wise because the CAA inspector required Anderson to complete ten additional hours of flying time in the trainer before certifying him as a secondary flight instructor. An official of the Chicago flying school, J. L. Patzolcl, assured Washington that the inspector's decision was "absolutely correct" because the Waco was "definitely different in flying characteristics" from the light aircraft used in elementary training. Patzolcl added that the additional training was certainly no reflection on Anderson's skill as a pilot. The staff at the school "had the highest regard for him that we can have for any instructor going through our school. He has shown himself more than willing to try to learn new procedures in spite of the fact that he has gotten all of his original time, flying on his own."[58]

With the trainer at his disposal Anderson was able to complete the additional flying time quickly, and on 1 August 1940 he left Chicago in the shiny new Waco, headed south for Tuskegee.[59] En route to

Alabama the cautious pilot avoided airports when he could, realizing that the arrival of a black pilot in a new airplane might attract undue attention. Midway on his flight to Tuskegee, he landed in a field in rural Tennessee and asked the white owner if he could stay overnight. The man willingly agreed and even invited Anderson to supper. At first the Tennessee man attempted to follow the southern rules of racial etiquette, explaining to his guest, "I've always heard that you don't eat with colored people. You sit over there at that table, and we'll sit over here at this table." Anderson politely complied, but his host's manners and common sense soon prevailed and he declared to the pilot, "You know, that sounds kind of foolish to me. . . . Come on over here and have a seat."[60] After his pleasant experience in Tennessee, Anderson had to refuel at the Birmingham airport, where he encountered open hostility from frustrated whites unable to obtain new advanced CPT trainers like the Waco, which were in short supply:

> I remember . . . them . . . saying, "That son of a bitch is taking that airplane to Tuskegee. These damn 'niggers' in Tuskegee can get the airplanes, and we can't get one of them. Can you imagine that?" I overheard the conversation, but I got fuel and left—glad to get out. Then I came on into Tuskegee with it.[61]

After the episode at the Birmingham airport, Anderson must have been grateful for the friendly welcome he received on his landing at Tuskegee. Once the pilot telegraphed his estimated arrival time, Washington alerted the entire Tuskegee Institute community that the school's first airplane, with its first flight instructor at the controls, was on its way. Anderson announced himself by swooping low over the campus as he headed south for Kennedy Field. On landing, he and the new Waco were greeted by Washington, President Patterson, B. M. Cornell, one of the Auburn ground school instructors, and the excited secondary students, anxious to begin their flight training. A large crowd quickly gathered, eager to take photographs of the new aircraft, which Washington later described as "the heaviest and fastest plane the area had ever claimed, and it sounded to us like a bomber."[62] After the welcoming ceremony was over and it was time to fly the plane to Auburn, the capabilities of the Waco caused an embarrassing moment for Washington, who was responsible for showing Anderson the way to the Auburn field: "I didn't realize the speed of the plane and was too excited to navigate accurately. So we wound up over Columbus, Georgia in no time at all. With apologies to Anderson I pointed to Auburn where we landed in a few minutes."[63]

Anderson began flight instruction the next day, flying seven days a week to make up for his late start.[64] Lewis Jackson completed his secondary instructor training on 13 August and accepted a position at Tuskegee; he could not report, however, until he completed the training for the summer CPT class at the Coffey School of Aeronautics at the end of September.[65] Thus for at least a month Anderson was Tuskegee's only flight instructor. Throughout August and September the Tuskegee secondary students conducted their flight training at the Auburn airfield; although their flying was closely monitored by local whites, Washington recounted that "not a single unpleasant experience was reported by the trainees."[66]

By early October 1940 the ten secondary students had completed their ground and flight training. On 10 October the CAA inspector came to administer the ground and flight examinations, and all ten students passed.[67] Once again Washington had organized a successful CPT program, but this time the flying operations had been conducted by black instructors working for Tuskegee Institute. The summer 1940 secondary course marked Tuskegee's entry into flight training operations. But before the secondary students had even flown their first training flight, Washington had begun to formulate plans for a complete takeover of flight training for all phases of the Tuskegee CPT Program.

From the outset of the CPT Program at Tuskegee, Washington and Patterson had intended for the institute ultimately to assume responsibility for flight training operations. Until late June 1940, when the CAA approached Washington about secondary training, they had almost certainly thought in terms of gaining experience by first conducting elementary flight training and then broadening their flight operations to include advanced training. Instead, they were overtaken by events and suddenly found themselves committed to conducting advanced flight training first, as they took advantage of an unanticipated opportunity to broaden their CPT Program. This unexpected turn of events meant that Washington had to build a flight training program from scratch in less than a month. That done, he immediately assessed the broader implications of Tuskegee's new role as an advanced flight training operator.

The financial aspects of the new flight training program were a crucial consideration. Washington knew that if Tuskegee was to have a permanent aviation program, it would have to be self-supporting. Indeed, in late June he had assured Patterson that secondary flight training would generate sufficient income to cover expenses: "As to [the] expense of becoming the operator, the money is in the operation of flight work and you can see from Mr. Nilson's figures that the money is there for the first year." But by the end of July, after he had prepared his own cost analysis, Washington had concluded that unless flight operations were expanded beyond the level of ten secondary students per training period, the aviation program would never be fiscally sound and might very well operate at a loss. He suggested three likely avenues for expanding Tuskegee's aviation operation and generating a self-sustaining profit: the addition of a CPT course to train secondary graduates as elementary flight instructors; an increase in the secondary quota from ten to fifteen; and the assumption of elementary flight instruction. He estimated that with the three new programs in operation, a yearly income of more than $24,000 could be realized, which could finance essential capital expenditures for equipment and facilities.[68]

Several days after Washington completed his financial analysis of the aviation program, C. Alfred Anderson arrived on campus to take responsibility for the secondary students and the exhausted CPT coordinator took a well-deserved vacation. Unfortunately, he became ill, perhaps drained by the pace of work associated with the aviation program.[69] By the end of August he had recovered and was once again devoting his full attention to the CPT Program.

Shortly after Washington returned to campus, Patterson sent a proposal to CAA headquarters embodying the recommendations for an expanded CPT Program that Washington had submitted at the end of July. The proposal called for larger elementary and secondary quotas, an instructor course, and refresher and recertification courses for black fliers who wanted to maintain or renew their licenses.[70] A week later, Washington presented a revised analysis of Tuskegee's aviation program and outlined a plan of action to put it on a firm financial footing.

On Monday, 9 September 1940, Washington submitted to President Patterson a six-page budget analysis of the aviation program together with a list of "Immediate Considerations" that had to be accomplished in order to meet the goals and objectives Washington envisioned for the coming fall and spring aviation programs. As in the assessment he prepared in late July, Washington analyzed the income, operating expenses, and capital expenditures for both the existing CPT Program and the expanded program. But in September the analysis of the existing program assumed that Tuskegee would have responsibility for elementary flight training as well as all advanced flight training. Washington emphasized that "*taking over immediately* flight operation of [the] primary program" was both "essential and required under the pending [2 September] proposal to [the] C.A.A. for additional programs." Simply by taking over elementary flight operations, with no increase in student quotas, Washington estimated that the aviation program would produce a $9,000 net profit, which could be used to amortize an outlay of $23,000 in capital expenditures required to become an elementary and advanced CPT flight operator. When the additional student quotas and new programs requested in the 2 September proposal were considered, the financial picture was even brighter. These programs would produce a net income of $17,000 with no additional outlay of capital.[71]

In short, the institute was on the verge of establishing a self-supporting aviation program, but only if steps were taken to implement the "Immediate Considerations" Washington had outlined; these items clearly show the scope of the operation Washington envisioned and the importance of prompt action:

Considering that the *Primary [i.e., elementary] Program is to start [the] middle of September* and that the *Secondary Program is to start October 1, 1940* there is no time to lose in accomplishing the following: (Listed in order of accomplishment importance)

1. Close deal with Mr. Joseph W. Allen of the Alabama Air Service by September 10, 1940, for purchase of *three Piper Cub Trainers.* Two to be delivered before October 1, 1940 and one delivered before November 1, 1940. . . .

   The fourth Cub trainer needed has been promised by Mr. Lewis A. Jackson. . . .

2. Secure agreement from Mr. Vaughn that the Tuskegee Airport [i.e., Kennedy Airfield] will be leased directly to Tuskegee Institute for twelve months from January 1, 1941. . . .

   Unless Mr. Kennedy's active interest in the Tuskegee Field is relinquished, we are handicapped under the present sub-lease in that no private flying is to be done by us.

3a. Begin now construction of two-plane and shop hangar at Tuskegee Airport to be completed by October 1, 1940. (The existing hangar will care for two of the above planes).

3b. Begin renovation of Metal Shed for shop-laboratory instruction under present programs and to meet C.A.A. requirement. In this laboratory, aviation mechanics will initiate.

3c. Place order for 6 parachutes and additional fire first aids [sic] to be delivered before October 1, 1940.

4. *Director's [i.e., Washington's] Concurrent Accomplishments* (deadline September 15–October 1, 1940)

   a) Immediate conference with Mr. Grove Webster on Tuskegee's proposal of September 2, 1940, for additional programs. . . .

   b) Secure two primary [i.e. elementary] instructors for flight [training]
      1. Two Negroes, first choice
      2. Two whites, second choice; Alabama Air Service and Alabama Institute of Aeronautics cooperation may be had in this connection.
      Agreements to be closed on this employment before October 1, 1940.

   c) Secure two C.A.A. certified Ground Instructors. Schedule for ground instruction will require both to have reported to Tuskegee between September 15th. and October 1st.

   d) Interview and secure an A&E rated mechanic for inspection and maintenance of Tuskegee aircraft.

   e) Compile (with aid of selected mechanic) tools, equipment and parts inventory to be secured for operations of aircraft maintenance at airport.[72]

Washington presented this ambitious plan of action on 9 September and within a week both he and Patterson had made significant progress in achieving its objectives. By 12 September Washington had met with Grove Webster at CAA headquarters, and he ebulliently reported to Patterson: "Got what we wanted at C.A.A." The instructor course would begin in October and the elementary quota was definitely to be increased from twenty to thirty. With the ten secondary students, a quota of ten apprentice instructor students, and thirty elementary students, an elated Washington predicted that

"we will have 50 students in three types of training at once beginning very soon (Fall Session) and on top of that a number doing 15
hours to renew license." Washington assured Patterson that he was
making progress in hiring the necessary personnel for the program,
reporting that he was interviewing mechanics as well as ground and
flight school instructors. The CAA's approval of Tuskegee's proposal
made it imperative, Washington told Patterson, that the institute
"*close out items* 1, 2, 3a, 3b, and 3c on 'Immediate Considerations' on
[the] yellow sheet exhibited in our conference of Monday. *There is
no time to lose.*" Especially concerned about obtaining aircraft for
elementary training, Washington reminded Patterson that "It is
absolutely necessary to clinch Allen's three Piper Cub planes and
Jackson's plane,—four in all."[73]

Washington scarcely needed to urge Patterson to action. On 10
September 1940, the day after their conference on the aviation
program, the Tuskegee president called a special meeting of the
Executive Committee (Southern Members) of the Board of Trustees,
seeking authorization to appropriate institute funds for the first
three items on Washington's list of "Immediate Considerations."[74]
When the committee met four days later, Patterson began by
explaining that the elementary program was presently "in the hands
of outsiders, [while] the secondary program or advanced flying was
handled entirely by Tuskegee Institute." When the Tuskegee president placed Washington's "Immediate Considerations" before the
committee, the vote was unanimous in favor of an emergency
allocation of $6,400 to purchase the four Piper Cubs and associated
equipment, improve hangar facilities, and lease Kennedy Field for
1941. Thus Patterson did move quickly and obtained the first
definite commitment by the trustees to establish an aviation program at Tuskegee Institute. The trustees' steps to provide emergency
funding for the aviation program, however, probably would not
have been taken without Washington's painstaking efforts to develop the framework of a self-supporting enterprise; the $6,400,
they warned, was not a grant but a loan that was to be refunded to
the board "from income based on United States government contracts."[75]

By the end of October 1940 Tuskegee Institute was in full control
of all phases of its CPT Program. No longer was it necessary to rely on
Alabama Air Service for flight training and API aeronautical engi-

neering professors to conduct the ground school, for Tuskegee had a staff of six aviation instructors, three blacks and three whites.[76] In early November, Washington named Charles Alfred Anderson the institute's chief instructor pilot, and he was soon known to all as "Chief" Anderson, an affectionate title he has carried ever since.[77] Tuskegee's CPT student quota for the fall 1940 session was thirty elementary, ten secondary, and ten apprentice instructor, with a similar number anticipated for the spring 1941 session. No other black college or private flight school could boast such numbers and such a variety of training. Moreover, the financial standing of Tuskegee's aviation program was excellent. Washington estimated the net profit for the year ending 31 May 1941 at $22,000, an amount sufficient to liquidate the institute's debts for the capital improvements associated with the aviation program. Washington, however, recommended that only a portion of the profit be used to amortize capital debt and that the remainder go toward an additional secondary training aircraft and improvement of ground school facilities; he predicted that an even greater profit would accrue in the 1941–42 budget year.[78]

One major element of Tuskegee's aviation program remained unfulfilled—the construction of an airfield. Nowhere in his budget analyses did Washington contemplate that profits from the aviation program would be used to build an institute airfield. He had not, however, forgotten this important matter in the months since the Alabama Aviation Commission provided estimates of construction costs. In mid-July he reminded Patterson that an airfield was still needed: "By the way,—we will have to have some money for the field." Then, somewhat flippantly, he quipped that he was "glad such matters are for the President to worry about," an apparent reference to a controversy that had arisen when Washington approached a member of the Board of Trustees directly on a matter relating to the aviation program.[79] In a mild letter of reprimand, Patterson had told Washington that at least one trustee was "embarrassed by your making representations to him on behalf of the aviation program instead of allowing this request to come through the President's office." He reminded Washington that he had no "permission whatever to make representations directly to the Trustees in regard to any program." Patterson conceded that Washington's motives were sincere, however, noting that he was certain that the CPT coordinator's "enthusiasm for the work and . . . desire to see activities go on as rapidly as possible are responsible for this action."[80]

Patterson apparently had no reservations about Washington's attempts to organize a fund-raising drive among Tuskegee alumni in Chicago, Detroit, and Cleveland. In early June, after Washington discussed the institute's accomplishments and future plans in aviation, the Chicago Tuskegee Club voted to embark on an airport fund-raising campaign and approach the Cleveland and Detroit Tuskegee Clubs about cooperating in such a venture.[81] Patterson met with the Chicago club in July and reported that Tuskegee was rapidly taking the lead in the field of aviation, having recently been approved for CPT advanced training, "which will establish Tuskegee Institute as a center for training instructors for the other Negro colleges. This is an additional reason why we must go forward as rapidly and effectively as possible with the drive for funds for our own air field."[82]

Patterson had no doubts about the enthusiasm of the alumni clubs for the venture, but he had serious reservations about their ability to raise the sort of money that was required to build an airfield. Consequently, the Tuskegee president approached a San Francisco public relations firm about managing the entire campaign, with the support and assistance of the alumni clubs.[83] The firm agreed to undertake the campaign, but the alumni clubs were apparently unwilling to cooperate and the initiative died.

In the fall active campaigns were launched in Chicago and Cleveland, but in Detroit there was little interest in the matter. By November 1940 a "Citizens Committee" for the campaign had been organized in both Chicago and Cleveland, each some thirty members strong.[84] When Washington first solicited the support of the alumni clubs, he had hoped that the Board of Trustees would provide partial funding for the airfield. But by the time the alumni campaigns were launched, it was obvious that such support would not be forthcoming. In late October, Washington explained the need for alumni support to Robert P. Morgan, the chairman of the Cleveland campaign. Such an effort was necessary because the institute's "funds for permanent construction have been depleted. Unless the Alumni and interested friends furnish Airport construction funds, it is likely that the project will not materialize." Although federal funding under the WPA might have been secured, the institute still would have been required to provide roughly forty percent of the funds and then deed the land "to the City, County, or State." Thus the "matter of control of the Airport and the deeding of so much land over to the State influenced our officials to forego any WPA consideration." Finally, Washington explained that even though Tuskegee's main objective was to train black aviation mechanics and technicians, "flight training is still necessary . . . as all technicians and mechanics in the best regulated [aviation] schools are required to be able to fly," just as automobile mechanics are required to be able to drive. Washington concluded that the emphasis on training mechanics and technicians was "by no means . . . intended to depreciate the program in flight training inasmuch as this is a permanent development at Tuskegee Institute irrespective of Government programs."[85]

Even though Washington had indicated to Morgan that federal funds had been declined because they would compromise the institute's autonomy, by September Patterson had so little confidence in the capacity of the alumni to raise funds that he was exploring several avenues for obtaining government support and was willing to deed land if necessary. When he learned from Edgar G. Brown that an appropriation bill for airport construction was pending before Congress, he wired Alabama senators Lister Hill and John H. Bankhead, asking that "part of special appropriation for this purpose may be earmarked for Tuskegee Institute."[86]

When Senator Hill advised Patterson that the funds for the pending legislation were to be administered by the CAA, the Tuskegee president wrote Grove Webster to solicit his support.[87] He explained that the Board of Trustees "would be willing to work in close cooperation with state and federal officials if a parcel of land could be accepted for airport development." Patterson indicated that he would be "willing to recommend the deeding of this land to whatever public agency would be acceptable to the federal government in order that we could take advantage of an allotment from this special appropriation. We are working hard to raise the money to develop our own field, but I am sure we shall not be able to do a job that would be as satisfactory as would be possible through a more generous allotment of funds."[88] At the 24 October meeting of the Board of Trustees, a resolution was passed at Patterson's request, empowering the Executive Committee (Southern Members) to deed land for airport development to Macon County or the state of

Alabama, if such action was necessary to obtain federal funds for development.[89]

By the end of 1940 Tuskegee Institute was no closer to having its own airport than it had been in April when the Alabama Aviation Commission provided cost estimates for its construction. The alumni fund-raising campaigns had been launched, but it became apparent that such efforts would only partially meet construction costs. While federal funding was still possible, no definite commitment had been received in this regard. Still, the Tuskegee aviation program had come a long way in a short time. The school had an ongoing flight training program that was being conducted by its own staff in its own aircraft. As the only black institution to offer advanced training, Tuskegee was regarded by many as the nation's emerging center of black aviation, a status that would be reinforced by the developments of the coming year. Washington summed it up best in mid-October 1940 when he observed:

The Civil Aeronautics Authority has been satisfied with [the] administration and operation of Flying School operations at Tuskegee Institute. Additional programs under government contract for civilians will be granted. Too, Tuskegee's interest and willingness to make aviation training a going, integral part of our program has brought forth unusual interest in us on [the] part of the Civil Aeronautics Authority. There are definitely bigger things ahead if Tuskegee will only continue and approve support in this National Defense Program.[90]

8

The Campaign for Air Corps Participation Broadens

As Tuskegee Institute was establishing and expanding its aviation program in 1939 and 1940, black spokespersons continued to insist that blacks be admitted to the Air Corps. By the summer of 1940 this campaign for participation in the Air Corps had become a key element in black America's crusade for full participation in the national defense effort. Throughout 1940, several factors kept blacks' attention focused on the question of their role in the nation's air arm. The outbreak of war in Europe after the invasion of Poland in September 1939 virtually guaranteed that the pace of the military build-up would quicken. As the war widened in the spring, defense became a top national priority; thus the campaign for black participation in the armed forces, launched by the *Pittsburgh Courier* in early 1938, quickly became a central issue in the minds of black Americans and their leaders. The Air Corps soon figured prominently in the campaign because its adamant refusal to admit blacks made it an easy target for black critics and an important one, given its status as an elite branch of service.

Another factor that kept the issue of Air Corps participation high on the civil rights agenda in 1940 was black involvement in the CAA's Civilian Pilot Training Program. Like the exploits of Banning, Anderson, Forsythe, and Robinson in the thirties, the success of black students in the CPT Program inspired black Americans and reassured them that the white majority did not hold a monopoly on flying aptitude. Equally important was the issue of compliance with the provisions of P.L. 18 that related to the training of black pilots. Most blacks and their congressional allies refused to accept the Air Corps' narrow interpretation of the law and indignantly accused the army of violating the will of Congress. Finally, in the fall of 1940, the passage of the Selective Training and Service Act and the approaching national election inspired many blacks and their leaders, especially the NAACP, to quicken the pace of the campaign for participation in the armed forces. Their efforts resulted in a number of concessions by the Roosevelt administration, including a promise that blacks would be admitted to the Air Corps.

In the waning months of 1939 the inauguration of the Civilian Pilot Training Program at Tuskegee Institute and four other black colleges—North Carolina Agricultural and Technical College, Howard University, Hampton Institute, and West Virginia State College—was widely covered in the black press. Throughout 1940—as students at these institutions soloed and completed their training, as other black colleges entered the CPT Program, and as blacks at predominantly white institutions completed the CPT curriculum—the black press gave full play to the news that Negro youth were active in the field of aviation.[1] At the conclusion of the 1939–40 academic year about one hundred blacks had completed the CPT Program. By the fall of 1940 the number of black colleges participating had grown to nine and more than 300 black students were expected to complete the program by mid-1941. The number of participating institutions would have been even higher, but several interested and eligible colleges were not allowed to participate because they lacked access to adequate airport facilities and were apparently unable to convince CAA authorities to grant the same waiver that had been accorded Tuskegee Institute in 1939. Especially gratifying to the black public, perhaps, were the black CPT students attending integrated schools who sometimes outperformed their white classmates. At the University of Minnesota the first student to solo, out of a class of fifty, was a black, as was the highest-scoring student in a Joliet, Illinois, ground school course.[2]

Despite the enthusiasm for the CPT Program, careful observers soon realized that pilot training programs in civil aviation were not a substitute for Air Corps flight training. While the *Pittsburgh Courier* acknowledged that the opening of the CPT Program to blacks was indeed a beginning, the paper continued to demand similar opportunities in military aviation: "The naming of five Negro schools where students may receive training as air pilots and mechanics is a step in the right direction but only a short step. No Negroes have been admitted to the Army Air Corps and there is not the slightest indication at this time that any will be."[3]

After the approval of P.L. 18 in April 1939, the Air Corps had come under increasing pressure to meet its provisions with regard to blacks. By late 1939 the Air Corps and the CAA jointly sought to establish a pilot training program for blacks that would meet the provisions of the law as interpreted by the Air Corps. Under the Air Corps' narrow reading of the law, one or more of its civilian contract schools was to be "designated by the Civil Aeronautics Authority for the training of any Negro air pilot."[4] Nothing in the legislation, Air Corps officials pointed out, required that blacks be admitted to the Air Corps, even though that had clearly been the intention of the senators who introduced the original amendment to the act. By the fall of 1939, with the expansion of the Air Corps well under way, neither the Air Corps nor the CAA had taken any steps to fulfill even the limited obligations they believed the law imposed.

Of the nine civilian contract schools providing Air Corps training in late 1939, the most likely candidate for designation as a school for training black pilots was the Chicago School of Aeronautics. It was the only school of the nine outside the South that was also situated near an important center of black population.[5] Moreover, the school's owner, Harold S. Darr, was on friendly terms with the black aviation community in Chicago, some of whom had learned to fly at Glenview in the early 1930s when it was the site of the Curtiss–Wright Flying School.[6]

In late December 1939 John C. Robinson discussed with Darr the matter of enrolling blacks as aviation cadets in the Chicago School of Aeronautics. According to Robinson, he knew Darr well and kept one of his planes at the Glenview flying field. Darr told Robinson that the issue of training blacks in the Air Corps was "red hot . . . [because NAACP executive secretary] Walter White was raising so much hell that the Army was being seriously embarrassed." When Robinson declared that blacks should be admitted to primary training, Darr disagreed, "afraid that the one social contact might be too much to swallow at one time." Robinson suggested that if social contact was the issue, then the cadets could be billeted on Chicago's Southside and driven to the Curtiss Airport daily, at army expense, for their training.[7]

Robinson's suggestions fell on deaf ears. By the beginning of 1940 the Air Corps and the CAA had settled on a plan to train black pilots, a plan which the Air Corps hoped would simultaneously satisfy the provisions of P.L. 18, silence critics like Walter White, and keep the Air Corps from being integrated. When Air Corps and CAA officials presented their plan at a 15 January meeting in Chicago, it immediately became apparent that the city's black aviation community believed that Congress had mandated the admission of blacks to the Air Corps with the passage of P.L. 18, and that they intended to see the will of Congress carried out.

On 15 January 1940 some twenty participants crowded into a conference room at the headquarters of the Rosenwald Foundation.[8] The site of the meeting had apparently been chosen by Dr. M. O. Bousefield, a prominent physician associated with the foundation, who had recently become the first black appointed to the Chicago School Board. The key participants were Maj. R. M. Webster, chief of the Air Corps' Training Section; Grove Webster, from CAA headquarters in Washington; and five representatives of the NAAA—Willa B. Brown, Cornelius Coffey, Earl Renfroe, Edgar G. Brown, and Enoch P. Waters, who was also managing editor of the *Chicago Defender*. Also in attendance were representatives of the Chicago school system, officials from the city of Chicago, regional CAA officials, the commander of Chicago's black National Guard regiment, Darr, and Robinson, who had recently severed his relationship with the NAAA.[9]

Bousefield opened the meeting by apologizing for his unfamiliarity with the situation at hand and expressed his interest in providing opportunities for young blacks who wanted to fly. Newspaperman Enoch Waters, speaking for the NAAA, showed little patience with the gradual approach advocated by Bousefield and demanded that "something should be done about Negroes in the Army right away; that there was a Constitution and that we as a race, were entitled to it."[10] The Air Corps representative, Maj. R. M. Webster, curtly dismissed Waters's complaint: "I know as much about the Constitution as you do and there is no need to bring that part of it in. We all know that Congress runs the Army and it does what it pleases about the Constitution. We have to get at this matter slowly, and we are here today to discuss the first step."[11]

As far as the representatives of the federal government—Grove Webster of the CAA and Major Webster of the Air Corps—were concerned, the first step was to arrange for the establishment in Chicago of CPT programs open to blacks. One of the courses would be conducted by the Chicago School of Aeronautics, which also conducted primary flight training for Air Corps cadets, with the Air Corps providing equipment for the ground school. Thus the provisions of Public Law 18, Section 4, would be fulfilled as far as the Air Corps was concerned. The other Chicago CPT course would be a noncollege unit sponsored by the NAAA, with flight training conducted by the Coffey School of Aeronautics.

Despite the best efforts of the CAA and Air Corps representatives to focus the discussion on CPT courses and avoid the issue of bona fide Air Corps training for blacks, Major Webster reported that the representatives of the National Airmen's Association took an

argumentative stand . . . [and] consistently attempted to broaden the discussion to include training such as that given by the Air Corps at civil flying schools. They stressed the fact that under the Air Corps program the students received not only transportation expenses, but pay during the period of their training, and that they later went forward for advanced flying at the Air Corps Training Center and subsequently were tendered commissions in the Reserve Corps. After considerable conversation the faction in question was impressed with the necessity of confining the deliberations to the specific points at issue, i.e., the completion of detailed plans for the functioning of the C.A.A. program for negro training in Chicago. Mr. [Edgar G.] Brown gave notable assistance in properly channeling later discussion.[12]

Brown had apparently concluded that it would be futile to press the issue of Air Corps participation at the conference and tried to serve as a mediator between the CAA and the Air Corps, on the one hand, and the indignant members of the NAAA, on the other. It was a natural role for him, because he had served as the NAAA's Washington lobbyist for almost a year and was probably acquainted with the CAA and Air Corps officials who were attending the meeting. He realized, no doubt, that the conference was not a suitable forum for negotiating the admission of blacks to the Air Corps; if Enoch Waters and the other blacks in attendance insisted on resolving the question then and there, then perhaps the Air Corps would cite their lack of cooperation as an excuse for disregarding P.L. 18 completely.

As a result of Brown's mediation, the fundamental issue—Air Corps participation—was tabled, and the details of the CPT Program were worked out. At the conclusion of the meeting it was agreed that two black CPT units would be established in the Chicago area, one to be conducted by the Chicago School of Aeronautics, which the CAA had "designated . . . for the training of twenty Negro air pilots in accordance with provisions of Public [Law] 18," and the other, under the sponsorship of the NAAA, to be conducted by the Coffey School of Aeronautics, with a quota of ten students.[13]

The twenty students to be trained at the Chicago School of Aeronautics were to have completed at least two years of college and would be selected by a committee headed by Dr. Bousefield. Those selected would begin their training with CPT elementary and those successfully passing that phase would be eligible for CAA secondary training once the course was inaugurated within the CPT Program and funds were appropriated, sometime after 1 July 1940. Students completing both courses could expect to accumulate 75–100 hours flying time and earn a limited commercial pilot license. The CPT Program sponsored by the National Airmen's Association was to be a noncollege elementary CPT Program with a quota of ten students, to be selected by Dr. Bousefield's committee, the same committee responsible for selecting the college students. Each flying school was to furnish the aircraft for the flight training and the Air Corps agreed to furnish equipment for the ground school.

A consolidated ground school, under the general supervision of Darr and Chicago School of Aeronautics, was to be taught in the basement of the Wendell Phillips High School, on Chicago's South-side, with classes held in the evenings. Darr planned to conduct the college unit's flying training not at his Glenview facility but at Ashburn Field on the Southside, using black instructors, if available, including his friend John C. Robinson, in whom he had "consider-able confidence."[14] The noncollege group planned to fly out of Harlem Airport, also on the Southside.

Once the details of the training program had been agreed upon, the NAAA spokesmen again raised the issue of admitting blacks to the Air Corps. At one point in the discussion, Major Webster reported, his CAA colleague, Grove Webster, characterized the Chicago School of Aeronautics CPT Program as "experimental," creating the "impression among the colored group that it would constitute an entering wedge for still further advancement towards the ultimate objective, which apparently visualizes training in the Air Corps establishment. This opinion was voiced by at least one of those present."[15] Apparently neither Grove Webster nor Major Webster was willing to disabuse the NAAA of the notion that a step was being made toward admission of blacks to the Air Corps. When the NAAA secretary, Willa B. Brown, reported the results of the conference, she used the term "demonstration unit" to describe the Chicago School of Aeronautics CPT Program, indicating that this group of students, who were to possess all the qualifications required of aviation cadets except skin color, would be expected to "demonstrate" their potential as Air Corps pilots.[16]

Despite the obvious differences between the Air Corps, the CAA, and the NAAA regarding the purposes of the black pilot training program in Chicago, all three parties began to arrange for the inauguration of the two flight training courses agreed upon at the conference. The Air Corps was especially anxious about the matter because scarcely a week after the conference it was once again charged with flagrant disregard of the law by a member of the Senate Military Affairs Committee.

In Senate debates on the supplemental military appropriations bill for fiscal year 1940, Republican senator H. Styles Bridges protested that the provisions of P.L. 18, Section 4, had not been carried out. He reminded his colleagues that when the bill was before the Senate he had "offered an amendment . . . to provide facilities for training Negro aviators; and under the able New Deal leadership on the other side, the Senator from Wyoming [H. H. Schwartz] offered a substitute which was approximately my amendment, which was carried, and provided for training Negro aviators." Nevertheless, the War Department had recently told a qualified black who applied for appointment as an aviation cadet: "It is regretted that the nonexistence of a colored Air Corps unit to which you could be assigned in the event of completion of flying training precludes your training to become a military pilot at this time." Such cavalier disregard of the will of Congress, Bridges asserted, "is a rather serious thing." Senator Schwartz's amendment had been passed, the New Hampshire senator continued, "in good faith to provide training for the colored men of this country who desire to participate and secure training as aviators in the United States Army; and apparently the law has been ignored." Bridges, an outspoken critic of the Roosevelt administration, concluded by demanding "some word as to why the administration here in Washington, headed by President Roosevelt, who claims to be so interested in these matters, has ignored the colored people of the country in that particular matter."[17]

Bridges's attack no doubt removed any lingering doubts the Air Corps might have harbored about the need to make some gesture that it could claim demonstrated an intent to comply with the provisions of P.L. 18 as it related to the training of black pilots. On 1 February 1940 General Arnold advised the Air Corps Training Center that the Chicago School of Aeronautics was to start training black students under the provisions of P.L. 18, Section 4.[18] By early February some preliminary ground school courses were begun with Willa B. Brown serving as instructor. Brown had received an appointment as an aviation instructor in the WPA Adult Education Program, an arrangement which had been discussed at the 15 January conference.[19]

As secretary of the NAAA, Brown was also responsible for publicity; in the second week of February she launched a recruiting campaign, hoping to attract enough students to the program so that it could begin without delay. Press releases were sent to the Associated Negro Press, *Chicago Defender, Chicago World,* the *Metropolitan Post,* the *Chicago Bee,* and *Pittsburgh Courier,*[20] and a thousand handbills describing the details of the training program were printed and distributed.[21] The announcements described both programs, with the NAAA-sponsored course labeled the "Non-collegiate Unit," and the Chicago School of Aeronautics course called the "Demonstration Unit."

The Non-collegiate Unit was clearly described as a course under the auspices of the CPT Program, following procedures similar to other noncollege programs across the nation. But the description of the Demonstration Unit was another matter, and it showed that the NAAA's vision of its purpose contrasted sharply with the Air Corps' vision. The opening lines declared that "Twenty young men will receive special aviation training preparatory to appointment as Army air corps flying cadets." To qualify, the announcement continued, applicants had to be unmarried American citizens, 18–25 years of age, with at least two years of college education, have had no previous solo flying experience, and be able to pass the army medical examination. All those who qualified would be enrolled in the ground school and the top twenty would be selected for flight training on the basis of their ground school examination scores. The flying phase was to consist of the "controlled private flying course as developed by the Civil Aeronautics Authority"—in other words, the CPT elementary course—but the connection with the CPT Program was not clearly stated. Those completing the controlled private flying course, the announcements continued, would "immediately enroll in an advanced . . . flying course," presumably the CPT secondary course. Finally, the announcement promised that completion of the prescribed course of study would lead to appointment as an aviation cadet in the Air Corps: "After successfully completing the advanced course pilots will then be inducted into the Army Air Corps Training Service."[22]

Nothing in the Air Corps or War Department records indicates that the promise of admission to the Air Corps had any official sanction. Indeed, in early April 1940, when the assistant chief of the Air Corps, Gen. Barton K. Yount, apprised Army Chief of Staff George C. Marshall of the status of pilot training for blacks, he made it very clear that the Air Corps had no intention of admitting blacks. Yount told Marshall that the War Department's only obligation under the law was to lend "such equipment as the Civil Aeronautics Authority deems necessary for one school for the training of Negro pilots, such school to be designated and operated by the Civil Aeronautics Authority." The CAA, Yount continued, had designated the Chicago School of Aeronautics as the school for the training of black pilots and the Air Corps had furnished the necessary equipment, "ground school equipment only—the school furnishes the planes." Yount related that General Arnold had sent a representative to several conferences in Chicago regarding the implementation of P.L. 18 and reassured the chief of staff that "the Negroes are very well pleased with the arrangements that have been made." According to Yount, P.L. 18 required the Air Corps to cooperate with the CAA, but it did

"not require Negro units in the Air Corps." Moreover, Yount emphasized, it was "not contemplated that any Negro units will be established, and it is not planned to take any of the Negro pilots into the
Army Air Corps."[23]

Thus there was a fundamental disagreement between the NAAA
and the Air Corps over the ultimate disposition of the black pilots
who were to be trained by the Chicago School of Aeronautics.
Certainly, there was some basis for the NAAA's assumption that
these young men were destined for more than simply the CAA pilot
license their counterparts in other CPT units would receive. The men
were to be trained in compliance with P.L. 18, the Air Corps Expansion Act, at a civilian school that also had a contract to conduct
primary training for aviation cadets. Furthermore, they were required to have the same entrance qualifications as aviation cadets. In
the end, however, the course they were to take was simply a CPT
course; it was not bona fide Air Corps flight training. In effect, it was
a noncollege CPT Program in that it was not affiliated with a college,
but its students were nevertheless required to possess at least two
years of college training. Thus there was little incentive, except the
illusory promise of admission to the Air Corps, for black college men
to leave school and apply for the training, especially if their institution was already participating in the CPT Program.

The subsequent history of the Demonstration Unit is not easily
traced. The selection committee found it difficult to attract applicants
who possessed the requisite physical and educational requirements,
and who also had the financial resources to meet their living
expenses while enrolled in the six-month training course. By mid-
March 1940 the quota of twenty still had not been filled; consequently, Dr. Bousefield extended the application deadline until 15
April and modified the sequence of instruction to allow concurrent
ground school and flight training.[24] Later in the month the Air Corps
shipped equipment for the ground school course to Wendell Phillips
High School, thus fulfilling the War Department's statutory obligations as far as the Air Staff was concerned.[25] By the summer of 1940
the Coffey School of Aeronautics' first CPT elementary course at
Harlem Airport had thirty students under instruction, which suggests that the Chicago School of Aeronautics' quota was reassigned
to the Coffey School.[26]

Noel Parrish, an Air Corps officer assigned to the Chicago School of

Aeronautics' primary school (and later the commander of Tuskegee Army Air Field), did not remember any blacks being trained by the Chicago school. Instead he was struck by the ambiguity of the relationship between the Air Corps and the black pilots in training in Chicago:

> In order to at least appear to comply with the law . . . the Air Force command decided to supply these schools with some ground school equipment and things of that sort, no military planes, and to assist them and to advise them and so forth. It didn't amount to much, there was no specific indication of what we should do, other than take an interest in it. So, because the Glenview School was only a few miles away, we were told to go over and take an interest in this school, find some way to help them and find out how well they were using the Air Force equipment. . . . My commander . . . had no interest in the project whatever. He told me to do all that. So, I went over there and talked to the people. I was very much impressed with them. They were very sincere, and seemed to be working hard, and a little puzzled by the arrangement as to what the Air Force was doing for them, just as we were puzzled by what we were supposed to do.[27]

By March 1940 a pattern of controversy between the Air Staff and blacks campaigning for admission to the Air Corps had emerged, and it prevailed for over six months. Black advocates and their allies attacked the War Department for failing to comply with P.L. 18, while the Air Corps and their allies asserted that the provisions of the law had been fulfilled and that a start was being made in the development of black personnel for the Air Corps.

In mid-March, when the ubiquitous Edgar G. Brown testified before the House Appropriations Committee, he asserted that the Air Corps had complied with the letter of Public Law 18 but not with its spirit. Brown acknowledged that the "demonstration unit in Chicago, under the C.A.A., has received some valuable assistance from the War Department, but Negroes do not receive equal opportunities." Reading from letters of disappointed black applicants, he emphasized that "All Negro citizens are denied an opportunity to enlist in the United States Army Air Corps as pilots or mechanics. . . . The truth is no Negroes have ever been admitted to the . . . Air Corps." Brown recounted the participation of blacks in the CPT Program, noting that many of the black students had shown an aptitude for flying and were eager to serve in the Air Corps. Especially frustrating to blacks were the recruiting posters prominently displayed across the nation that proclaimed the nation's need for military pilots while "Negroes who request aviation training are unable to get in the Air Corps."[28]

When the military appropriations bill for 1941 was debated before the House, the bill's sponsor, Democrat J. Buell Snyder of Pennsylvania, was questioned on the subject of black pilot training by two of his colleagues, Louis Ludlow and Everett M. Dirksen, who had actively supported such measures in 1939. Ludlow, a fellow Democrat, was careful not to question Snyder too closely on the matter. When he asked about the Air Corps' policy on enlisting blacks, he seemed satisfied with Snyder's response that "Negroes are enlisted in the Army, and there is no law that prohibits their assignment to the Air Corps. Whether or not any have been so assigned, I am not advised. But I want to say to the gentleman that I am satisfied we are making headway."[29] But Dirksen, a Republican, refused to accept such an equivocal response and asked Snyder if the Air Corps was refusing to enlist blacks. Snyder told the Illinois congressman that "Thus far there has been no occasion to enlist them." He went on to explain that as soon as "the school has trained a sufficient number of students to warrant the creation of a special unit, I am advised that the War Department will then take such steps as may be necessary looking to the utilization of this additional trained personnel." When Dirksen pressed the matter and asked Snyder if he had "any idea when that will be," the congressman from Pennsylvania answered optimistically "I should think in the very near future. These civilian schools are turning out pilots in from six months to a year."[30]

In May 1940 the Committee on the Participation of Negroes in the National Defense Program joined the fight to admit blacks to the Air Corps. Led by Howard University historian Rayford W. Logan, the committee was concerned with all aspects of black participation in the armed forces and paid particular attention to exclusionary practices of the Air Corps. An umbrella organization for a number of groups, the committee was sponsored by the *Pittsburgh Courier* and sought to end racial discrimination in the armed forces and defense industries.[31] Although the committee opposed racial segregation, it was prepared to accept separation if it led to opportunities that would otherwise be denied. Logan clearly outlined the position of the committee on this point in one of his first appearances before Congress, when he announced that his organization "protests the continuation of separate Negro units. We deplore segregation in any form, especially when it is practiced by the Federal Government. But in accepting these separate units which are forced on us, we do so only because of the hope that these units will be commanded by Negro officers."[32] This pragmatic stance on segregation, however, cost the committee the participation of the nation's most outspoken civil rights organization, the NAACP, which declined the *Pittsburgh Courier*'s invitation to join the committee but agreed to cooperate on specific issues.[33]

In May 1940 two representatives of the Committee on the Participation of Negroes in the National Defense Program, Logan and Louis Lautier, appeared before the Senate Appropriations Committee. In his prepared statement, Logan raised the issue of Air Corps participation, observing that "no Negro as of today is being specifically trained for service in the Army Air Corps as either a flying cadet or an enlisted mechanic."[34] Republican senator H. Styles Bridges interrupted Logan to renew his attack on the War Department and the Roosevelt administration:

> Senator Bridges: Just a minute. When the last appropriation bill was under consideration a little over a year ago, I introduced an amendment which provided for the training of Negro aviators. My amendment, through some maneuvering in the Senate, was substituted by one of Senator Schwartz here, of Wyoming. That was passed, specifically authorizing and directing the War Department [to provide for the training of Negro aviators]. Has that been done? . . .
>
> Mr. Lautier: Our information, Mr. Senator, is to the effect that the Secretary of War designated a school at Glenfield [*sic*], Ill., and the War Department accepted that, but the War Department has refused to accept a Negro in that corps, and they have no separate corps for service. . . .
>
> Senator Bridges: The amendment which I offered and the Schwartz amendment which was substituted . . . has been carried out, but in effect even though they are trained, the War Department is not accepting them. Is that correct?
>
> Mr. Lautier: They are not being given military training. The school has been designated, the equipment has been loaned, but no Negroes

are being trained at that institution. Just complying to that extent with the Schwartz amendment.

Senator Bridges: Well, personally, Mr. Chairman, I think that this committee or somebody ought to talk with the War Department on that particular matter, because that certainly was the sense of Congress.[35]

Logan no doubt concurred with the senator and concluded his remarks on the Air Corps by noting that "Of the $165,762,162 provided for the Air Corps in . . . H.R. 9209, not one penny would be spent in developing Negro personnel, either enlisted or commissioned, for service in the Army Air Corps." He urged, therefore, that ten percent of the $2.1 million "allocated for the tuition of flying cadets and enlisted mechanics . . . be earmarked for the training of Negro flying cadets and enlisted mechanics for the Army Air Corps."[36]

Despite the indignation of Senator Bridges and the reassurances of Congressman Snyder, the Air Corps stood fast in its policy against admitting blacks. By May 1940 the War Department, under growing pressure to increase the number of blacks in the army, had begun to seriously consider the creation of additional black units.[37] Consequently, the General Staff asked the chiefs of the various branches to comment on how blacks might best be used in their particular specialty. General Arnold adamantly asserted that there were no Air Corps units "combat or service, for which it is recommended that Negro personnel be used for mobilization planning purposes or as peacetime augmentation in the Regular Army." He noted that P.L. 18 "was so worded that the training of Negro pilots was allotted to the Civil Aeronautics Authority." Consequently, the secretary of war approved a policy that stipulated "that Negro pilots would not be trained for the Army Air Corps but that their training would be given by the Civil Aeronautics Authority at one of the civil schools used by the Army Air Corps. This method of training is being carried out at the present time at Chicago, Illinois." The black pilots being trained could not, however, be used in existing Air Corps units, "since this would result in having Negro officers serving over white enlisted men," creating "an impossible social situation." He rejected out of hand the idea of organizing an all-black squadron, since "it would take several years to train the enlisted men to become competent mechanics." Thus, Arnold concluded, it was not "feasible to train Negro pilots for the Air Corps or to organize Negro units for augmentation of the Regular Army or for mobilization planning purposes."[38]

Although some on the General Staff objected to the attitude of the Air Corps, arguing that all arms and services should have a proportionate share of black recruits, Arnold's policy held throughout the summer of 1940.[39] It was, however, totally at odds with the reality of the political situation that unfolded before the upcoming presidential election. By autumn, the exclusion of African Americans from the elite Air Corps had become something of a cause célèbre among black Americans, seen as tangible evidence of the army's intention to relegate blacks to labor battalions and fatigue duty, just as in World War I. The issue figured prominently in congressional debates over an antidiscrimination provision in the Burke–Wadsworth Bill, passed as the Selective Training and Service Act, the first peacetime draft in the nation's history. As the election approached, perceptive black leaders demanded that Roosevelt, running for an unprecedented third term, end racial discrimination within the armed forces. Even the Air Corps could not withstand such pressure; by October the Air Staff was reluctantly, and secretly, drafting plans for the establishment of a token black squadron.

By June 1940, when the NAACP opened its annual conference, Denmark and Norway had fallen before German forces, the Low Countries had been unable to resist the Nazi invaders, and Hitler's Panzer divisions thrust resolutely toward Paris; two days into the conference, France capitulated and Germany's forces seemed invincible. The NAACP's theme for the meeting—the role of black Americans in the national defense effort—suddenly took on a new significance as the delegates realized with the rest of the nation that only Britain stood between the United States and Germany. On the first day of the conference Rayford W. Logan reported on the efforts of the Committee on the Participation of Negroes in the National Defense program; he noted that although the War Department had no plans for increasing the black presence in the peacetime army, in the event of a national mobilization blacks could expect to be drafted in numbers sufficient to comprise approximately 10 percent of the infantry, artillery, and quartermaster corps. But even in the event of a national emergency, Logan continued, blacks would still be excluded from the Air Corps.[40] Logan went on to comment on P.L. 18 and the training in Chicago, explaining that the Air Corps cited the lack of segregated units as the rationale for refusing to admit blacks.[41]

Several other speakers at the NAACP conference made special reference to the issue of Air Corps participation. William H. Hastie, dean of the Howard University Law School, reminded his listeners that they had recently heard the Roosevelt administration call for "ten thousand airplanes, and we know that there have got to be pilots, skilled mechanics, and a tremendously expanded force." The NAACP, Hastie declared, "must insist that Negroes are given every opportunity to participate in every phase of this expansion."[42] Walter White's closing address also focused on the issue of black participation in the armed forces, and like those before him, he found the Air Corps' policy of racial exclusion particularly insulting. If the leadership of the nation wanted the loyalty of black Americans in time of war, he warned, then they

> had best take a page from the book of experience of England. The British Empire, with but little difference in her methods of exploitation of her black and brown subjects, was forced in desperation to abolish race prejudice in her army by removing the color bar and admitting non-European British subjects to the Royal Air Force of the British Empire. But our own country, faced with danger almost as great as that of England, still clings tenaciously to the stupid and vicious prejudice of the past. . . .
> Negroes are barred from the Air Corps, from Annapolis, and can get into West Point only at the expense of almost overwhelming difficulty. Not only are Negro Americans of education, training and courage barred from the more skilled posts in the armed forces of America but they are even being banned from enlistment as Privates.[43]

By the close of the NAACP's annual conference for 1940 the association had adopted a number of resolutions relating to black Americans and the national defense program, including one which condemned the fact that "Negroes are barred from the air corps and most other branches of the service," a discriminatory practice that was "dangerous and treasonable."[44]

While the efforts of organizations such as the Committee on the Participation of Negroes in the National Defense program and the NAACP received the lion's share of attention in the black press, they were by no means the only voices of protest denouncing the Air

Corps' refusal to admit blacks. By June 1940 local and regional civil rights organizations were also calling for an end to exclusion from the air arm. The Commission on Interracial Cooperation (CIC), a conservative organization of southern blacks and whites which usually avoided criticism of segregation, was openly critical of the racial policies of the armed forces. A special CIC committee on national defense recommended that blacks be admitted to all branches of the army and navy, including "all phases of the air program." It noted that the Air Corps "has refused to accept a single Negro flying cadet." The committee was not, however, concerned with "pilots only, but we have equally in mind the employment of Negroes in ground crews in connection with aviation such as mechanics and the like."[45] In New York City a community council wrote President Roosevelt criticizing the racial policies of the Air Corps and urging that blacks be trained as military pilots.[46] One disheartened black who wrote the president denouncing the administration's policies toward blacks observed that African Americans "can take no pride in our armed forces as we are openly barred from the Marine Corps and regular air force."[47] In mid-July black cabineteer Mary McLeod Bethune forwarded a "Memorandum on Negro Participation in the Armed Forces," from an unidentified source, to the White House. The three-page memorandum noted specifically the exclusion of blacks from the Air Corps: "Negroes have not to date been accepted as flying cadets, despite the fact that Congress specified that one air school should be designated for the training of colored military pilots."[48] Indeed, the fact that the Air Corps refused to accept blacks was widely known among black Americans, and most suspected that the real motivation behind the refusal was not a lack of separate units but the prejudice of army officers who thought blacks lacked the technical aptitude to become competent military pilots and aviation mechanics.

Their suspicions were confirmed in late July by the remarks of the army chief of staff, George C. Marshall. At congressional hearings on military appropriations, Indiana congressman Louis Ludlow questioned Marshall on the participation of blacks in the army. When Ludlow asked about the Air Corps, Marshall declared "there is no such thing as colored aviation at the present time. A start has been made, and I think the C.A.A. is the proper place for that start or beginning to be made."[49] Although Marshall doubtless meant no offense by his remark, the black press was incensed at his assertion that blacks had not participated in aviation. The *Pittsburgh Courier* printed the full text of his testimony under front-page headlines which announced that the army chief of staff "Implies Race Men Are Not Capable of Being Army Pilots."[50]

Marshall's pronouncement that "there is no such thing as colored aviation" also caught the attention of the *Chicago Tribune*, a journal not known for its strong stand on civil rights but rather for its outspoken opposition to the Roosevelt administration. Under the guiding hand of Col. Robert R. McCormick, the *Tribune* had emerged by 1940 as the leading newspaper of the Midwest, a steadfast critic of the New Deal and a strong advocate of isolationism.[51]

Sensing another opportunity to embarrass the administration as the presidential campaign was under way, the *Tribune* editorialized on the subject of "Negro Pilots" in its 8 August 1940 issue. After briefly describing the exchange between Congressman Ludlow and Marshall which led to the chief of staff's unfortunate comment on black aviation, the *Tribune* urged Ludlow to continue his fight against the exclusion of blacks from the Air Corps, "not only in protest against discrimination but also in recognition of the very large contribution to national defense which may be expected of Negro pilots." The qualities that made a good military pilot, the editorial maintained, were the same as those found in a top athlete—good reactions, coordination, balance, and overall sound health. Additionally, a military pilot required "physical and moral courage." Blacks, the paper concluded, possessed these attributes in abundance:

A race which has produced, in the span of a few years, Joe Louis, Henry Armstrong, Jefferson of Northwestern, Ozzie Simmons of Iowa, a substantial number of golden gloves champions, and a score of other absolutely top-notch athletes provides a rich resource which ought not to be lost to the country through prejudice. In the face of this roster of world champions the physical fitness and courage of their race cannot be questioned by any reasonable man.

The record suggests that the country would lose less by refusing to train Harvard, Yale, and Princeton men for the flying corps than by refusing to train Negroes.[52]

Although the editorial based its conclusions on a racial stereotype which presumed that all blacks were natural athletes, it was widely and enthusiastically reprinted in the black press. The *Pittsburgh*

Courier ran it next to Marshall's testimony.[53] The *Chicago Bee* prefaced its reprint with an editor's note: "The Chicago Tribune of Thursday morning carried the following editorial 'Negro Pilots,' which we are printing in its entirety, without comment—it speaks for itself."[54] The *Washington Tribune* hailed it as a "thunderous editorial [by] one of the most influential white dailies in the country, [which] lambasted the color line in the U.S. Air Corps and urged that Negroes be given equal treatment."[55]

The *Chicago Tribune's* editorial was part of the struggle for the black vote, which was to become more intense as the presidential election approached. When the Republican National Convention closed on 28 June, GOP candidate Wendell Wilkie, a dark horse whose nomination had taken the old guard by surprise, spoke with black reporters and declared: "I want your support. I need it. But irrespective of whether Negroes go down the line with me or not, they can expect every consideration."[56] Wilkie could claim a long-standing concern over civil rights—he had been an outspoken opponent of the Ku Klux Klan in the 1920s—which dovetailed conveniently with the Republican party's aim of recapturing the black voters who had defected to the Democrats in 1936. To this end the GOP platform included a strong civil rights plank, which declared that "Discrimination in the civil service, army, navy, and all other branches of the government must cease."[57] The black vote was equally important to the Democrats, especially after the Democratic National Convention nominated Roosevelt for a third term, an unprecedented development in American politics which, many believed, might well make the election exceedingly close.

It was in this politically charged atmosphere that legislation to enact the first peacetime draft in the nation's history came before Congress. Roosevelt had initially opposed a draft, fearful that its defeat might encourage further fascist aggression, but by early August his new secretary of war, Henry L. Stimson, had persuaded him to support the measure.[58] Shortly afterwards, the NAACP and the Committee on the Participation of Negroes in the National Defense Program sought to influence Congress to amend the Burke–Wadsworth Bill to include a proviso prohibiting racial discrimination. In mid-August Rayford W. Logan and Charles Houston represented the Committee on the Participation of Negroes in hearings before the House Committee on Military Affairs and urged that safeguards be written into the law to preclude circumvention of measures intended to provide opportunities for blacks, as had occurred in the case of P.L. 18.[59] Walter White of the NAACP turned to Sen. Robert F. Wagner, a Democrat from New York who had been a strong supporter of the campaign for antilynching legislation, and urged him to sponsor an amendment "to guarantee to Negroes the right to serve in every branch of the armed forces without any discrimination because of race or color."[60] Such an amendment was necessary, White argued, because of navy and army policies that excluded blacks from various specialties, including aviation.

On 26 August 1940 Senator Wagner introduced an amendment to the Burke–Wadsworth Bill that modified the section on voluntary enlistments, stipulating that men subject to the draft could voluntarily enlist, "regardless of race, creed, or color," in any branch of the land and naval forces "(including aviation units)."[61] The issue of Air Corps participation figured prominently in the ensuing debate. Wagner noted that he had received communications from various sources protesting against "a rule adopted by the Army, particularly in the aviation units, that no matter how well fitted they may be physically, mentally, or in any other way, they will not take certain American citizens because of their color."[62] Southern Democrats were virtually unanimous in their opposition to the measure, with John H. Overton, Allen J. Ellender, both of Louisiana, and Tom Connally of Texas leading the debate against the Wagner amendment. A coalition of Republicans and nonsouthern Democrats supported the measure, with two Republicans—W. Warren Barbour of New Jersey and Arthur H. Vandenberg of Michigan—speaking on behalf of the amendment, along with Democrat H. H. Schwartz of Wyoming.

The southern senators attempted to discredit Wagner by implying that his proposal would force racial integration on the army. Realizing that any measure whose purpose was to overturn segregation would be filibustered, Wagner asserted that there was "no question of whether they are to be integrated or not. The complaint is against the refusal to permit them to serve. That is the only point I am making." Overton responded if there were "mixed units in the Army, it would be subversive of discipline, subversive of morale, and would not be of benefit either to the colored race or to the white race." Thus the War Department had created separate units, not out of a "desire to discriminate against the colored race; on the contrary, they encourage them [to enlist]." Wagner challenged Overton's assertion: "The Senator from Louisiana may make that statement, but I think if he will inquire he will find that in the aviation units no colored enlistments at all are accepted. The colored American citizen cannot enlist there; they will not accept him; which is quite a different thing from the question of segregation. That question I am not considering at all. . . . I am talking about units from which they are excluded because they are colored, and for no other reason; and it is frankly stated by enlistment officials."[63]

Overton adamantly defended the Air Corps, claiming that if blacks were "excluded from the air forces it is because the Army is not ready yet to have separate units," and thus the ban did not constitute discrimination. Sen. W. Warren Barbour of New Jersey supported Wagner, observing that "it cannot be doubted—I know it of my own personal knowledge—that there is and has been and will continue to be this tragic, unfair discrimination of which the Senator from New York speaks. . . . These are not just intimations. They are justified complaints, and they are legion."[64]

Sen. H. H. Schwartz entered the debate to outline, as he understood it, the War Department's position with regard to accepting blacks in the Air Corps, an account which hardly corresponded to the position General Arnold had taken at the end of May when he told the General Staff in no uncertain terms there was no place in the Air Corps for blacks:

> About a year ago we incorporated in a bill before us [i.e., P.L. 18] a provision that an opportunity should be given to the Army to train colored pilots. The Army has not been able to work out that provision. I do not understand that they would be particularly adverse to it if they could find a place, after they educated the pilot, where he could perform service to his country and not of necessity be within a mixed social situation whereby he would have to be associated in his flights with white officers, and possibly the colored man and his family would have to be part of the white organization at the different posts. I think most of the colored men understand that; but it was the idea of the colored men that with the present increase of the Army there would be a place for colored pilots connected with colored regiments or at other places where they would not have to be working with white pilots. The War Department have [sic] told me very frankly that they have not

created what they call the social situation in the South and in the Army. It is there, and it exists, and they have to meet that situation, and they are trying to do it.[65]

Sen. Allen J. Ellender of Louisiana challenged Wagner's claim that the army had discriminated against blacks, maintaining that the purpose of the exclusion from the Air Corps was to maintain segregation. "Was that not the reason," Ellender asked, "why the discrimination was made?" Wagner responded: "I do not think so. It was just the feeling of the Army. It excluded them altogether from the aviation units. In some of the regiments they have included them, but in aviation they have excluded them entirely. I cannot believe that has the support of the people of the United States, who are tolerant, and want all Americans treated alike, so long as they are loyal to our country, and are willing to fight for it."[66]

When Senator Overton asserted that no evidence had been offered to support the charge that the army discriminated against blacks who tried to enlist, Sen. Arthur H. Vandenberg of Michigan spoke on behalf of the amendment, citing a recent letter from the secretary of war stating that because of the limited openings in the black regiments, only a few openings existed at any one time. Thus, Vandenberg concluded that

> while the discrimination may not run against the colored citizen as an individual when he appears for enlistment; the discrimination actually results, and is equally effective, due to the failure to provide a unit to which he is eligible under the terms of the War Department rule. . . . The units are not provided in sufficient quantity to assimilate the applicants for enlistment. Then the colored citizen is rejected. . . . It is indirect discrimination, it is covert discrimination, but it is no less discrimination. We might even say it is unwitting discrimination as far as the Army is concerned. But whatever its character or motive, it is discrimination and so far as possible it ought to be removed. The record of the Negro soldier is the complete recommendation for its removal. The constitutional rights of the Negro citizen complete the case in behalf of the pending amendment.[67]

Senator Tom Connally of Texas strongly opposed the Wagner amendment and was especially critical of the provision that required the army to provide an opportunity to enlist in aviation units, noting sarcastically that the Army now does not have any colored aviation units, but under the amendment of the Senator from New York it will be obliged to have one, because if three colored men enlist, the War Department, under this amendment, must establish a colored aviation unit—even if only three colored men enlist, or perhaps four—there might be four. But if any number at all go to a recruiting office and say, "We want to enlist in the aviation unit," the Army must take them; it must establish an aviation unit, or else it must put them in with the other trainees, the other members of the Army, white men.[68]

When Wagner asked if qualified American citizens ought not to have the right to enlist in the Air Corps, Connally agreed they should, "If there is a corps to enlist in; but I do not think, if a half dozen of them want to enlist, that is any reason why the Army should change its whole basis of organization to accommodate a half dozen and to establish an aviation unit for them."[69]

Connally's comments reflected the position of the Air Corps, which was based on the dual assumptions that blacks must only serve in the Air Corps on a segregated basis and that only a handful of blacks could qualify for duty in the Air Corps. Thus the establishment of a separate black unit was impractical and would serve no useful military purpose. Like his fellow southerners, Connally raised the specter of integrated units, charging that Wagner's real purpose was to "put colored men and white men in the same company, make them sleep together in the same tent, [and] make them eat together at the same table."[70] Why, Connally asked, did Wagner's amendment specifically mention aviation? "Because," he concluded, "the Army has no colored aviation units, and the Senator from New York would make the Government of the United States establish an aviation unit to accommodate a little handful of colored men who may possibly want to enlist in the [Air Corps]."[71] Connally declared that, as a southerner, he knew "as much about the colored race as . . . the Senator from New York," noting that he was "raised with colored people, and played with colored boys when I was a boy, and I worked with them side by side in the cottonfields and other places." Thus he realized "better perhaps than [Wagner] that constitutional and legal rights are one thing, and the right to select one's associates socially is another thing." Indeed, Connally asserted—to snickers in the galleries—"I am ready to fight for the right of the colored man under the laws and the Constitution, not simply during election time, as the Senator from New York is." The Texan concluded his tirade against the Wagner amendment by commending the "great mass of the American colored people," who "want to serve their country if their country needs them." But he warned against the few who "want continually to agitate, disturb, stir up dissension, and raise the devil about what they speak of as their political and social rights." Such sentiments, he claimed, had caused black soldiers to riot and mutiny earlier in the century at Houston and Brownsville, Texas.[72]

Despite the racebaiting of the southerners, the Wagner amendment carried by a wide margin, 53 to 21. Only three nonsouthern senators opposed the measure—Republican Warren R. Austin of Vermont, and two Democrats, Elbert D. Thomas of Utah and Millard E. Tydings from the border state of Maryland. The lone southerner to support the amendment was James F. Byrnes of South Carolina, a close political ally of President Roosevelt but usually at odds with him on civil rights issues.[73] In the final version of the bill, approved 16 September 1940 as the Selective Training and Service Act of 1940, the section on volunteering for induction into the "land and naval forces" included Wagner's prohibition on racial discrimination but not the parenthetical phrase "(including aviation units)." The effect was the same, however, because the final version of the bill also specified that the term "'land and naval forces' shall be deemed to include aviation units of such forces."[74]

The inclusion of a modified version of the Wagner amendment in the Burke–Wadsworth Bill was the turning point in the campaign for full participation in the Air Corps. By the summer of 1940, when the draft bill came before Congress, it was obvious that the first legislative initiative—P.L. 18—had failed. But the summer of 1940 also marked a tremendous surge of popular interest by the black public on the issue of Air Corps participation. Two major organizations, the NAACP and the Committee on the Participation of Negroes in the National Defense Program, had thrown their influence behind the cause, supplanting Edgar G. Brown and the National Airmen's Association to become leading critics of the Air Corps' proscription on admitting blacks. Moreover, a host of smaller organizations and individuals had likewise come to regard the ban from the elite air arm as an ominous portent, an indication that the army high command saw even highly qualified black Americans as nothing more than hewers of wood and drawers of water. The fortunate confluence of this surge of interest in the Air Corps, the impending presidential election, and the introduction of the draft bill provided an opportunity to once again pass legislation intended to force the Air Corps to admit blacks. The consequences of the Burke–Wadsworth Act were markedly different from those of P.L. 18: at last, black spokesmen did not have to base their arguments for Air Corps admission on a law that had been carefully worded in favor of the Air Corps, or on broad concepts such as constitutional rights, fairness, and a record of loyal service in past wars. Congress had acted and the law was clear, even though the connection between voluntary enlistments and admission to the Air Corps was not as obvious as it had been in the amendment as originally proposed by Wagner.

While Congress had enacted legislation that put the War Department and the Air Corps on notice that total exclusion of blacks from the air arm would not be tolerated, the debates over the Wagner amendment also sent a strong signal that blacks should be admitted to the Air Corps only on a segregated basis, and that the tradition of segregated units would prevail. Once the southern senators forced the supporters of the Wagner amendment to go on record as opposing racial integration, they acquiesced and did not attempt a filibuster, allowing the measure to come before the Senate, where it won by a wide margin, as they no doubt knew it would. Thus the effect of the modified Wagner amendment that found its way into the Burke–Wadsworth Act was to open ever so slightly the door leading to black admission to the Air Corps, but only after those guarding the door were assured that a policy of strict segregation would be followed.

Fruits of the Campaign

The Ninety-ninth Pursuit Squadron

The approval of the Burke–Wadsworth Bill, or the Selective Training and Service Act, on 16 September 1940 was a milestone in African Americans' campaign for full participation in the armed forces; moreover, it had a special significance with regard to the Air Corps, since it prohibited racial restrictions on voluntary enlistments in any branch of the land or naval forces, including the aviation units. The significance of the Burke–Wadsworth Act was apparent to Howard Williams, a young black man who had been trying to enter the army's air arm since early 1939. The day the draft law was approved, Williams wrote Secretary of War Henry L. Stimson, President Roosevelt, and First Lady Eleanor Roosevelt to advise them that he had previously been rejected by the Air Corps because of his race and to assert his right to serve in it without discrimination, as guaranteed by the Burke–Wadsworth Bill.[1] Williams's letter was turned over to the adjutant general, who communicated the standard response: blacks were not accepted into the Air Corps because there were no black units to which they could be assigned. The adjutant general also reiterated the claim that black initiatives in aviation were nonexistent and that efforts were under way in Chicago and elsewhere to foster the development of aviation among Negroes: "There being no colored aviation development in this country, the War Department arranged with the Civil Aeronautics Board . . . to make a beginning by conducting an experimental aviation school at Glenview, Illinois, and furnished equipment to this school. In addition the Civil Aeronautics Board is conducting colored aviation schools at Hampton College, Delaware State College for Colored Students, Tuskegee Institute, West Virginia College Institute, and Chicago."[2]

Although the adjutant general's response to Williams did not acknowledge the provisions of the Burke–Wadsworth Act, the law's implications had begun to influence the thinking and actions of the White House and the War Department even before the act was approved. During the first week of September 1940, President Roosevelt asked the War Department to prepare and hold in readiness "a statement to the effect that colored men will have equal opportunity with white men in all departments of the Army." The statement was to be released "in the event that its issuance becomes desirable."[3] In his Cabinet meeting of 13 September President Roosevelt expressed concern regarding the complaints of blacks who feared they would only see service in labor battalions if drafted into the armed forces. To counter such fears the president recommended that the army publicize its tentative plans to give blacks "proportionate shares in all branches of the Army, in the proper ratio to their population—approximately 10%." After the meeting, Secretary of War Stimson asked Chief of Staff George C. Marshall for an "exact statement of the facts in the case, and as to how far we can go in the matter."[4]

On 16 September 1940, the day President Roosevelt signed the Burke–Wadsworth Act into law, the White House issued a press release that addressed the effect of the act on black Americans. The release concluded with a statement on black participation in the Air Corps: "The Civil Aeronautics Authority, in cooperation with the Army, is making a start in the development of colored personnel for aviation service. Pilots, mechanics and other specialists must first be trained as a nucleus for the formation of colored aviation units."[5] This statement certainly ran counter to the position of the Air Corps, articulated by both General Arnold and General Yount several months earlier, opposing the formation of black flying units. Thus to claim that the army and the CAA were "making a start in the development of colored personnel for aviation service" was a somewhat disingenuous attempt by the White House to put the best possible face on the Air Corps' activities with regard to P.L. 18 and the training program at the Chicago School of Aeronautics. The statement that personnel "must first be trained as a nucleus for the formation of colored aviation units" is, in retrospect, significant: it constitutes the first official statement by the executive branch of the federal government which suggests that blacks might eventually be admitted to the Air Corps, albeit on a segregated basis.

Before the end of September 1940, the civilian and military leadership of the army had indicated, privately, that they would make good on their obligations to open the Air Corps to blacks. Less than two weeks after the approval of the Selective Training and Service Act, General Marshall directed the General Staff to recommend a general plan for training a cadre of black aviation mechanics pursuant to the establishment of a Negro combat aviation unit.[6] The chief of staff had the reluctant concurrence of the secretary of war in taking preliminary steps to establish a black flying unit. On 27 September Secretary Stimson set down in his diary his misgivings about admitting blacks to the Air Corps, along with his views on the importance of maintaining segregated units: "In the draft we are preparing to give the negroes a fair shot in every service, however, even to aviation where I doubt very much if they will not produce disaster there. Nevertheless they are going to have a try but I hope for Heaven's sake they won't mix the white and the colored troops together in the same units for then we shall certainly have trouble."[7]

Stimson's comment clearly shows the connection between the decision to admit blacks to all the arms and services of the army, including aviation, and the enactment of the conscription act. The black public and its leadership, however, remained skeptical about the army's willingness to comply with the provisions of the Selective Training and Service Act. Walter White, executive secretary of the NAACP, was especially apprehensive about the matter. He prevailed upon Mrs. Roosevelt to arrange for him and two other black leaders—A. Philip Randolph, president of the Brotherhood of Sleeping Car Porters, and T. Arnold Hill, an NYA official and acting head of the National Urban League—to meet with the president on 27 September 1940 to discuss the matter of blacks in the armed forces.[8]

When White, Hill, and Randolph were ushered in to see the president, he was accompanied by the secretary of the navy, Frank

Secretary of War Henry L. Stimson was skeptical about proposals for training blacks as pilots. Nevertheless, Stimson (with cane) subsequently visited Tuskegee Army Air Field (TAAF), where he was greeted by Lieutenant General Barton K. Yount (center). Although the photograph includes Benjamin O. Davis, Jr., Stimson did not meet Davis during his visit to Tuskegee. To satisfy a request from the Pentagon, TAAF officials made a photograph of Davis and superimposed the image on this photograph of Stimson's arrival, thus giving the impression that Davis had met the secretary while he was at Tuskegee. The incident is related in Davis's autobiography *American: An Autobiography* (Washington: Smithsonian Institution Press, 1991). (Courtesy United States Air Force Historical Research Center, 222.01, June–December 1943, vol. 6)

Knox. Roosevelt had hoped to have the secretary of war at the meeting as well, but Stimson had excused himself and sent his deputy, Assistant Secretary of War Robert P. Patterson, noting in his diary that "I really had so much else that I had to do that I couldn't take it in." Perhaps the pressure of his office did indeed keep Stimson from the meeting, but he nevertheless had little incentive to participate because he considered it nothing more than an effort "to satisfy the negro politicians who are trying to get the Army committed to colored officers and various other things which they ought not to do."[9]

The tone of the meeting was positive, according to Walter White. The president "listened attentively and apparently sympathetically, and assured us that he would look into possible methods of lessening, if not destroying, discrimination and segregation against Negroes."[10] Both the president and Assistant Secretary Patterson emphasized that the War Department was planning to accept blacks into all branches of the service and that a number of black reserve officers would be recalled to active duty. At the conclusion of the meeting, the black spokesmen presented a memorandum outlining several important points that they considered essential to the "integration of the Negro into military aspects of the national defense program."[11] With respect to aviation, they recommended: "Immediate designation of centers where Negroes may be trained for work in all branches of the aviation corps. It is not enough to train pilots alone, but in addition navigators, bombers [sic], gunners, radio-men, and mechanics must be trained in order to facilitate full Negro participation in the air service."[12]

Twelve days later President Roosevelt approved a War Department policy statement on the use of black personnel, and it was released to the press.[13] In what was to become the standing policy of the War Department for the duration of World War II, the statement called for proportional representation in the army in all branches of the service, but on a segregated basis. The issue of blacks in aviation service was specifically addressed and carried the implication that the Air Corps was currently training pilots and mechanics: "Negroes are being given aviation training as pilots, mechanics, and technical specialists. This training will be accelerated. Negro aviation units will be formed as soon as the necessary personnel has been trained."[14] The statement differed from earlier statements on blacks in the Air

Corps, apparently a reflection of the studies currently under way in the War Department regarding the formation of a black squadron. But to outsiders the difference seemed inconsequential.

The release of the War Department's policy disturbed the three men who had met with the president on the twenty-seventh. The president's attitude had encouraged them to believe that some relief from the rigid segregation that had prevailed in the past might be forthcoming, but the policy contained no hint of any concession: "The policy of the War Department is not to intermingle colored and white enlisted personnel in the same regimental organizations. This policy has proven satisfactory over a long period of years, and to make changes would produce situations destructive to morale and detrimental to the preparation for national defense."[15] Equally disturbing to White, Hill, and Randolph was the implication in the White House press release that the policy was the result of the 27 September meeting and had their blessing, a development which threatened to seriously damage their credibility with the black public. Realizing the precarious position in which they found themselves, the three launched a vigorous counterattack in the press. A casualty of this episode was the hint of goodwill and compromise that the meeting of 27 September had engendered. An 11 October NAACP press release denounced the White House statement of 9 October and emphatically denied that Hill, White, and Randolph had approved a policy of segregation in the armed forces. It also challenged the claim that blacks were being trained for aviation duty: "We ask proof that even one Negro is now being given aviation training as [a] pilot in [the] army air corps. As recently as October first nineteen forty the Adjutant General of the War Department wrote 'applications from colored persons for flying cadet appointments or for enlistment in the air corps are not being accepted.'"[16]

This episode reconfirmed the traditional adversarial relationship between the NAACP and the War Department, overturning the tentative steps toward cooperative action that had been taken at the 27 September meeting. The association's long-standing opposition to segregation could not be sacrificed, and the War Department was not to be trusted. Consequently, the NAACP renewed its belligerent attacks on the War Department and in particular the Air Corps, which was especially vulnerable because it rejected blacks completely and refused to provide even token segregated opportunities.

By October 1940 sentiment among the black public was so strong against the Air Corps that the NAACP's campaign encouraged other black advocates and organizations to step up their criticism of the Air Corps. Ironically, this mounting criticism came just as General Marshall directed his staff to begin planning for a black air unit, a decision that was kept confidential and only hinted at in the War Department's public statements.

By failing to announce the decision, the War Department and the Air Corps unwittingly put themselves under attack both from conservative black organizations, which would have grudgingly accepted segregation as inevitable, as well as from organizations like the NAACP, which sought to end segregation completely. A firm, unequivocal statement by the secretary of war or the chief of staff in late September or early October that a black unit would definitely be established, that its personnel would receive standard Air Corps training, and that planning for the unit had begun might have won approval from segments of the black public that were willing to endure segregation a while longer in exchange for some sign of progress. Instead, the staff work that was under way throughout October, November, and December was shrouded under a veil of secrecy while black aviation advocates and their allies waxed indignant over the issue of admission to the Air Corps.

On 16 October 1940 the policy approved by the president on 8 October and released to the press the following day was codified within the War Department via a letter from the adjutant general, "Subject: War Department policy in regard to Negroes," which was addressed to the "Commanding Generals of all Armies, Corps Areas and Departments, and Chiefs of Arms and Services."[17] The same day the *New York Times* carried a story with the headlines that announced "Negro Air Force Planned."[18] Citing sources at the War Department, the article explained that segregated units would be established as part of the army expansion program, once personnel had been trained. Indeed, by mid-October General Arnold had agreed to the formation of a single black squadron, but plans for training its personnel were still pending.[19] But the story did not ring true to careful readers, for it claimed that men for the segregated unit were already being trained, that pilots were receiving instruction at the Chicago School of Aeronautics, and that ground crews were being trained by the NYA. If the information about War Department plans to form a black unit was released with the intent to placate the black public, it was rendered ineffective by the inaccurate claim that efforts to train blacks for Air Corps duty had already begun. Air Corps officials were apparently unwilling to acknowledge that their attitude toward the training in Chicago had been duplicitous and persisted in characterizing their so-called cooperation with the CAA under P.L. 18 as a sincere effort to ultimately admit blacks to the Air Corps. Past experience had made the black public dubious of any claim that the War Department made, and blacks were doubly suspicious of claims that segregated air units would be established when in the same breath it was asserted that training under the CAA had already begun.

The next issue of the *Pittsburgh Courier* attacked the proposition that black pilots should be trained for Air Corps duty by the CAA, with an article that announced "One Type of Training for Whites: Another for Us!"[20] Without challenging the accuracy of the assertion that blacks were actually being trained by the Chicago School of Aeronautics, the article pointed out such students were of a "different category from that of white youths enrolled in accredited civilian aviation schools under the same program." The article explained that the training at Glenview was part of the CPT Program, such as that being offered at black colleges such as Tuskegee Institute, and noted that whites being trained for the Air Corps at civilian primary schools—like the Chicago School of Aeronautics—received the pay of a flying cadet and would be commissioned in the Air Corps Reserve on completion of the prescribed course of study.[21]

Throughout November 1940, as the Air Corps was formulating plans for the establishment and training of a black squadron, no facts were released and the War Department persistently maintained that a beginning was being made through the CAA and NYA. By the end of November John C. Robinson characterized the Chicago initiative as a sham, "set up to save the face of the Army and as an excuse not to admit Negroes into the Army Air Corps."[22] Others petitioned the War Department to admit blacks. One of the recommendations from a conference held at Hampton Institute on participation of Negroes in national defense was that the War Department immediately implement P.L. 18 and comply with the recent "statement of policy released by the White House as to the training of Negroes as military pilots."[23] In mid-November the National Negro Council—organized by the United Government Employees, African Methodist Episcopal Church, and the National Baptist Convention and claiming to speak "in the name of 12,000,000 loyal, patriotic Negro Americans"— called for an "Appropriation of $25,000,000 to provide similar facilities at Tuskegee Institute, Tuskegee, Alabama, as those now at Kelly and Randolph Fields, Texas, and another under construction at Corpus Christi, Texas in order to provide training of Negro air pilots." The resolution included a demand for a separate training base for black naval aviators in the Virgin Islands, similar to the facilities at Pensacola, Florida.[24]

The agitation over Air Corps participation reached fever pitch in December 1940, when the NAACP's monthly magazine *Crisis* showed an Army airplane in flight over Randolph Field with the caption "FOR WHITES ONLY—A U.S. Army air corps training plane over the 'West Point of the air'—Randolph Field, Texas. Negroes are not being accepted and trained by the Army air corps at *any* field in the nation, despite all the talk of national unity and of the urgency of every group serving in national defense."[25] The issue's feature article was "When Do *We* Fly?" by James L. H. Peck, a young black aviation writer whose work had appeared in a number of national magazines.[26] Peck's book, *Armies With Wings*, recently published by a major New York publishing house, had been reviewed favorably and he reputedly had served as a pursuit pilot with Loyalist forces in the Spanish Civil War.[27]

A skilled writer and a knowledgeable observer of military aviation, Peck had little difficulty exposing the inconsistencies in the Air Corps' public statements regarding the enlistment and training of Negroes. He quoted the *New York Times* article of 16 October 1940 in full, along with a letter to Walter White from President Roosevelt's secretary, Col. Edwin M. Watson, explaining that they formed the basis for answering the question "When Do *We* Fly?" Both sources claimed that black air units would be formed once the CAA and NYA trained the personnel, a premise which Peck proposed "to tear . . . to bits." Such a claim, he asserted, does "*not* hold water in the light of [the] past, present, or contemplated Army Air Corps policy of procurement of personnel."[28]

Peck dismissed the notion that the CAA's Civilian Pilot Training Program could provide black pilots for the Air Corps, noting that not until September 1940 "was there even an indirect connection" between the CPT and military aviation: "At the beginning of the new school term, [all CPT] candidates were asked to pledge their services to the flying services *if called*." Indeed, many black CPT Program graduates, he continued, had not waited to be called but had volunteered for duty in the Air Corps, only to be rejected. Peck noted that thirty black CPT students were in training at the Harlem Airport on Chicago's Southside and commended them for their efforts. But, he asked rhetorically, was the "picture of 30 students flying three or four Taylor Cub 'put-puts' around a small field [the] beautiful picture of the 'Negro training center' the [War] Department would have the race believe is 'already in operation?'" Equally absurd was the idea of the NYA-trained mechanics taking responsibility for servicing sophisticated military aircraft:

> So the N.Y.A. is training mechanics and technical specialists. I would like to ask the office from which that statement emanated whether or not any of the enlisted personnel charged with the maintenance of the $200,000 bombers and pursuit craft of the Air Corps tactical units was "trained" on any N.Y.A. project.
>   . . . I would like to ask any Army pilot—in fact I have, for I know a great many—how he would like to make a "blind," or instrument flight with gadgets calibrated by an N.Y.A. graduate; or how he would like to meet an enemy in combat with guns attended by an N.Y.A. armorer. This innocent query . . . had the whole hangar gang, at a certain field, in an uproar and I'm still laughing.[29]

Peck admitted that the thought of CAA-trained pilots flying combat aircraft, maintained by NYA mechanics, might be laughable in the abstract, but it was "not a laughing matter apropos this discussion." Instead, it showed precisely what the War Department and the Air Corps "think of the intelligence of the Negro people."[30]

Despite the claims of the War Department, Peck asserted, blacks would not enter the Air Corps as mechanics and pilots so long as they were denied the opportunity to enlist and train as aviation cadets at one of the civilian primary flight training schools or as support personnel at one of the Air Corps technical schools, Chanute Field near Rantoul, Illinois, and Lowry Field near Denver, Colorado. Peck concluded by outlining four steps that the War Department should take to implement its announced policy of training personnel for black air units:

When, first of all, several of the many competent instructors of our race are sent to one of the Air Corps Training Centers for what is known as a "familiarization course," the War Department's declaration of good intentions toward the Negro flyer will just begin to make sense. . . .

When, secondly, provision is made for the entrance of our boys into an Army-supervised commercial school, or into the technical schools of the Air Corps, the Department may truthfully announce that it plans to "provide for the acceleration of such training."

When, thirdly, the applications of enlisted personnel and flying cadets are no longer relegated to the waste basket, but are referred, through the proper military channels, to whom ever the Department shall appoint as Personnel Officer of the Negro aviation unit, or units, they can, in truth, say, "Negro aviation units will be established as soon as trained personnel can be obtained."

When, fourthly, the Department requests the presence of two or three aviation experts of our race to sit in meeting to discuss the hundred and one details involved in the setting up of a colored flying unit; or, as an alternative, calls into conference two or three of our high-ranking Army officers to actually map organization of such a unit, we will be beginning to get our just representation as taxpayers.[31]

To ensure that the information in Peck's article was widely disseminated among black Americans, the NAACP issued a press release late in November under the headline "Army Air Officials Fool Negro Citizens About Training For Flight, Says Negro Aviator."[32] Several black newspapers carried the story and it prompted the leading black journal, the *Pittsburgh Courier*, to publish a strident editorial that declared:

When it comes to FLYING military airplanes for the United States, the Negro aviators of this country are "grounded."

Other pilots are grounded TEMPORARILY by bad weather, but Negro pilots are grounded PERMANENTLY regardless of weather.

With white pilots it is a matter of CLIMATE; with colored pilots it is a matter of PIGMENT.

Echoing Peck's arguments, the *Courier* admitted that "Negro students ARE BEING TAUGHT TO FLY under the auspices of the Civil Aeronautics Board, but this has NOTHING to do with military aviation." Likewise, the "National Youth Administration is SUPPOSED to be training some Negroes as aviation mechanics, but every informed person knows that the training given WILL NOT FIT

ANYBODY for the exacting duties of a military aircraft mechanic." These programs, the editorial concluded, are simply "sops to Negroes, designed to keep them BELIEVING that they are actually making some progress." Claims that black flying units will be organized when sufficient personnel have been trained, the *Courier* went on, were patently false:

> The government is just KIDDING us.
> The government knows that NO Negroes are being TRAINED as military aviation mechanics.
> So, even if there WERE a sufficient number of trained Negro military aviators, there would NEVER be any segregated aviation units until EIGHT TIMES their number of aviation mechanics were trained.
> But there are NOT enough trained Negro military aviators because no Negroes have EVER been given such training in this country.

And even though black Americans were barred from flying, servicing, and building American warplanes, the editorial concluded, they were nevertheless taxed to support the expansion of the Air Corps: "When it comes to paying the bill, the order is 'Ceiling Zenith.' When it comes to getting opportunities in either flying, making or repairing airplanes, the order is 'Ceiling Zero.'" The black American, the *Courier* warned, "is TIRED OF PAYING to make the world safe for HYPOCRISY [and] will NO LONGER PRETEND that he is satisfied to be 'Grounded.'"[33]

By the end of 1940, with the War Department still silent on its plans to establish segregated air units, many black Americans had concluded that the army was indeed playing them for fools and flooded Washington with letters protesting their continued exclusion from the Air Corps. The beleaguered War Department was becoming impatient with the apparent intransigence of the Air Corps about accepting black recruits. In late December Assistant Secretary of War Robert P. Patterson upbraided General Arnold for the Air Corps' lack of cooperation:

> Complaints come in regarding the delay in making use of negroes in the Air Corps.
> It was announced in September or October that negroes would be placed in every arm of the service, including the Air Corps.
> I appreciate that delays have been inevitable, but every effort must now be made to expedite the training of negroes as pilots.[34]

Assistant Secretary Patterson was apparently as unaware as the black public of the plans that the Air Corps had been formulating since October to establish a segregated pursuit squadron.

The planning had begun in late September, after the Burke–Wadsworth Act was approved, when General Marshall had asked the Operations and Training Division of the General Staff, G–3, to prepare, in coordination with the Air Corps, recommendations for training black aviation mechanics who would ultimately form the maintenance cadre of an air combat unit. The General Staff proposed that the black mechanics be trained at the Chicago School of Aeronautics, assuming it the logical site for such training since it was the school designated for training blacks under the provisions of P.L. 18. When the Air Staff explained that the Chicago School of Aeronautics trained only pilots, not mechanics, G–3 suggested the Aeronautical University of Chicago; the Air Corps also found this unacceptable, because the students lived at the YMCA, making it "manifestly impossible to assign colored students under the existing arrangements."[35]

Even though the Air Corps rejected G–3's proposals, there was little agreement among the various divisions of the Air Staff on exactly how to proceed in the matter of training the black mechanics. The Air Training and Operations Division opposed the use of civilian schools on grounds that securing room and board for black students would impose an undue burden on the civilian contractors. Training and Operations, therefore, urged that the black recruits be trained as aviation mechanics at the Air Corps Technical School at Chanute Field, where military quarters and messing facilities could be used.[36]

The Air Plans Division took issue with the recommendations of Training and Operations, warning that it was exceedingly dangerous to assign black trainees to military facilities because "disturbances and riots will probably ensue both at Chanute Field and the nearby communities." To preclude such occurrences, the Plans Division recommended that black schools be used to train black Air Corps personnel, suggesting Tuskegee Institute as an appropriate site to begin the instruction: "If colored units are to be formed, colored schools should be provided for their training [and] separate schools for colored pilot training likewise should be organized." But Training and Operations refused to accede to the recommendation of the Air

Plans Division, arguing that such a course was impractical and inefficient, because only a few black trainees were to be trained.[37]

When it became apparent that the Air Staff was at an impasse, the matter was brought before General Arnold for resolution. Irritated that his staff could not agree on a plan for training a few black technicians, Arnold scribbled a note to Gen. Davenport Johnson, chief of the Training and Operations Division, asking "How should we go about training the colored mechanics for *1 squadron* with the least trouble and effort?"[38]

Apparently Arnold agreed that it was inadvisable to train blacks at existing Air Corps installations, because the plan the Air Corps submitted to the General Staff a week later conformed to the position of the Air Plans Division. The proposal recommended that black personnel for one corps and division observation squadron should be trained at "a recognized colored school, such as Tuskegee," so as to preclude the likelihood of racial disturbances that would inevitably occur if the training was conducted at other sites. The Air Staff was careful to disassociate itself from the idea that a black squadron be established at all, urging that the plan be implemented only "if it is imperative that negro tactical units should be formed," and cautioning the General Staff that "no country in the world has been able to organize a satisfactory air unit with colored personnel." Moreover, the establishment of such a unit would necessarily be a slow process, primarily because of the time required to train the maintenance supervisors—three years for a crew chief, five years for a hangar chief, and ten years for a line chief. Thus, the Air Corps admonished, there was little chance that an effective black air unit could be formed in time to meet the immediate national crisis.[39]

The Air Corps' belief that it was ill-advised to expend resources on training blacks for duty in air units during a time of impending national crisis was shared by the secretary of war. On 22 October 1940, the same day that the Air Staff forwarded its proposal for a segregated observation squadron to the General Staff, Stimson confided to his diary that his department planned on "training colored pilots for the Air Corps. . . . I have very grave doubts as to their efficiency. I don't think they will have the initiative to do well in the air."[40]

Two days later, the General Staff rejected the proposal for a separate observation squadron and instead asked the Air Corps to submit a plan for the establishment and training of a black single engine pursuit squadron.[41] Since a principle function of observation units was control of artillery fire, perhaps the Air Corps had advocated the formation of an observation squadron on the theory that it could be attached to one of the black field artillery regiments which were planned as part of the army expansion.[42] The General Staff, however, was apparently anxious to avert any criticism that blacks were only being admitted to the Air Corps in a support role and mandated the establishment of a separate combat pursuit squadron. Thus on 24 October 1940 the Air Corps was directed to abandon its long-standing refusal to admit blacks and was obliged to develop a detailed plan for the establishment of the unit that would ultimately be known as the Ninety-ninth Pursuit Squadron.

By the first week of November, General Johnson had approved a preliminary proposal for the training and establishment of a separate pursuit squadron to be permanently based in the vicinity of Tuskegee, Alabama.[43] The plan was essentially the same as the one developed for the observation squadron, except for the training of the initial cadre of enlisted personnel. The General Staff's rejection of the proposed black observation squadron, which would fly unsophisticated light aircraft, in favor of a pursuit squadron, equipped with modern high-performance aircraft, apparently forced the Air Corps to abandon its plans to concentrate all training at Tuskegee. Evidently, the shift from observation to pursuit aircraft made it imperative that the initial complement of maintenance personnel be trained at an Air Corps school where there were adequate facilities to teach them to maintain high-performance combat aircraft. Thus Training and Operations' earlier proposal to train the maintenance personnel at the Air Corps Technical School at Chanute Field prevailed.

On 8 November 1940 Johnson wrote Gen. Walter Weaver, commander of the Southeast Air Corps Training Center at Maxwell Field in Montgomery, regarding the formation of the Tuskegee pursuit squadron. He noted that the three "major problems involved are . . . the technical training of enlisted personnel, the training of pilots and the establishment of the necessary facilities at Tuskegee required for the housing, maintenance, operation and training of this tactical unit." Weaver was directed to "submit a general plan at the earliest practicable date covering the establishment of this pursuit squadron at Tuskegee and the accomplishment of the flying training." All enlisted men, Johnson advised, would be "sent to Chanute Field initially, where the necessary technical training will be accomplished."[44]

The development of the unit, Johnson told Weaver, would consist of three distinct phases. During the first phase the enlisted technicians would be trained at Chanute Field, while facilities at Tuskegee were being constructed. In the second phase, the pilot training phase, some thirty-three black pursuit pilots would be trained at Tuskegee, supported by the black mechanics trained at Chanute. At the completion of this phase a permanent tactical unit would be established, consisting of almost five hundred black men and officers permanently stationed at the Tuskegee field.[45]

Racial considerations pervaded the planning for all three phases. In order to minimize racial mixing, the flying training phase could not begin until the black enlisted men completed their training at Chanute, a minimum of twenty-two weeks: "The necessity of adhering to present policies concerning segregation will make it necessary to delay the initiation of flying training until such time as a sufficient number of colored mechanics are trained at Chanute Field. It will probably be necessary, however to utilize a few white non-commissioned officers at Tuskegee during the pilot training phase."[46] The second and third stages of the pilot training course—basic and advanced training—would be conducted at the Tuskegee field under Air Corps instructors. But the first stage, primary training, presented a problem; it had been given over to civilian contractors and the Air Corps was unwilling to require these private concerns to accept blacks. Thus, Johnson concluded, because no primary training "is being conducted at Air Corps flying schools, it will probably be necessary to have that training done by the Civil Aeronautics Administration." The CAA was graduating fifteen potential candidates for basic training from Tuskegee Institute's CPT Program every four months, Johnson continued. Consequently, he suggested that General Weaver contact Grove Webster at the CAA, who was "thoroughly conversant with the situation at Tuskegee, and . . . has offered to go to Tuskegee or to Maxwell Field for the purpose of rendering such assistance as he can in connection with this project."[47]

Weaver's instructions regarding the third phase, the establishment of a permanent tactical unit, clearly show that the Air Corps envisioned the squadron not as a contribution to the expansion of the nation's air arm, but as simply a token gesture intended to show compliance with the requirement to admit blacks to all arms and services of the army. Weaver was instructed to plan for facilities to accommodate 476 black personnel, 429 enlisted men, and 47 officers. The bulk of the personnel were to be assigned to the pursuit squadron, 210 enlisted men and 33 commissioned pilots. The remainder would be distributed among the support detachments, with 160 enlisted men and 10 officers assigned to the Base Group Detachment, 20 enlisted men and 2 officers to the Weather and Communications Detachment, and 39 enlisted men and 2 officers to the Services Detachment. The most important consideration, Johnson concluded, had nothing to do with military matters and everything to do with racial matters. Care must be taken, he warned Weaver, to ensure that the goal of establishing a separate but equal air unit was met: "It is desired that this entire matter be approached with the purpose of providing a set-up that will be fully equivalent, with respect to the character of the living conditions, facilities, equipment and training, to that provided for white personnel under similar   conditions."[48]

The Air Corps' plans for training black pilots did not go beyond the thirty-three required to staff the proposed pursuit squadron. Once a black unit was established, the Air Corps leadership hoped that no further action would be necessary to refute charges of racial discrimination. Its position in the matter of black personnel initially caused some confusion to planners on the General Staff, who assumed that once the Air Corps began to accept blacks for pilot training, there would be an ongoing requirement for black aviation cadets. When the Mobilization Branch projected an annual requirement for sixty—out of a total of 15,000 aviation cadet recruits—an officer from another branch, fully cognizant of the Air Corps plans, urged that the requirement be deleted because there was no need for a steady flow of black aviation cadets into the Air Corps:

In showing 60 of those 15,000 [aviation cadet recruits] as negroes, it is not clear that that number is intended to produce some 36 pilots in one nine-month course, and that it is not intended at this time to continue so large an output of negro pilots. Since detailed plans have not yet been established on the training of colored pilots and the activation of colored units (General Andrews and General Arnold have instructed General Johnson . . . to work up this project), it is recommended that all reference to [Negro] cadets be deleted.[49]

Significantly, General Johnson's letter to Weaver, outlining the considerations the Southeast Air Corps Training Center was to use in developing its plan for the pursuit squadron, did not mention the training program in Chicago, which for almost a year had been the basis for the Air Corps' claim of cooperation with the CAA to "begin" the training of black pilots. The CPT Program at the Coffey School of Aeronautics apparently did not figure into Air Corps plans at all because the only training being offered there was at the elementary level; it was clearly inadequate to prepare prospective black aviation cadets to bypass primary flight training and enter basic directly. The CAA equivalent of Air Corps primary was a combination of CPT elementary and secondary training, and the only black CPT Program offering both courses in the fall of 1940 was at Tuskegee Institute. Thus the Air Corps sought to abandon the pretense that the Chicago CPT Program constituted the first step toward admitting blacks, as it no longer served as a convenient subterfuge for excluding black Americans.

The Chicago initiative was not, however, so readily dismissed by others, both within the War Department and without. While the Air Corps proceeded to develop plans to establish a black pursuit unit at Tuskegee, General Marshall maintained that the unit should be situated in Chicago, apparently concerned that the army should follow through after insisting for almost a year that the Air Corps had been cooperating with the CAA to train black pilots in Chicago. In mid-November the General Staff warned one of Arnold's key staff officers, Maj. Gen. Frank M. Andrews, that Marshall did not support the plan to base the unit at Tuskegee:

> Reference the Air Corps' plans to put negro units (pursuit) in at Tuskegee, Eaker said that General Arnold, in talking with the Chief of Staff, got a directive from the Chief of Staff that enlisted men were to be trained at Chanute and the unit was to be formed in the vicinity of Chicago, for reasons best known [to] himself. This happened during your last absence.
>
> I have notified General [Davenport] Johnson . . . that this is the Chief of Staff's desire, in so far as we understand it.[50]

Arnold, however, directed his staff to "make the study as to the most efficacious carrying out of the plan and should the Chief of Staff then decide Chicago to be the location we will do so but I want the best location shown."[51]

Despite the Air Corps' best efforts to keep the planning for the black pursuit squadron a secret, by mid-November the National Airmen's Association had learned of the proposal to concentrate the training at Tuskegee and immediately began an effort to forestall the plan. On 15 November 1940 NAAA spokesmen Enoch Waters and Howard Gould protested to the NAACP and to William H. Hastie, who had just been appointed as the racial advisor to the War Department, with the official title of civilian aide to the secretary of war. Waters and Gould couched their protest in terms calculated to appeal to the NAACP. They argued that the Air Corps was rejecting Chicago in favor of Tuskegee in order to ensure complete segregation: "Urgently request your aid. We are disturbed over plans to develop air base only at Tuskegee rather than Chicago—where [black] aviation received its impetus and is not Jim-Crow. We protest vigorously against plans to develop complete Jim-Crow setup in the South. [We protest] vigorously against Army refusal to really expedite [the black training] program by extending subsistence of $75 monthly, flying togs, and [aircraft] to Negro trainees in Chicago."[52]

Howard Gould, an official of the Chicago Urban League, outlined the complaints of the NAAA in virtually identical letters to Hastie and the NAACP. He began by noting that Chicago blacks had been active in aviation for over ten years, principally at a field in Oaklawn on Harlem Avenue. According to Gould, who was not himself an aviator, the Harlem Airport was a model of racial cooperation and harmony:

The field *is not* and *never has been a jimcrow airport*. The overwhelming majority of pilots using the field are white. Fortunately for us perhaps, the owner of the field and the aviators using the field were happy to have Negroes come out to fly and Negroes have been integrated into the flying program. In fact, one of the key men at the airport is a Negro, Cornelius Coffey, and as a flight instructor has taught both white and colored people how to fly. This, we believe, is as it should be and we are sure you will agree that Negroes can best advance in aviation where they participate equally with, and where they have the same interests and same opportunities as white citizens.[53]

Gould also recounted for White and Hastie the details of the NAAA's efforts to see blacks admitted to the Air Corps. He noted that "after considerable political pressure by the Nat'l Airmen's Ass'n and newspapers," CAA and Air Corps officials called the conference of 15 January 1940, which led to the establishment of a noncollege pilot training program under the supervision of Willa Brown.[54] By Gould's secondhand account of the meeting (there is no record that he attended), the Air Corps agreed to give full consideration to the admission of blacks after the training was completed. Once the training program was established, Gould emphasized, it was not a segregated endeavor: "Although Miss Brown is a Negro and the Nat'l Ass'n of Airmen [sic] were Negroes they did not operate a jim crow program. White students were enrolled in the ground school and received flight instruction."[55]

Having laid out the facts regarding Air Corps training as understood by the NAAA, Gould outlined their grievances against Tuskegee Institute. He implied that Tuskegee had used its considerable influence to ride roughshod over the efforts of the NAAA, obtaining its advanced CPT courses at the expense of the Chicago program. He asserted that Tuskegee Institute president F. D. Patterson, in effect, had stolen two black instructors after they had completed their refresher training in Chicago.[56] He also suggested that the institute had unfairly recruited students from the CPT programs of other black colleges, observing that "Tuskegee arranged to bring the two best . . . students from Hampton, Howard, W. Va. etc. in order to have ten men available for the [secondary] course at Tuskegee."[57]

The plan to develop an army air field in the vicinity of Tuskegee, Gould continued, was an ill-advised scheme to silence black criticism of the Air Corps by abandoning efforts to train black pilots in Chicago and instituting a rigidly segregated program in a racially hostile environment:

> We do not wish to imply in anyway that we are opposed to the development of air training at Tuskegee. However, we protest against and we resent deeply any plans to sacrifice or limit the program here in Chicago, and to answer the demand for Negro admission to the air corps by a jimcrow unit in the South. We insist that [the] C.A.A. be instructed to cooperate in every way in the development of the aviation program here in Chicago. We ask, too, that the War Department give real cooperation to our flyers in this area.[58]

Continuation of the training in Chicago was vital, according to Gould, for several important reasons:

> In Chicago our men are not flying in [a] complete jimcrow environment. In Chicago our men have the advantage of being close to the facilities of the big army schools. In Chicago we can bring political pressure to bear for the integration of our men into the army setup at [Chanute] Field. In Chicago we can move forward and push for opportunities for our advanced students to fly as observers on the big air lines, for Chicago is a great commercial air base.[59]

Gould concluded by soliciting the assistance of White and Hastie to keep the Chicago initiative alive and inviting them to Chicago for discussions on the matter.

Immediately after Gould and Waters contacted the NAACP, Assistant Secretary Roy Wilkins communicated with Hastie on the matter and explained that the NAACP was unaware of any War Department plan to concentrate the training of black Air Corps pilots at Tuskegee. Wilkins, whose position on segregated military units was more pragmatic than that of Executive Secretary Walter White, cautioned against either the NAACP or the Civilian Aide entering a "fight between Tuskegee and Chicago, if it is true that the War Department is actually going to start the training of Negro pilots. As I see it, the main idea is to train flyers." As for the site of the training, Wilkins noted that the "great aviation centers for pilots already are in the South, and it does not seem that we would get very far objecting to the location. Of course, I would rather see our men trained in Chicago than anywhere in Dixie, but if it comes to a choice between being trained in Alabama or not being trained at all, I would be for Alabama."[60] Wilkins's attitude prevailed and the NAACP did not commit itself publicly on the establishment of a segregated squadron at Tuskegee until after the War Department announced its plan in mid-January 1941.

Perhaps part of the NAACP's reluctance to become embroiled in a dispute between Tuskegee and Chicago over the site for the training of black pilots was due to its attempts to bring suit against the War Department for refusing to admit blacks to the Air Corps. By early October 1940 the NAACP legal staff had begun to explore the possibilities of challenging the Air Corps policy of racial exclusion in federal court.[61] The impetus for legal action had come the month before, on 9 September 1940, when the Board of Directors authorized the Legal Defense Committee to "give aid to any American citizen desiring to enlist in the Army or Navy, who is refused the privilege of enlisting on account of race or color."[62] Litigation had always been an important aspect of the NAACP's fight for equality and racial justice, so the board's decision to provide legal assistance to blacks who were refused enlistment on account of race was consistent with the overall policies of the association.

Some NAACP members considered legal action particularly appropriate in the case of the Air Corps, which could not, like the rest of the army, offer the defense that it accepted blacks on a segregated basis and therefore did not discriminate. In early September, shortly before the board authorized legal aid, the secretary of the NAACP's Boston branch made this point to Executive Secretary Walter White. Throughout August, Boston branch secretary Alfred Baker Lewis had warned White that NAACP members in Boston were becoming increasingly concerned over the exclusion of blacks from the Air Corps and he urged White to pursue the matter from the national office.[63] On 5 September Lewis told White that he questioned whether the Air Corps could legally exclude blacks simply because there were no segregated air units, suggesting that "we can cause some embarrassment by taking the position that no military official has the right to refuse Negroes enlistment merely because there is not a segregated unit for them in the particular branch of service to which they apply, unless there is a specific law authorizing such practice."[64]

By late November 1940 Thurgood Marshall, special counsel to the NAACP, had been apprised of Yancey Williams's pending application to enter the Air Corps as an aviation cadet and had concluded that there might well be grounds for a suit if Williams was indeed rejected because of his race. Except for the color of his skin, Williams met or exceeded all the qualifications for appointment as an aviation cadet: he was a twenty-four-year-old senior in mechanical engineering at Howard University, unmarried, and had passed an army flying physical. Moreover, he had demonstrated an aptitude for flying, having completed CPT elementary training at Howard and CPT secondary at Tuskegee. Williams had submitted his application for appointment as a flying cadet on 20 November, which he had carefully prepared and filed according to army regulations with the assistance of Washington attorney W. Robert Ming, Jr., an associate of Thurgood Marshall and a member of the faculty of the Howard University Law School.[65]

Williams received an immediate response to his application: on 23 November 1940 he was advised that "appropriate Air Corps units are not available at this time, at which colored applicants can be given flying cadet training." In a departure from past procedure, Williams's application papers were not returned; instead, he was told that his application was being held "for future consideration in the event facilities become available for training of colored flying cadet applicants."[66] When Ming learned of the reply, he notified Marshall, speculating that it was perhaps "an indication that the Air Corps actually is contemplating the establishment of Negro training facilities."[67] Ming also noted that another black cadet applicant, George S. Roberts, had recently been approved by a flying cadet examining board in Charleston, West Virginia, observing that apparently "the only thing left for Roberts is for the Chief of the air corps to appoint him as a flying cadet." Despite these developments, which apparently reflected plans to form a black squadron which were then being developed in the Air Corps, no action was taken on Williams's application. Thus Marshall and Ming resolved to bring suit on his behalf and force the War Department to admit blacks to the Air Corps. Throughout December 1940 they sought to exhaust all appeals and administrative remedies open to Williams so that when they filed the complaint in federal court they could argue that he had been denied admission to the Air Corps only because of his race.[68]

While the NAACP Legal Committee prepared its suit against the War Department, the Air Corps continued to formulate its plans for the establishment of a black pursuit squadron. By the end of November, almost three weeks after General Weaver and his staff at Maxwell Field were directed to submit a plan for training and basing a pursuit squadron at Tuskegee, Arnold's office wired Weaver to ask when the plan could be expected.[69] On 3 December Weaver responded that it was complete except for the exact location and cost of the field, "now being studied by representatives of Tuskegee Institute, Municipality of Tuskegee and this office." The general reported that the entire plan would be ready the following week and offered to send a partial report outlining all the details except for the airfield data.[70] The Air Staff, no doubt anxious to come to some conclusion on the matter in the face of mounting criticism, asked that the "report excluding airdrome data be forwarded without delay and airdrome data with least practicable delay."[71] Weaver complied immediately, forwarding the plan to Washington on 6 December 1940.

The particulars of the plan, prepared by Maj. L. S. Smith, provide several important details regarding the Air Corps' approach to the incorporation of blacks into its ranks. In contrast to black ground units, which were really mixed units with whites holding the officer positions and Negroes serving as enlisted men, the proposed pursuit squadron and support detachments would be virtually all-black organizations, once the training phase was completed and the white instructors departed. Important exceptions were the three key positions at the installation, which would be reserved for white officers—the post commander, the executive officer, and the pursuit squadron commander. The post commander was designated as a white position because the "Commanding Officer of the Post must be a white officer as required by Army Regulation 95–60." No justification was given for the requirement that the other two key positions be held by whites, but it was noted that the pursuit squadron commander "must be a white officer of considerable experience." Although the stipulation that the flying squadron be commanded by a white pilot was not absolute, the plan observed that the pursuit squadron commander "will necessarily have to be a white officer for an indefinite period of time."[72]

The plan also reveals how the Air Corps intended to respond if the establishment of one Negro squadron failed to silence black criticism. Anticipating that the role of blacks in the Air Corps might inevitably have to be enlarged beyond a single squadron, the staff at Maxwell recommended that a building plan for the field be developed that would allow its facilities to be expanded to accommodate a pursuit group, which typically consisted of three squadrons. Having settled on Tuskegee as the site for its first black unit, the Air Corps hoped that it would have sufficient capacity for any other black units that might have to be established. In short, Tuskegee was emerging in the minds of the Air Corps as the headquarters of the "Negro Air Corps."[73]

The plan also noted that the "city of Tuskegee has insufficient facilities to afford any of the necessary accommodations for the

Brigadier General Walter R. Weaver (front row center, looking right) and staff at Maxwell Field, August 1941. Major L. S. Smith is on Weaver's immediate left and Captain Noel Parrish is in back row, far right. All three played important roles in the establishment of the flight training program at Tuskegee. (Courtesy United States Air Force Historical Research Center, 222.01, January 1939–7 December 1941, vol. 3)

personnel required." Thus, if the unit was to be located in the vicinity of Tuskegee, it would be "necessary to construct at the site all the municipal facilities which may be needed." Consequently, the building requirements outlined in the plan were "in excess of those normally required." One of the "excess" requirements in the building plan were separate living quarters and eating facilities for fourteen white officers and seventeen white enlisted men.[74]

Despite the shortage of facilities in the environs of Tuskegee, General Weaver and his staff concluded that the proposal to establish and train a black pursuit squadron at Tuskegee, Alabama, was feasible and practical. They recommended that their plan be implemented and the installation be constructed at an estimated cost of just under $1.1 million.[75]

The Maxwell Field staff took care to point out to their counterparts in Washington that the Tuskegee Institute administration fully endorsed the proposal to establish a segregated pursuit squadron and base it near the campus. A key finding reported in their plan was that "Tuskegee Institute is most anxious to have this unit located within its vicinity and will secure options for the purchase by the Government of a site meeting the approval of the War Department." Indeed, President Patterson was working closely with local officials and the county agricultural agent to locate a suitable site for the base.[76] By 17 December 1940 he had identified a likely parcel of land and executed a four-month purchase option, in the name of Tuskegee Institute, on some 800 acres at forty dollars per acre.[77]

Patterson's support for the project, however, went far beyond cooperating with Maxwell Field officials to locate a suitable site and obtain a purchase option. Once he learned that the Air Corps was indeed going to establish a black squadron and was seriously considering a plan to base it at Tuskegee, he contacted various federal and local officials and urged them to support the proposal. When he learned that the Chicago aviation community was opposing the plan, he redoubled his efforts to get the War Department's final approval of the Air Corps' proposal to base a black squadron in the vicinity of Tuskegee Institute.

Shortly after General Weaver forwarded his plan for the Tuskegee squadron to Washington, D.C., Patterson sought congressional support. Realizing that the acquiescence of Alabama senators Lister Hill and John H. Bankhead and Alabama representative Henry B.

Steagall, whose congressional district encompassed Macon County, might well be crucial, Patterson wired all three to advise them that a black squadron might be based near Tuskegee and to solicit their support: "Tuskegee ideal base for Army pursuit squadron. True because of year round training possibilities and cooperation possible between training programs and facilities of the Institute. Will greatly appreciate your endorsement to General Arnold."[78] Patterson probably did not expect an enthusiastic endorsement from the Alabama senators or Tuskegee's congressman, but he was a perceptive student of race relations in Alabama. He knew that some local whites would object to the presence of a black military unit and would no doubt lodge a strident protest with their elected representatives. Thus he took care to ensure that Bankhead, Hill, and Steagall were well armed with the facts before they received a letter or a call from an irate constituent.

When Senator Bankhead responded promptly, declaring that he was always willing to seek federal projects for Alabama and asking for more information, Patterson quickly provided him with all the pertinent facts.[79] He realized that Bankhead, a strong supporter of the New Deal with significant influence in Alabama and Washington, D.C., would be a valuable ally in his effort to ensure that the War Department followed through on the proposal to base the squadron at Tuskegee. He wrote the senator on 17 December 1940, explaining that he was "confidentially advised that the army has definite plans for the establishment of a pursuit squadron of Negroes somewhere in the United States" and that a "strong effort is being made to have this unit located in Chicago."[80] A decision to base the unit in Chicago, Patterson explained, would be "an unfortunate move, particularly because of the short flight season which would prevail there." Tuskegee, he asserted, was the logical site for a black air unit, not only because of favorable flying weather, but also because "Tuskegee Institute has most of the programs being administered by the Civil Aeronautics Authority; in fact, [it] has more such programs than any other Negro institution in the country." Moreover, he concluded, there was strong support for a black flying unit near the Institute campus from Maxwell Field and municipal authorities: "Our close proximity to Maxwell Field, along with the good wishes and hearty cooperation of the officers there will insure the sympathetic supervision of a unit located in Macon County. I am also advised by the Mayor that the City Council of Tuskegee is heartily in favor of this unit and that they will do all in their power to help bring it to this area."[81]

Despite his efforts to inform the Alabama congressional delegation of the proposed plan and solicit their endorsement, Patterson knew that their support would not be forthcoming if a majority of the local whites opposed the measure. Thus he had taken care to keep local officials informed of his contacts with the Air Corps and his efforts to select a suitable site for the proposed army flying field. By 3 January 1941 Patterson had succeeded in winning the full support of the white city officials, convincing them to send a wire to Bankhead, Hill, and Steagall that urged them to use their influence to have the pursuit squadron brought to Tuskegee:

> Please do every thing you can to have Government locate Regular Army Pursuit Squadron Air Post, for the training of colored aviators, in Macon County. An ideal site has been selected half[way] between Tuskegee and Union Springs on paved highway and main line Rail Road.
>
> At the last regular meeting of the City Council a resolution was unanimously adopted calling on you to assist us in locating this Squadron in Macon County.
>
> Would like to call your attention to the fact that Alabama, especially this section of the state, east Alabama, is sharing very little in the preparedness program.
>
> Do all you can to help us in the matter, because you know no better place could be selected from every view point.[82]

Patterson also attempted to head off opposition from the NAACP, which he knew would not be pleased over the prospect of a separate Air Corps unit, given its consistent opposition to any form of segregation, regardless of the purpose. On 11 December, the same day that he had sent his initial telegrams to the Alabama congressmen, Patterson also wired NAACP executive secretary Walter White, informing him of the advantages of basing the unit near the institute and asking for his endorsement of the plan.[83] White apparently did not respond, and almost certainly did not endorse the project. Such an action would have compromised the association's litigation on behalf of Yancey Williams and would have also been inconsistent with White's response when Patterson approached him again on the matter several weeks later.

In early January 1941 President Patterson advised White that he would be in Washington, D.C., on the eleventh and suggested they meet with William Hastie for "a conference on aviation."[84] Initially, White was receptive, wired Hastie, and asked him to arrange the conference.[85] The following day, however, White had second thoughts, perhaps fearful of another episode like that of the previous September after his conference with Roosevelt, and he told Patterson that "Before attempting to arrange conference wish you [to] understand I must oppose any proposal for segregated training of Negroes. Please advise."[86] Patterson explained that he sought the meeting to make "available [the] largest possible opportunity for participation of Negroes in all branches of [the] service," and that he regarded the "question of segregation as academic and unnecessary to be considered in [the] conference." Acknowledging the NAACP's position on the segregation issue, however, he assured White that his "endorsement need not be given if same will jeopardize your position."[87] Two days later, after he had consulted with Hastie,[88] White told the Tuskegee president that he understood the "purpose [of the] War Department conference would be to seek Army air unit for Negroes in Tuskegee area and that you consider racial segregation in this connection academic. In such circumstances N.A.A.C.P. participation [is] inappropriate. If I have misconceived [the] purpose or scope of [the] conference please advise."[89] After this exchange Patterson no doubt concluded that any further effort to head off NAACP criticism was pointless, and he abandoned the initiative.

While Patterson was seeking support from federal and local officials and the acquiescence of the NAACP, General Arnold's staff in Washington, D.C., considered the plan the officers at Maxwell Field had drawn up and submitted on 6 December 1940. By 10 December the Air Staff had adopted the Maxwell plan virtually unchanged and had also developed a separate plan for conducting the technical training of the support personnel at Chanute Field, Illinois.[90] Several days later General Arnold reviewed the two plans and pronounced them "admirable," except that "the training of crew, line and hangar chiefs seems to be ignored. For years we have made the statement that it takes from 5 to 7 years to train a good crew chief, and from 5 to 9 years to train a good line or hangar chief. With this background we must take a definite stand relative to how we are going to train personnel for these jobs, or superimpose well trained white soldiers within the organization for that purpose."[91]

On 18 December 1940 the Air Corps sent the plans for the training and establishment of the black pursuit squadron, based at Tuskegee, forward to the adjutant general for final approval by the General Staff and the War Department.[92] With slight modification, Arnold's comments appeared in the covering letter, which observed that although completion of the training programs outlined in the plans would take approximately eighteen months, *"at least from three to five years is normally required before enlisted personnel are qualified as competent crew chiefs, line chiefs, and hangar chiefs."* Therefore, "white non-commissioned officers must continue to act as inspectors, supervisors, and instructors for an indefinite period of time."[93]

The Air Corps emphasized to the War Department that the flight instruction at Tuskegee would begin with the second phase of instruction, basic flight training. Black aviation cadets, it was explained, could not complete the first phase of instruction, primary training, at one of the civilian contract schools used by their white counterparts because it was "impracticable." Thus the only alternative was "to obtain trainees by the enlistment of graduates of the C.A.A. secondary phase, as flying cadets. A number of qualified candidates are now available and approximately fifteen additional potential candidates for basic military flying instruction will become available each four months."[94]

Finally, the Air Corps rejected the proposal that the unit be based near Chicago, citing the high cost of land, dense air traffic, and poor weather as justification for preferring a southern location:

> An investigation has been made of the sites available in the Chicago area and no existing installations have been found suitable for the purpose. Lansing Airport, located twenty miles southeast of Chicago, is the only airdrome considered desirable in any respect. This field, however, is now leased to the University of Chicago and the fact that it is located on a civil airway with heavy air traffic renders the selection of this site inadvisable, if not entirely impracticable. Although satisfactory sites can be procured some distance south of Chicago, the cost of this real estate will probably run as high as $400.00 or $500.00 per acre. Additional disadvantages affecting the continuity of flying training in this area would occur as a result of climatic conditions.[95]

After the Air Corps sent its plan for establishing and training the Tuskegee pursuit squadron to the adjutant general, the civilian and military leadership of the War Department deliberated on the recommendations of the Air Corps for three weeks. In late December Assistant Secretary of War Robert Patterson reviewed the plan, and then asked William Hastie, the civilian aide, for his comments; Hastie responded in a highly critical three-page memorandum. He began by acknowledging that the "type of unit to be established, its station and its tactical relation to other units in the Air Corps are, of course, matters of military strategy." Nevertheless, he urged that the rationale for the decision be disclosed, that it be "made clear that there is some reason which justifies the establishment of the proposed post and the separation of the proposed squadron from normal relations in a pursuit group at a field already existing or planned in the general expansion of the air defenses."[96]

Hastie was openly critical of the aspects of the plan that dealt with training, declaring that he could "see no reason whatever for setting up a separate [training] program for Negroes in the Air Corps." He outlined four "persuasive reasons for training Negroes in existing institutions of the Air Corps." First, there was the matter of morale. If black and white pilots were trained together at one of the existing flight training facilities, the "acquaintance, understanding and mutual respect established between blacks and whites" in that setting could be "the most important single factor in bringing about harmonious racial attitudes essential to high morale." The nature of flying duty in the Air Corps, Hastie warned, made it inevitable that the racial "contacts which the Air Corps seems to fear cannot be avoided. Such contacts should be established normally in the training centers."[97]

Second, the civilian aide charged that the Air Corps' plan to train black pilots separately was "an anomalous exception to existing and contemplated procedures for training [blacks] in other arms and services," citing the interracial training of infantry officers at Fort Benning and plans to train blacks in other specialties at existing army schools. The Air Corps, he declared, "proposed to pursue a different and abnormal course . . . where there is the greatest reason for following the normal procedure." Third, Hastie pointed out that establishing a separate training center for such a small group of cadets was uneconomical.[98]

Judge Hastie's final criticism of the plan concerned the reaction of black Americans. He warned that, despite the cooperative attitude of the Tuskegee Institute administration, the black public would not be enthusiastic about the Air Corps' plan:

> Whatever the attitude of Tuskegee may be, there would unquestionably be very great public protest if the proposed plan should be adopted. In addition to objections on principle, charges of different and inferior training for Negroes would be widespread. The fact that other arms and services do not have separate schools for the training of Negroes, would increase public objection to the Air Corps plan.
>
> In brief, the plan as submitted creates problems and aggravates difficulties without offering any compensating advantages. If the Army will only begin the training of Negroes for the Air Corps in the normal manner, I am confident that immediate and long-range results will be gratifying.[99]

Hastie concluded by raising the issue of recruiting, an issue that the plan did not address, but nevertheless an important aspect of the program. Specifically, he was concerned that the number of blacks qualified for pilot training might very well exceed the number needed for one pursuit squadron: "In view of the difficulty being experienced in obtaining enough suitable [white] men for pilots, it would be unfortunate to refuse any Negro who is qualified. It may be that the number of Negroes accepted by the examining boards will be sufficient to provide flying personnel for more than one squadron. It seems important that the acceptance of qualified men not be restricted by arbitrary limitation to the number required for a single squadron."[100]

Significantly, Hastie did not criticize the establishment of a segregated air unit, even though there can be little doubt that he was personally opposed to it. He no doubt realized that, as a member of the secretary of war's staff, it would be impolitic for him to openly oppose the policy of segregated units which had been clearly articulated in the adjutant general's memorandum of 16 October 1940.[101] The War Department had not prohibited integrated training; indeed, as Hastie pointed out, at least one integrated training program had been instituted and there were plans for more. Thus Hastie appears to have accepted the inevitability of segregated units in the Air Corps but not segregated training, especially segregated training at an isolated post established for that purpose alone.

Hastie's arguments had little effect on the Air Corps leadership. On

6 January 1941 General Arnold explained to the assistant secretary of war for air, Robert A. Lovett, why blacks could only be trained at Tuskegee. Black cadets, he asserted, could not be trained at any of the existing Air Corps facilities because their "fields are so congested with their existing load it would be unwise to tax their facilities with additional students."[102] Moreover, Arnold continued, supply channels might become clogged if additional students were suddenly added to the programs at these training bases. Tuskegee was an ideal location, the Air Corps chief concluded, because it was a readily available site, in an uncongested area, near a major air installation:

> Tuskegee was selected because it would be possible at that place to start a negro training school in the shortest possible space of time. The majority of the facilities are available and there is no question of air congestion at Tuskegee. The school would be under the direct supervision of the Commanding General, Maxwell Field Training Center. Thus it affords a means of starting the school with a minimum delay, avoids air congestion and is close enough for control and supervision by the Commanding General, Maxwell Field.[103]

After he received Arnold's memorandum, Secretary Lovett discussed the matter with Judge Hastie, citing the policy statement of 16 October 1940 as yet another factor that precluded training black pilots at existing Air Corps facilities. The civilian aide refused to endorse the Tuskegee plan but agreed to withhold his objections, still maintaining that blacks should be trained at one of the existing training centers. When Arnold learned that Hastie had withdrawn his objections, he ordered that the plan for establishing the Ninety-ninth Pursuit Squadron be implemented without delay, "in view of the fact that Judge Hastie will not object to this and of the absolute impracticability of training these negroes at Randolph Field or Maxwell Field."[104] The following day, 9 January 1941, the plan received the formal approval of the secretary of war: the era of the all-white air force had ended, and the day of the segregated air force had arrived.[105]

10

The Reaction
to the Ninety-ninth

Most, if not all, of the leaders of the army and the Air Corps considered racial issues a matter for the society at large to address and not within the scope of their responsibility. They viewed racial questions as a bothersome distraction from the overwhelming task at hand, the challenge of creating, virtually overnight, ground, naval, and air forces to match the vast Axis military machines in Europe and Asia. General Arnold and his lieutenants no doubt hoped that, once the black squadron which the General Staff had directed that they establish was formed, the nettlesome issue of black participation in the Air Corps would be settled once and for all. But the black public's criticism of the Air Corps was not abated by the news that a segregated squadron would be formed and based at Tuskegee; Judge Hastie's warning that "there would unquestionably be very great public protest" if the Tuskegee plan was adopted proved prophetic indeed. The circumstances of the announcement, the squadron's isolation from the rest of the Air Corps, and its relationship with Tuskegee Institute all precipitated a storm of black protest, and the Air Corps continued to be the target of criticism from various segments of the black public.

Nothing in the record indicates that either the War Department or the Air Corps had considered either the timing or the forum for announcing the Tuskegee plan, a curious circumstance considering the fact that the War Department had mandated the development of the plan in response to the intense public pressure campaign for black participation in the Air Corps. If the leadership of the War Department and the Air Corps had no definite time or place for announcing the decision to establish a separate air unit, Tuskegee Institute president F. D. Patterson did. In late December 1940 he agreed to host a conference of representatives from black schools and colleges who were interested in broadening opportunities in aviation for their students, both through the CPT Program and, ultimately, in the Air Corps. After the conference was scheduled for mid-January 1941, Patterson learned that the Tuskegee plan had received final approval from the War Department, and he apparently persuaded the Air Corps to make the plan public while the conference was in session.

The meeting had been proposed by Hampton Institute president Malcolm S. MacLean, who wired the Tuskegee president in late December to urge that he call an aviation conference at Tuskegee. MacLean noted that all of the nation's black colleges were "deeply aware of [the] importance of training Negro aviators with speed and efficiency towards ultimately forming [an] Army air unit." He acknowledged Tuskegee's recent efforts in the field of aviation and urged that the efforts of black colleges be carefully coordinated: "We recognize and are grateful for Tuskegee leaders[hip] which Hampton wil[l] follow if you make clear what we [ought] to do. Have we not gone far enough now to make mandatory full cooperative planning of [the] next steps and discussion as to functions each institution should perform?" MacLean concluded by "strongly" urging his counterpart at Tuskegee to call a conference at the "earliest possible date . . . to work out [a] strong[,] careful[,] total plan to instruct us all in our specific duties in further developing flight training plus aviation mechanics and maintenance," and to develop a plan for "united pressure" to force the federal government to admit blacks to the Air Corps.[1]

Patterson responded immediately, declaring the idea of an aviation conference an "excellent" suggestion and predicting that a "comprehensive program including all interested institutions" would yield positive results. The Tuskegee president suggested 16 January 1941 as a tentative date and arranged to meet with MacLean on his upcoming trip to Washington, D.C., during the second week of January.[2] Patterson's enthusiasm for President MacLean's proposal was due largely to his inside knowledge of the plan that Maxwell Field's Southeast Air Corps Training Center had submitted to the chief of the Air Corps during the first week of December. Under it, completion of CPT secondary training would be a prerequisite for blacks seeking appointment as aviation cadets. As the only black college offering CPT secondary training, it was incumbent upon Tuskegee to ensure that the pool of black secondary graduates was large enough to meet the needs of the new black air unit. Thus Patterson and his CPT coordinator, G. L. Washington, needed the cooperation of their sister institutions, because, as Washington later recalled, "Tuskegee would be hard pressed to keep the Secondary Course supplied with trainees unless an ample supply of Primary [i.e., elementary] Course graduates were available" from the other black colleges that conducted CPT courses.[3]

By early January Tuskegee had settled on the sixteenth as the date of the conference, and President Patterson had invited over thirty black institutions to send representatives to the meeting. The Tuskegee president's letter of invitation indicated that the purpose of the conference was to "consider a broad program of aviation training to be participated in by our several institutions, which will be pointed toward supplying the necessary personnel for an army program in aviation."[4] Most of the institutions Patterson contacted supported the purposes of the conference and fifteen sent representatives to the meeting, a surprising response considering that the invitations were sent out less than two weeks before the date of the conference.

By 4 January 1941 plans for the aviation conference had been finalized and the invitations had been mailed, although the final decision on the Air Corps' plan for a segregated squadron was still pending. Patterson knew he could not freely discuss the details of the plan until it was approved and announced to the public; thus it seems likely that he intended to urge the War Department to approve the plan and release it to the public when he visited Washington, D.C., several days later. His trip fortuitously coincided with the final approval of the plan by the secretary of war and another important, but unanticipated development—the premature disclosure of the NAACP's plans to file suit on behalf of Yancey

Williams against the War Department for the exclusion of blacks from the Air Corps.

During the first week of January 1941, as the leadership in the War Department considered Judge Hastie's objections to the Tuskegee plan, the NAACP's Legal Defense Committee had redoubled its efforts to challenge the exclusion of blacks from the Air Corps. Thurgood Marshall's interest in the Yancey Williams case took on a special urgency which strongly suggests that he knew of the Tuskegee plan and suspected it would soon be approved and announced to the public.[5] On Wednesday, 8 January, he wired his Washington colleague, Robert Ming, to inquire about the Williams case: "How is the Air Corps case coming? Can we get it filed by Friday of this week? If filed on Friday [it] will make our regular press release. If we wait another week we will lose effect. Send rough draft air mail special."[6] Ming's return wire advised Marshall that the complaint was still being drafted and could not be filed on Friday, but that a rough draft would be sent to New York by Thursday evening. Nothing should be released to the press, Ming warned, until the complaint was actually filed in federal court.[7]

The following day, as Ming sent the draft to NAACP offices in New York City and once again admonished Marshall to defer publicity until the suit was filed,[8] news of the case inexplicably appeared in the black press and at least one national news magazine contacted NAACP headquarters for further details. Marshall immediately apprised his colleague in Washington of the development and suggested decisive action to control the damage to the case that might be caused by the premature disclosure:

> In some manner unknown to us [the] Air Corps story has broken in [Baltimore] *Afro[-American]*. Other Negro newspapers will be on our necks claiming we gave [the] *Afro* a scoop. *Time* magazine likewise interested and has just telephoned. In order to keep faith with other Negro newspapers seems to me we should release some sort [of] story tomorrow announcing that a case will be filed. This will be followed by press release to local press in Washington by you and follow up story in our press releases next week. Would also like [to] include story in bulletin to branch members going to press this Saturday afternoon.[9]

Once word of the Williams case leaked to the press, the NAACP had lost the element of surprise, and it finally became obvious to the Air Corps that its silence on the Tuskegee plan was counterproductive. The effect of the case on the War Department was apparent to at least one member of the secretary of war's staff, Truman Gibson, a black Chicago attorney who had recently been appointed assistant to Civilian Aide William Hastie. On 11 January, shortly after the news of the NAACP suit against the Air Corps appeared in the *Afro-American*, Gibson confided to his good friend Claude Barnett that "the air corps situation is about to break wide open," because "a very scared military" had learned of the NAACP's plans to "seek an injunction against high War Department officials" on behalf of Yancey Williams and other qualified blacks seeking appointment as aviation cadets.[10]

Gibson could not comment further on the matter without compromising his position in the War Department, but he advised Barnett to "keep in close touch with Patterson." The Tuskegee Institute president, Gibson continued, "has been in conference and I don't think I need say more. I know you will not indicate that I suggested you call him."[11] Although no record of the conference has been found, Gibson's cryptic comments suggest that Patterson had learned of the Tuskegee plan's approval and that the Tuskegee president was attempting to influence the timing of the announcement to coincide with the upcoming aviation conference. After Gibson's letter, Barnett realized that Patterson was the best source of information on the Air Corps' plans, and he appealed to his friend and fellow Tuskegee Institute board member for news: "What happened when you were in Washington? Don't forget that our business here is publishing news and we naturally want it before some one else breaks with it."[12]

Barnett's appeal went unanswered, which suggests that Patterson returned from Washington, D.C., without authorization to disclose the details of the Air Corps' plans for the new black squadron. The uncertainty over the release of the Air Corps plans for admitting blacks was revealed clearly in the agenda that G. L. Washington proposed for the aviation conference. Only hours before the conference began, Washington submitted a proposed program to Patterson and noted that all the session topics had been finalized except for the evening dinner session; he suggested that this session focus on "Cadet Training in the Army Air Corps and the Negro," with President Patterson presiding. The evening session, Washington explained, "is the place proposed for giving [the] status of [the] Army plans for training Negro cadets."[13]

Patterson, however, knew it would be imprudent to discuss the Air Corps' plan for training black pilots until he received definite word of its approval and public release; to do otherwise would alienate the Air Corps officials whose goodwill he considered crucial to the success of the project and, equally important, could prove embarrassing if the plan was disapproved. When the day of the aviation conference arrived and he still had received no further word on the status of the plan beyond what he had learned on his trip to Washington, D.C., the previous week, Patterson sent an urgent wire to War Department officials, reminding them of his interest in the matter: "Having conference on aviation today involving . . . Negro schools interested in program. Can War Department give me definite statement to be read to conference relative to development of pursuit squadron[?]"[14] General Arnold replied by providing him with the text of a statement the under-secretary of war was to make on the subject at an afternoon press conference:

The War Department will establish a Negro pursuit squadron. The plan will begin by the enlistment of approximately four hundred thirty Negro high school graduates to undertake technical training and other specialized instruction at Chanute Field[,] Rantoul[,] Illinois in courses varying from twelve to thirty weeks. Approximately six months after training is begun at Chanute Field a nucleus of trained technicians will be available for transfer to Tuskegee[,] Alabama to start organization of [the] squadron. A site for installations at Tuskegee already has been selected. Pilot trainees will be obtained from those completing Civil[ian] Pilot Training Program Secondary course and will be enlisted as flying cadets subject to present standards. They will be sent to Tuskegee for Basic and Advanced flying training and unit training in pursuit types of aircraft. The squadron will be organized at Tuskegee as soon as fully trained pilots become available. Instruction will proceed as soon as funds are made available for this purpose by the Congress.[15]

General Arnold's wire cleared the way for a full discussion of the Air Corps' plans for incorporating blacks into its ranks. Patterson acceded to Washington's plans for an evening session on "Cadet Training in Army Air Corps and the Negro," but did not give the main address himself. Instead, he briefly reported on the "cooperative efforts between Tuskegee Institute and the Army Air Corps" and then introduced Maj. L. S. Smith, director of training at Maxwell

Field, the officer who had drafted the original plan for establishing a pursuit squadron at Tuskegee. Major Smith spoke at length, giving "the Conference a complete report on the Air Corps' plans as regards Negroes" and then fielding questions from the audience.[16] G. L. Washington concluded the evening session with discussions on the requirements for nonflying officer positions, in the meteorology, communications, and engineering specialties. Washington also urged those black schools conducting elementary courses in the fall session to provide their best candidates for Tuskegee's spring secondary course immediately, because the first black aviation cadets would be selected from the graduates of this course.[17]

Judging from the "findings and recommendations" adopted by the conference delegates, it is clear that they accepted Tuskegee's leadership and intended to support the school's efforts to provide the Air Corps with black aviation cadets through its CPT secondary course. Acknowledging Tuskegee's leadership in the field of aviation, those attending commended "the pioneering work which Tuskegee Institute has done in offering advanced training" and urged support for the "development of a strong center of aeronautical training at Tuskegee Institute." They recommended that Tuskegee take the lead in urging the CAA to "expand Civilian Pilot Training in Negro institutions to more nearly equal a quota of trainees in proportion to the Negro population ratio," noting that only 100 blacks were trained out of the 10,000 students who participated in the 1939–40 CPT programs. In the interim, the conferees recommended that a "cooperative arrangement with Tuskegee Institute in the matter of secondary training be supported," and that Tuskegee place high priority on training the necessary black flight instructors.[18]

Finally, the conference participants endorsed Tuskegee Institute's collaboration with the Air Corps and promised to support the plan for establishing a segregated squadron near the Tuskegee campus. Before the conference began, G. L. Washington had consulted the delegates regarding their views on the issue of a separate unit, for he knew that the institute's role in the Air Corps plan would inevitably provoke criticism from the NAACP and other groups who denounced segregation. At an informal gathering of the delegates the evening before the conference, Washington led a discussion on the issue of *amalgamation* vs *separate unit* as [the] only two Army Air Corps approaches to training Negroes."[19] The next morning he advised Patterson that the "consensus of opinion was for [a] separate unit and [we] likened this unto [the] government [Tus-kegee Veterans] Hospital case some years ago." Washington told Patterson that the delegates "would accept a resolution backing your stand," and he tactfully urged the Tuskegee president to solicit the support of the key participants in introducing such a resolution at the evening session: "Two or three of the stronger delegates including President [F. D.] Bluford [of North Carolina A. and T. College] and [the] Hampton president I am sure (if approached during the day) would draft any resolution you suggest regarding the Cadet Training Stand."[20] Apparently, Patterson took Washington's advice, for the final conference report included a recommendation "That full cooperation and backing be pledged to Tuskegee Institute and mobilization of the forces necessary for presenting complete unity be made regard[ing] plans for a complete training program for Negro cadets in the Army Air Corps."[21]

In the week between the untimely disclosure of the NAACP's suit against the War Department and the announcement of the plan to establish a segregated air unit, NAACP attorneys Thurgood Marshall and Robert Ming completed the final details of the Yancey Williams case. By 13 January they had secured the cooperation of local attorney Wendell L. McConnell, a member of the District of Columbia Bar, who agreed to serve as Williams's counsel.[22] Three days later, on the same day that the War Department announced plans to establish a segregated pursuit squadron, McConnell filed suit in the D.C. District Court on behalf of Williams and "other qualified Negroes."[23] The complaint named five codefendants: Secretary of War Henry L. Stimson; Gen. George C. Marshall, the army chief of staff; Maj. Gen. E. S. Adams, the adjutant general; Maj. Gen. H. H. Arnold, chief of the Air Corps; and Maj. Gen. Walter S. Grant, Third Corps Area commander.[24]

Although the Williams case was filed on the same day that the Air Corps' plan for a black squadron was released to the public, no evidence has been found to suggest a causal relationship between the two events. The fact that the two closely coincided, however, has led to speculation that the segregated unit was established in response to the Williams suit. An account of the case in the NAACP Annual Report for 1941 explained that after Williams lodged his complaint in federal court, "plans were announced and put into effect for the establishment of a Negro air corps unit at Tuskegee, Alabama."[25] In his autobiography, Roy Wilkins, NAACP assistant secretary and editor of the *Crisis* at the time the Williams suit was filed, implies that the Williams case was the key factor in the War Department's decision to establish a segregated air unit:

> The most flagrant example of discrimination came . . . in the army air force. From New York, we sent a wire to Washington asking if Negroes were to be trained as pilots. We received a one-sentence reply saying the War Department didn't think so. Then an intrepid, twenty-four-year-old engineering student from Howard University sought our help in breaking this color line in the sky. His name was Yancey Williams, he had a private pilot's license, and when the army turned him down he sued. Thurgood Marshall did what he could to help in the case, and under pressure, the army made a small concession. It agreed to train thirty-three black pilots for a pursuit squadron of twenty-seven planes, the crack 99th Pursuit Squadron, an outfit that subsequently attracted some of the most promising young leaders of the race.[26]

Wilkins's brief account of the Williams case overlooks the broad campaign for Air Corps participation that had been under way for several years. Although the NAACP had unquestionably played a crucial role in focusing public pressure and attention on the exclusion of blacks from the Air Corps, especially in the latter half of 1940, Wilkins's assertion that the source of that pressure was the Williams case is incorrect and an oversimplification. The inclusion of the Wagner amendment in the Selective Training and Service Act, together with the politics of the 1940 presidential election were the key factors which prompted the War Department to insist that the Air Corps establish at least one black unit. By the time the Williams case came to public attention and was subsequently filed in federal court, the decision to establish a black pursuit squadron and base it at Tuskegee had already been made. The role of the Williams case was to accelerate, and perhaps force, the public release of the Air Corps' plans, a development which, ironically, benefited Tuskegee Institute. If news of the Williams case had not been prematurely released, the War Department might have postponed announcing the plan, a delay which would have made it difficult—if not impossible—for Patterson to have obtained the cooperation and support of the black colleges attending the aviation conference.

G. L. Washington had been correct in his conclusion that Tuskegee should seek the endorsement of other black colleges. Once the plan for establishing a segregated unit and basing it at Tuskegee was released to the public, Patterson and Tuskegee Institute came under strong criticism from several sources, most notably the NAACP, and the black aviators in Chicago and their supporters who comprised the leadership of the National Airmen's Association of America. Two days after the War Department's announcement, Claude Barnett sent a confidential letter to President Patterson, which began with a warning: "It appears that some misguided people are preparing to attack the aviation program at Tuskegee on the basis of its being segregated."[27]

Barnett assured Patterson that the challenge would not emanate from the office of the civilian aide: "Any program which Negroes develop will be segregated," Barnett asserted, a fact that Judge Hastie had recognized when he "objected to a special training school [at Tuskegee] and set forth that it should be set up in Illinois, California, etc." Hastie's office had condemned the Tuskegee plan, Barnett explained, "for principle's sake since they are objecting against discrimination and segregation on so many fronts[;] . . . while they expected to be defeated and were not at heart opposed to the Tuskegee designation, they felt in all fairness that they had to impose a paper objection."[28]

Instead, Barnett warned, the NAAA, under the influence of Howard Gould and Enoch Waters, was planning to "bring the NAACP in to stop Tuskegee [from] being designated as an army training center." Barnett reminded Patterson of an article that Waters, city editor for the *Chicago Defender*, had written earlier "in which he threatened along with Gould and Miss [Willa] Brown to send for Walter White to address a mass meeting to oppose the formation of an air squadron at Tuskegee." Barnett implied that Waters's criticism of the segregated unit at Tuskegee was opportunistic and hypocritical, since he had played a key role in securing the Chicago CPT Program. Moreover, he continued, the Chicago program was "entirely segregated in the north where it need not be," a characterization of the racial policies at the Harlem Airport, which contrasts sharply with the glowing descriptions Waters and Gould provided the NAACP the previous November, portraying the flying field as a model of racial harmony and cooperation. Nevertheless, Barnett warned, "Waters has let it be known that he plans to lead an attack on the Tuskegee program," and "Gould, whose intimates despair of his ever being practical, is urging them on." The threat of a controversy between the NAAA and Tuskegee Institute did not bode well for the future of blacks in the Air Corps, Barnett concluded, because Waters, Gould, and the NAAA "can do some harm if they start a lot of mud slinging. Here we have what we have been yelling for, a chance to get into the army air corps and they want to give the army a chance to go back on its program by starting opposition to it."[29]

The NAACP's reaction to the announcement of the segregated squadron was predictable and no doubt encouraged the Chicago opposition to proceed with their plans to attack Tuskegee and the Air Corps. The NAACP had responded immediately to the War Department's public statement that a black squadron would be formed. The day after the announcement, Walter White wired Hastie to ask if the "*New York Times* report is correct that a segregated Air Corps squadron is to be established by the War Department at Tuskegee Institute." If accurate, White concluded, the NAACP "vigorously protests [the] surrender of the War Department to the segregation pattern."[30]

Hastie responded by summarizing the statement issued the previous day. He confirmed that a black pursuit squadron would be formed; that flying cadets would be selected from Tuskegee CPT secondary course graduates and then enter military flight training at a field to be constructed in the vicinity of Tuskegee; that ground crews would be trained at Chanute Field and then transferred to Tuskegee; and that the number of pilots called for training would exceed the thirty-three required to staff the squadron, to allow for normal attrition during the training phase.[31]

In a 28 January response, which constituted the official NAACP position on the Air Corps program at Tuskegee, White attacked the plan as "wholly unsatisfactory, not only from the point of view of the Negro but . . . that of national defense." By instituting separate training and separate units in the Air Corps, where none had previously existed, the War Department was "fixing the pattern of segregation more firmly than has ever before been the case." The NAACP secretary urged, as a "more effective and courageous procedure," the integration of blacks into training courses in states where, unlike Alabama, segregation was not an entrenched practice. "Such action by the War Department," White maintained, would indicate "some intention on its part to begin abolishment of the undemocratic and un-American practice of segregation of the Negro." Moreover, White concluded, the plan to isolate the black air unit at Tuskegee amounted to an open invitation to exclude the squadron from training exercises with larger Air Corps units in preparation for combat duty.[32]

White and Hastie communicated in their official capacity—White as executive secretary of the NAACP and Hastie as civilian aide to the secretary of war—but NAACP special counsel Thurgood Marshall contacted Hastie informally the day after the Tuskegee plan was announced. He explained that James L. H. Peck, the black aviation writer who had written the December 1940 *Crisis* article that accused the Air Corps of playing black Americans for fools, had just called to express his concern over the news of "the Jim Crow air corps to be set up at Tuskegee." Peck argued that it was "impossible to set up a separate unit that will be equal to a regular unit."[33] Marshall was so impressed with Peck's arguments against the Tuskegee plan that he proposed to send Peck to Washington, at NAACP expense, for "an *off the record* conference with you and Gibson to give you the material to attack this set up on a basis of inequality as well as the question of the evil of segregated units. This fellow really has the dope and could act as an expert advisor."[34] Whether Hastie accepted Marshall's suggestion is not clear, although he asked the NAACP to arrange a meeting with Peck the following month when he visited New York.[35]

Walter White was also impressed with Peck's critique of the Tuskegee plan. When he was preparing his 28 January letter to Hastie outlining the official NAACP position on the segregated pursuit squadron, he asked George B. Murphy, the NAACP's director of publicity and promotion, to prepare a "paragraph based on your conversation with Peck about the necessity of a pilot being trained in a larger unit than the one set up at Tuskegee in order to make him an efficient flier."[36] Murphy complied and provided the following sentence, which was included verbatim in White's 28 January letter: "As far as information available to us is concerned, there can be no such thing as properly training Negro aviators in a Jim Crow squadron because, on the basis of the War Department's past performances, there is no assurance whatever that members of this segregated squadron will take part in the necessary tactical training with larger units, a situation which must obtain if these men are to be fitted for actual combat flying."[37]

The NAAA and its supporters in Chicago were also quick to react to the War Department's 16 January statement announcing plans for a black squadron. According to a press release prepared by Enoch Waters, the NAAA had adopted a "strong resolution condemning the War Department's plan to establish an all-Negro pursuit squadron" the following day.[38] Noting that the unit was to be commanded by a white officer, the release included a statement by NAAA president Cornelius R. Coffey:

> Our fight for entrance into the air corps has been long. We don't intend to compromise now. Both the army and the navy have stressed tradition in arguing against the abolition of segregated units.
>
> In the air corps there is no tradition either favorable or unfavorable to complete interrgation [sic]. If we permit the establishment of a Negro unit, it will be establishing a precedent which will be hard to break down.
>
> We'd rather be excluded than to be segregated. There's no constitutional support for segregated units and the only traditions existing in aviation as I know it are ones which would make complete integration sane and logical.[39]

NAAA spokesmen claimed privately that they were not opposed to the development of aviation at Tuskegee but rather "the establishment of a jim-crow unit anywhere as the answer to requests for enrollment of Negroes in the army air corps."[40] Nevertheless, they believed that the Air Corps had decided to train and base its black unit at Tuskegee at the behest of President Patterson, undermining their two-year struggle to force the Air Corps to open its ranks to blacks and establish a flight training program in the Chicago area.[41] Thus their condemnations of Air Corps racial policies usually characterized the black unit as the "Jim-Crow 99th Pursuit Squadron to be established at Tuskegee Institute," implying that the only reason that a segregated unit existed was because the school had cooperated with the Air Corps.

The NAAA became even more suspicious of Tuskegee's motives when the Air Corps abandoned its plan to enter black CPT secondary graduates directly into the second phase of instruction, basic training, bypassing the standard army primary training at civilian contract schools, which provided the first phase of flying instruction to white cadets. The Coffey School of Aeronautics had begun CPT secondary in mid-December 1940, using two army primary training aircraft, the only secondary course at the time so equipped.[42] Thus the NAAA had reason to hope that if it developed an effective advanced CPT training program, it could compete with Tuskegee and serve as a feeder program to provide black aviation cadets to enter military flight training. Only a month after the plan to establish a black squadron was announced, however, the Air Corps dropped the idea of using CPT secondary in lieu of contract primary training and awarded a contract for primary flight training to Tuskegee Institute.

The genesis of the decision to use a primary contract school rather than CPT secondary for the first phase of flight instruction is difficult to determine. G. L. Washington maintains that the War Department adopted the plan "as a result of a decision at Headquarters to equalize training throughout for Negro youth."[43] Probably no such change would have been made had not Judge Hastie and his assistant, Truman Gibson, become concerned that black cadets might not receive adequate training under the plan to use CPT secondary as a substitute for primary training at a civilian contract school. In early February Claude Barnett discussed the status of plans for training black pilots with Hastie and Gibson and learned that they suspected the Air Corps of setting the black aviation cadets up for failure. They told him that "the army heads say constantly that the CAA trained people are no good and yet they expect the Negroes to jump from CAA secondary to army [training]." They urged Barnett to advise Tuskegee to seek a contract for a primary flying school, noting that the money was available; the War Department had just awarded eleven new contracts, bringing the total number of civilian primary flight schools to twenty-eight: "There is no reason why Tuskegee should not have this they say. The men would get the correct training and not be failing all over the place as CAA trained men have done everywhere."[44]

Hastie and Gibson had another reason for urging that the plan to use CPT secondary in lieu of primary training be dropped. They hoped that if the Air Corps abandoned the idea of using the CAA to begin the training of black pilots, they might eventually be persuaded to train blacks as pilots at other Air Corps schools. If black enlisted men could be trained at Chanute Field, then the Air Corps could, "in addition to Tuskegee's training, send men to . . . Glenview, Illinois, or to California."[45] This hope proved illusory, however, once Tuskegee received approval to establish a primary school. In May Hastie noted that plans were under way for a school at Tuskegee and commented to Robert A. Lovett, the assistant secretary of war for air, "that the contract primary training schools in Arizona and California are located in areas where the admission of Negro candidates should not create any substantial problem." He suggested to Lovett that the "assignment of Negro cadets to such schools would avoid a bottleneck which may be created by the limited capacity of the proposed Tuskegee primary school and, also, would go far toward removing the widespread public criticism of limiting Negro cadets to attendance at Tuskegee."[46] Lovett summarily rejected any suggestion that blacks be trained outside of Tuskegee: "I am assured by the Chief of the Air Corps that the establishment of this school will provide adequate facilities for the elementary flying training that is required in connection with the approved War Department program for the activation of the 99th Pursuit Squadron."[47]

G. L. Washington also suggested another factor that may have influenced the decision to use a primary contract school instead of CPT secondary. In late January 1941 he recalled that Grove Webster, the CAA official in charge of the CPT Program, visited Tuskegee Institute as a representative of the Air Corps and advised Patterson that it had been decided to establish a primary school for black aviation cadets. He had been sent "to feel out Tuskegee Institute on the idea of the Primary Flying School being located at Chicago, by way of appeasement, since Chicago had lost out on the Army Airfield project."[48] Patterson refused to go along with the idea and insisted that the primary school should be at Tuskegee Institute "because of the close proximity to the Army Airfield, favorable flying climate, and other considerations."[49] Whether the leaders of the NAAA ever learned of Webster's trip to Tuskegee and Patterson's refusal to go along with a primary school in Chicago is not clear. If they did, it no doubt confirmed their suspicions that the NAAA was a victim of Tuskegee's influence and reputation among white Americans, and convinced them that their earlier efforts to pressure the War Department to admit blacks counted for nothing.

In mid-February 1941 G. L. Washington met with Maj. L. S. Smith, director of training at Maxwell Field, to arrange for the establishment of Tuskegee Institute's primary flying school. The

Southeast Air Corps Training Center had received a letter from General Arnold directing that a preliminary contract for a primary school at Tuskegee be drawn up immediately so that the first group of cadets could begin training in July.[50] By the end of the month officials at Maxwell Field and Tuskegee Institute had come to an agreement and the proposal was forwarded to the War Department for approval; two weeks later Gen. Walter Weaver, the commander at Maxwell, was directed to negotiate a contract with Tuskegee Institute.[51]

The decision to establish a primary school at Tuskegee, of course, did not curb the criticisms of the NAACP or the NAAA. During the spring, as Tuskegee and the Air Corps were preparing to begin the training, both groups used various tactics to undermine the program and force the War Department to abandon the plan to segregate blacks in the Air Corps. Working independently, both organizations tried to influence the recruitment of blacks for the unit in such a way that the Air Corps would find it expedient to reconsider the overall plan for establishing and training a black air unit.

By May Claude Barnett had become so concerned over the attacks on the Ninety-ninth Pursuit Squadron by the NAAA and the *Chicago Defender* that he sent a confidential report to President Patterson on their activities with regard to enlistment.[52] By Barnett's assessment, "[t]here has been every possible effort made to sabotage" the enlistment of men for the ground crew training at Chanute Field. The attacks of the NAAA, mainly in the pages of the *Chicago Defender*, had kept many likely recruits in Chicago from enlisting: "You have seen last week's editorial and the illustrated 'Jim Crow' story in the *Defender*. It and their other activity had much greater effect than I anticipated. The better element of young men who might have enlisted from Chicago were confused and Chicago fell down altogether on its quota. . . . Those they are getting are not of the highest calibre and most of them come from about [the] state, very few if any from Chicago."[53] The army recruiters had discussed their difficulties with Barnett, describing how, despite their best efforts, they had been unable to interest Chicago blacks in enlisting for training and a tour of duty with the new black air unit:

> Last year when they opened enlistments for the regular army corps, they merely made an announcement and they were over enlisted in the first day or two. They made this announcement on the air squadron and

got no results. They put a captain on the job and sent a lieutenant from [Chanute] Field to conduct the examinations right here. Still no applicants. The captains and the lieutenant called on the *Defender* and the Urban League which they soon learned were the influences which were keeping people out of the unit. Waters at the *Defender* and Gould at the League are the fomenters. The army men were shocked. They thought, of course, that colored people had an interest and were amazed that leaders among us had not recognized the formation of the unit as an opening opportunity and sought to secure men who could make the best showing possible.[54]

Barnett warned Patterson "that this sort of thing is going to continue," and urged him to send the report to the CAA, because "these fomenters who are rooted in the local CAA program are actually interfering with the army." Such action, he predicted, might convince the CAA to do a "little tightening up and it would force the hands of Waters and Gould" as well as Willa Brown, who "works hand in glove with them."[55] Patterson took Barnett's advice and sent copies of the report to Grove Webster at the CAA and also to Chief of Staff George C. Marshall.

He explained to Webster that an effort was under way in Chicago to "nullify the development of the 99th Pursuit Squadron through slanderous and derogatory publicity," and noted that the report was "entirely accurate as regards the persons mentioned who are participating in this." Acknowledging Webster's long-standing support for Tuskegee's aviation program, Patterson explained that he thought it appropriate to apprise CAA officials of the situation in order to "bring some pressure to bear on those persons who, while enjoying the benefits of the CAA program, are attempting to undo this next significant step which has been taken." He assured Webster that most black Americans supported the efforts of the federal government to open up new opportunities in aviation for their race: "I know that the attitude of Negroes in this country at large is not only favorable but enthusiastic about the 99th Pursuit Squadron. If Chicago falls down on its quota, all that will be necessary will be to re-allocate [it] to some other part of the country, and with proper publicity we shall have no difficulty whatever in securing the necessary enlistments." Patterson concluded by soliciting Webster's cooperation in assuring "the military authorities that this is largely a scheme emanating from those who are selfishly interested in the development of aviation rather than a reflection on Negroes throughout the nation."[56]

Patterson's letter to General Marshall noted that Barnett's report outlined the details of "the situation in Chicago in which concerted attempts are being made to embarrass the military program in aviation which has been set up for Negroes." He likewise assured the chief of staff that the majority of black Americans supported the plan to establish a black air unit and that the necessary enlistments would be forthcoming. Nevertheless, Patterson explained that he considered it "wise to take certain steps at once to overcome the unfavorable publicity which the Chicago group is sending out, and would like to discuss this with you at your convenience"; he suggested that the chief of staff authorize him to fly to Washington, D.C., on a military aircraft for a conference. He concluded by assuring General Marshall that Tuskegee Institute was "deeply appreciative of the action which has been taken by the Army, and that we intend to leave no stone unturned in seeing that the plans are satisfactorily executed."[57]

Marshall apparently did not consider the matter serious enough to warrant flying Patterson to Washington, but he did refer his letter to General Arnold, who assessed the situation for the chief of staff and provided a suggested response to the Tuskegee president. Arnold explained that, according to sources at the CAA, the "northern negroes are trying to sabotage the Tuskegee plan" for four reasons:

(1) The faculty and staff at Tuskegee will not be equal to that at the regular Air Corps centers.
(2) The racial attitude on the Tuskegee campus and environs is pleasant but it is located in the general area where they feel racial prejudice is high. These difficulties would be very much less likely, say in western Texas.
(3) They are convinced that the army must ultimately take over qualified aviators of that race. They state that if the country needs flyers badly enough, like the R.A.F. they will take the negroes regardless of anything else.
(4) These people are willing to take a chance on losing the whole Tuskegee opportunity in order to gamble on obtaining training on different circumstances which they claim will give them a more even break.[58]

Arnold's determination to proceed with the plan for a segregated unit at Tuskegee, however, remained firm. None of these reasons, he declared to Marshall, "hold water," and he refuted each charge in turn:

a. We are going out of our way to get the best possible instructors and equipment for this project.

b. It is not believed that any change of environs will change the racial feeling if such exists.

c. The R.A.F. has not taken any negro pilots and as a matter of fact, the Air Chief Marshall of the R.A.F. expressed his pity to me in that we had to in the United States.

d. It looks as if it is a case of the whole or nothing that this group of people are waiting for.[59]

Marshall accepted Arnold's report and rewrote the letter that the chief of the Air Corps had proposed he send to Patterson. He told the Tuskegee president that the CAA had found that the campaign against Tuskegee was "not confined solely to the airport or the airport operators, but extends beyond that and they cannot see any steps which they might take at this time to provide a solution." He expressed concern that "Tuskegee is laboring under such unfortunate disadvantages," but saw little more that the army could do "other than to give you the best equipment and the best personnel possible, in order to provide the same standards of training at Tuskegee as at other training centers."[60]

Ultimately, the threat from the Chicago forces did not seriously damage the operations at Tuskegee. In fact, it may have been beneficial in the long run, because it forced Arnold to state in writing to the chief of staff that the Air Corps was making a concerted effort to "get the best possible instructors and equipment for this project" in order to refute charges that the black cadets were being set up for failure.

The NAACP's efforts to undermine the Tuskegee plan in the months that followed the initial announcement differed substantially from that of the NAAA and their supporters at the *Chicago Defender* and the Chicago branch of the Urban League. Instead of attacking Tuskegee Institute directly, they sought to swamp the War Department with applications from young blacks who aspired to become pilots, hoping that if an overwhelming number of qualified black applicants were placed on the waiting list for training at Tuskegee, enough public pressure could be brought to bear to force the Air Corps to expand opportunities for flight training beyond Tuskegee.

The origins of the NAACP's tactic of flooding the War Department with applicants are unclear, but it had a precedent in World War I when an organization of undergraduates at Howard University called for "overwhelming numbers" of qualified applicants to seek appointments to the officer candidate school for blacks at Fort Des Moines, Iowa.[61] By the second week in February 1941 the decision had been made to proceed, and Walter White wired the District of Columbia branch secretary, John Lovell, Jr., to ask for his help in obtaining the names and addresses of all blacks who had completed CPT courses or who held pilot's licenses.[62] White shortly had at hand the September 1940 issue of the Bureau of Census publication *Negro Aviators*, Negro Statistical Bulletin Number 3, and letters from black colleges participating in the CPT Program—as well as from several predominantly white northern colleges that had trained blacks under the CPT Program—providing names and addresses of graduates.[63] These sources provided perhaps a hundred names, and on 21 February 1941 White sent a form letter to these men asking them to respond if "you would be willing to join the Air Corps . . . if given an opportunity and if you come within the general requirements as to age, freedom from dependents, and other qualifications given in the enclosed memorandum." White explained that the NAACP had "certain plans based upon replies from yourself and others who have licenses or who have taken C.A.A. courses, which we believe will have considerable bearing upon the question of discrimination against Negroes in the Air Corps."[64]

The NAACP subsequently broadened its campaign and solicited the names of all qualified black men who were interested in becoming Air Corps pilots. At the end of February Hastie provided Thurgood Marshall with the five general qualifications for aviation cadets: unmarried male citizens; aged 20–26; completion of at least one-half the credits for a bachelor's degree or capable of passing a qualifying examination; excellent character; good health.[65] The NAACP used this information to prepare the first of a series of press releases that outlined the qualifications for appointment as an aviation cadet and asked young men "willing to enter this training" to contact their offices in New York so that information could be "assembled for a special inquiry and possible action."[66]

By April almost three hundred men had responded.[67] The following month the NAACP's main offices began mailing blank aviation cadet application forms to those who had responded. The cover letter urged each applicant to "request that you be given training at the training school nearest your home. Our reason for making this suggestion is because the present plan is to send such Negroes as are accepted for training to a segregated training school at Tuskegee Institute."[68] A press release was also prepared announcing that application forms were available at NAACP headquarters and including the suggestion that applicants request training at the nearest school.[69]

The NAACP's method of countering the Air Corps' plan for segregated training at Tuskegee was certainly a more reasonable approach than the smear campaign and posturing that characterized the Chicago opposition's tactics, which was not constructive and only served to alienate the decision makers in the War Department. Indeed, the NAACP could well argue that their efforts were simply well-intentioned efforts to aid the national defense by encouraging those qualified to fly to apply for duty in the Air Corps. Moreover, the NAACP's approach conformed to that of the civilian aide's office, which had suggested that the Associated Negro Press sponsor a campaign in the black press urging qualified blacks to apply for flight training so that the "Air Corps [would] be swamped with applications" from Negroes.[70] In the final analysis, however, the NAACP's recruitment campaign was more of an aid to the segregated program at Tuskegee than a hindrance, for it ensured, at least at the outset, that a backlog of qualified candidates were available for appointment as aviation cadets.

Another unintended effect of the NAACP's recruitment campaign was to raise the hopes of young black men who aspired to become military aviators. Eager for an opportunity in an era of rapidly rising expectations, many of these hopefuls saw the NAACP's interest as proof positive that they had a real chance to fly and fight in their nation's air arm, and they responded enthusiastically. Typical was the reaction of Charles W. Dryden, of the Bronx, who answered Walter White's query "would you be willing to join the Air Corps if given the opportunity" with a resounding "Yes, definitely! I would not only join but would pursue a lifetime career of service in the [Air] Corps."[71] Others did not realize that the NAACP opposed the Tuskegee plan and assumed that the recruitment campaign amounted to an endorsement of the program. Another young black man from New York, Oscar Oliver, wrote the NAACP offices to request "an application blank, for the newly organized 99th pursuit squadron, which is at present stationed at Tuskegee Institution [*sic*]." Oliver went on to declare his enthusiastic endorsement of the black air unit: "I think this branch of the service, is a wonderful opportunity for the Negro youth of America too [*sic*] show the world, that we too are equally as efficient as any man in the country, when we are given an opportunity too [*sic*] prove ourselves."[72]

The enthusiasm and excitement of young Americans like Dryden and Oliver was more often than not replaced by disappointment and frustration as they found themselves caught in the middle of a power struggle between the Air Corps, Tuskegee Institute, the NAACP, and the NAAA. Some of these high-minded young men only dimly perceived the positions and tactics of each group and were simply grasping for an opportunity to serve their country during the struggle against fascism in the glamorous field of aviation. Others understood only too well the conflicts dividing the black community. One of these perceptive young men was Alexander Anderson, a Tuskegee student who had learned to fly in Tuskegee's first CPT class; he had achieved the second highest score of his class in the written examinations and had completed the secondary and apprentice instructor courses as well.[73]

Anderson wrote to Walter White three times in 1941 regarding his plans to enter the Air Corps. He responded to White's initial letter of February 21 by declaring that he "would be very glad to join the Air Corps . . . if given an opportunity, and will and plan on doing all within my power to do so." Although he clearly understood the NAACP's opposition to the Tuskegee plan, he refused to sacrifice his chances to fly in order to take a stand against segregation: "But I can not and will not turn down this opportunity here at Tuskegee, if given the chance, because a bird in the hand is worth ten in the bush. We feel that we have this program, now what we must do is to fly these pursuit ships like nobody's business, fly them so that they can't turn other applications down. Otherwise I'm with your program. I understand."[74]

Several months later Anderson received aviation cadet applications as part of the NAACP's mass mailing campaign, and the young pilot wrote back to thank White for the gesture; he explained, "I've already applied and also passed the physical examination at Maxwell Field." He told White of his record of training in the CPT Program at

Tuskegee and noted that he was the only CPT student "white or black to hold [a] parachute license and have made parachute jumps before several thousands here at Tuskegee." The letter also contained hints of Anderson's frustration over the wait to begin Air Corps training. He asked White for information "at once" on flying opportunities in the Royal Air Force or "any other information you can give us connected with flying anyplace, Canada, England, Africa or anywhere." Still confident in his ability as a pilot, he declared to White, "I can do the flying if I can get a break."[75]

Six months later Anderson wrote White again, an impassioned and touching twelve-page letter in which he outlined his frustrations at not being called to training in the Air Corps. No longer self-assured and enthusiastic over the limited opportunity offered by the one unit in training at Tuskegee, he reminded White of his earlier correspondence "concerning this segregated Air Corps" and explained that it had been over six months since he, "along with many other young men took the physical examinations at Maxwell Field . . . and we are still waiting to receive our appointments to the Army Air Corps." Anderson went on to recount how he and others at Tuskegee had, at great sacrifice, struggled to complete CPT advanced training and gain valuable flying experience; yet it afforded no priority in obtaining appointment as an aviation cadet and entering training at Tuskegee. Only a trickle of blacks entered training at Tuskegee—ten every five weeks—and these were taken in order of application approval, without regard to previous flying experience, which worked against the Tuskegee CPT students: "We fellows flying were about the last to get a chance to take the [physical] examination, because (as you know) a place for Negroes to take any type of examination in the South was harder to get than in the North or East. When the first Army training program at Tuskegee for Negroes [was] started, to our disgust, the trainees were selected according to the order in which . . . we took our physical examinations, and to top that, only one from each Corps Area every five weeks."[76] Anderson concluded with an impassioned litany of complaints that clearly showed the frustrations and bitterness of many young black men who were forced to wait for months on end to be called for training at Tuskegee while they heard countless pleas in the national media for whites to enter flight training:

I see where there has been an appropriation to train five hundred thousand air cadets. . . . I see where they are building many fields to train these white boys. At Randolph Field . . . alone they are training over 2,500 white boys every 5 weeks. . . . This is only at one field, yet they can only train ten Negro boys. Why Mr. White, there aren't two thousand five hundred of us [blacks who have qualified for flight training] in all. . . . Yet I sit day after day and listen to advertisements over the radio, begging young men to take the United States Army, Navy, and Marine Air Corp [*sic*] training, and here we are begging for an opportunity to be trained for a government that claims to have a Democracy.[77]

The problems and frustrations Anderson faced as he sought to avail himself of the limited opportunities open to qualified blacks who aspired to serve in the Air Corps underscore the limitations in the approach of both the black conservatives, who were willing to accept segregated air units, and the black progressives, who opposed any extension of segregation. Patterson's policy of "segregated opportunity" was in the tradition of Booker T. Washington and it had indeed led to new opportunities in the Air Corps. The segregated opportunities were, however, extremely limited; consequently, the rising expectations of young men like Anderson were often frustrated because of the limited number of spaces open to blacks. The integrationists pointed to the bottleneck at Tuskegee as the natural result of accepting a segregated program, arguing that if blacks had held out for integration such problems might have been avoided. Given the official racial policy of the War Department—that segregated units would be maintained—and the opposition of the Air Corps leadership to accepting blacks even on a segregated basis, it seems likely that even if black Americans had insisted on integrated air units, the Air Corps would have remained all-white. Consequently, even the limited opportunities possible with segregated units not would have been forthcoming. In short, the racial climate of the period precluded any tidy solution that would have shielded the dreams of patriotic young men like Anderson from the realities of racial prejudice.

11

Military Flight
Training Begins

Although President Patterson and G. L. Washington were disturbed by the criticism of the Air Corps' plan to establish a segregated air unit following the announcement of 16 January 1941, they could not let the complaints of the NAACP and the NAAA paralyze them, for Tuskegee Institute figured prominently in that plan. Patterson and Washington concluded that the best response to their critics was to make certain the Air Corps followed through on its commitments. In the months that followed, both men carefully monitored the Air Corps' efforts to establish a segregated unit and worked diligently to develop an aviation program at Tuskegee Institute capable of supporting the army's plans for training black pilots.

Under the initial proposal developed by Maxwell Field's Southeast Air Corps Training Center (SEACTC), Tuskegee was to expand its CPT secondary course and train some thirty prospective aviation cadets, who would be fed directly into basic flying training when they graduated from secondary. In mid-February the Air Corps abandoned this arrangement and asked Tuskegee Institute to build a new airfield and establish a primary flying school for black aviation cadets under a contract with the United States Army. Several months later, the institute was also asked to administer the preparatory preflight training for the first class of aviation cadets. By fall

1941 Tuskegee had established a contract flying school, built an airfield, and completed primary training for the first class of black aviation cadets.

Tuskegee began to prepare for the expansion of its secondary flying operations in late 1940, just as the Air Corps was finalizing its plans to establish a segregated squadron. In December 1940, the institute launched an effort to shift its CPT secondary flying operations from the airfield at Auburn to Kennedy Field in Tuskegee, where the institute conducted elementary flight training. Commuting the twenty miles to Auburn became a burden to secondary students who were also enrolled in classes at Tuskegee; when Chief Anderson assured Washington that Kennedy Field could handle the heavier aircraft used in secondary training, Washington sought CAA approval to use the field for advanced CPT flying operations. The CAA immediately sent an inspector, who advised Washington that Kennedy Field would be acceptable for CPT secondary, if certain improvements were made, including additional grading, additional runway markers, and lighting for night flying.[1] Washington quickly went to work completing the improvements. By early January the CAA inspector had approved Kennedy Field for advanced CPT flying, and all of the institute's flying operations were transferred to Tuskegee.[2]

The transfer of CPT secondary flying operations from Auburn to Kennedy Field marked the consolidation of Tuskegee's CPT flying operations. The institute had made remarkable progress in the field of aviation; in little more than a year Tuskegee's CPT Program had been transformed from a shoestring affair, dependent on a helpful white pilot and access to flying fields inconveniently located in Montgomery and Auburn, to one staffed and controlled by the institute, operating from a field near the campus. Thus when the Air Corps announced the plan to use Tuskegee's secondary course as a feeder for Air Corps basic training, the school's future as the center of black aviation seemed certain.

By the end of January 1941 some thirty students had entered Tuskegee's CPT spring secondary course, "earmarked as the first Negro students to be trained in the Air Corps."[3] Tuskegee's original quota of ten for the secondary course had been tripled in order to provide an adequate pool of black aviation cadet candidates for basic flight training. Although these men were not actually inducted into the army prior to enrolling in secondary, the plan called for their enlistment as aviation cadets at the completion of the training; thus they had to pass the same physical examination required of aviation cadets and meet the approval of the War Department before they were admitted to the secondary course. Moreover, the Air Corps provided tangible evidence of its support when it supplied Tuskegee Institute with two Waco trainers to use in training the thirty secondary students.[4]

Shortly after the secondary students entered training, the Air Corps dropped its plan to use CPT secondary as an alternative to primary training for prospective black pilots.[5] In mid-February 1941 G. L. Washington met with Major Smith at Maxwell Field to arrange for the establishment of Tuskegee Institute's primary flying school. The conference was called in response to a letter from the chief of the Air Corps, which directed officials at Maxwell to negotiate a contract for the school with Tuskegee Institute. The commanding general of the SEACTC, Walter R. Weaver, was present for part of the meeting and admonished Washington that the "success of Negro youth in the Army Air Corps hinged upon every step being taken by Tuskegee Institute to insure [a] safe and creditable operation, that [Tuskegee owes] such to the Negro populace, and that . . . [a] contract would not be issued until . . . [the institute obtained] facilities which would insure this." Tuskegee's task would not be easy, Weaver concluded, because "the facility would have to be fully equipped and staffed and ready for receiving cadets [by] July 15, 1941."[6]

After the general left, Major Smith outlined the details of the plan to establish a primary flying school under contract with Tuskegee Institute. The quota for each incoming class was to be set at thirty, with the course of training lasting for ten weeks. The classes would overlap, with a new group of cadets entering training every five weeks; this meant that Tuskegee would have to develop facilities to handle sixty cadets. The quota of thirty was based on a goal of producing fifteen black pilots from each class that entered training. Using the normal graduation rates for each phase of flight training—sixty percent for primary, ninety-four percent for basic, and ninety-nine percent for advanced—Smith had calculated that twenty-seven cadets would have to enter primary training in order to produce fifteen pilots per class. Concerned that the graduation rates for blacks might be somewhat lower than the standard rate, Washington persuaded Major Smith to raise the quota for each entering class to an even thirty.[7]

At the 15 February conference Smith also outlined the terms of a standard contract between a civilian flying school and the Air Corps for primary flight instruction. The Air Corps provided the aircraft, one for each three cadets assigned, and furnished the cadets with textbooks, flying clothes, helmets, goggles, parachutes, and mechanics' suits. The civilian contractor received $1,050 for each cadet who graduated from the ten weeks of primary training and $17.50 per flying hour for those not graduating. Compensation for the cadets' room and board amounted to $1.67 per day per cadet. Three Air Corps officers were assigned to supervise the operation and ensure that the training met Air Corps standards. As the contractor, Tuskegee Institute was expected to provide one civilian flight instructor for each five cadets in training. Each instructor was required to complete a six-week Air Corps primary instructor training course to qualify as a primary flight instructor. Finally, primary contractors were required to carry adequate insurance, perform routine maintenance on the army aircraft (major overhauls were the responsibility of the Air Corps), and provide the cadets with transportation to the flying field if they were housed more than one mile from the field.[8]

As General Weaver had pointed out, the first class was scheduled to begin training in mid-July, which allowed Tuskegee only a few months to obtain financing and construct suitable facilities for the primary school. In addition to a flying field, Tuskegee Institute was required to provide full facilities for aircraft and personnel: quarters and mess for the cadets; hangars and maintenance shops; and offices for Air Corps personnel, flying instructors, ground school instructors, and mechanics. Both Weaver and Smith urged Tuskegee to develop the primary flying field on a site that allowed for expansion, and they suggested that it could be financed through the Reconstruction Finance Corporation (RFC), which had underwritten the construction of a number of other civilian primary schools.[9]

General Weaver suggested that Washington visit Darr Aero Tech, a new primary flying school under construction near Albany, Georgia, as an aid to developing preliminary cost figures. Shortly after his conference at Maxwell Field, Washington inspected the Albany school, which was, ironically, a project of Harold S. Darr, owner of the Chicago School of Aeronautics at Glenview Field, the school the

Air Corps had originally designated for the training of black pilots under the provisions of P.L. 18; Darr was in the process of transferring his operations to the South because of a navy takeover at Curtiss Field.[10] Washington examined the plans for the school, interviewed key personnel, and obtained a general idea of the type of facility Tuskegee should construct. From his inspection of the Albany school, he estimated that between $300,000 and $400,000 would be required to develop a first-class field with the capacity for future expansion, and $150,000 to develop a minimally acceptable facility.[11]

Such a project dwarfed the modest plans for an institute airport that Washington had proposed the previous year. Although the Tuskegee Institute Airport Fund Campaign was still under way in Chicago and Cleveland, its success had been limited.[12] If Tuskegee was to obtain the contract for the primary school, funds would have to be obtained immediately so that work could begin at once. Throughout March and into the early part of April, Washington and Patterson sought funding from a variety of sources.

They began their search for funds by approaching the RFC. As General Weaver had indicated, the RFC was familiar with the Air Corps program of contracting out the primary phase of flight training to civilian schools and was willing to finance the project. Washington and Patterson, however, were not pleased with the repayment terms: the RFC would carry the loan for no more than three years. The Tuskegee administrators doubted that the operation would generate sufficient income to repay the loan in such a short time; it would be subject to federal taxes if financed through the RFC, which President Patterson believed would "literally eat the thing up."[13]

At the end of February 1941 Patterson and Washington met in New York with the chairman of the Board of Trustees, William Jay Schiefflin, and the members of the finance committee to consider alternatives to the RFC.[14] A variety of funding options were discussed, including a proposal for a ten-year loan, which most of the committee supported. Alexander B. Siegel, however, the "most conservative member, did not feel Tuskegee Institute's trustees should obligate the Board for such a large outlay with little assurance the project would run ten years."[15] In an effort to counter Siegel's objections, Patterson called Air Corps headquarters for information on the estimated duration of the contract and to confirm the quota of thirty cadets per class. When he was advised that the War Department was considering a drastic reduction in the class size, Patterson discreetly adjourned the meeting and traveled to Washington, D.C., to meet with Air Corps officials to discuss the changes under consideration. After they learned of plans to reduce the quota for the first two classes to fifteen and that quotas for subsequent classes might be subject to similar reductions, Patterson and Washington prepared a less ambitious financing proposal and met again with the finance committee in early March.[16] This proposal called for a more conservative capital investment of $200,000, which could be retired in only a few years.[17] The finance committee agreed to provide a loan of $25,000 from institute funds and authorized Patterson to raise the balance from private sources.[18]

Armed with the endorsement of the finance committee, Patterson contacted several foundations with a record of supporting self-help programs for black Americans—the Carnegie Corporation, the General Education Board, and the Rosenwald Foundation.[19] He requested an outright grant of $150,000 from the Carnegie Corporation, an appeal that was rejected.[20] The General Education Board was also unwilling to support the construction of Tuskegee's primary school.[21] The Julius Rosenwald Fund, however, expressed an interest in the project, and its chairman met with the Northern Executive Committee of the Tuskegee Board of Trustees to discuss the matter.[22] Prospects for Rosenwald Fund support were perhaps enhanced by the fact that its board of trustees was to hold its annual meeting at Tuskegee at the end of March, only a week before the institute's trustees were to gather at the campus for their annual spring meeting.

One of the Rosenwald Fund trustees who attended the meeting at Tuskegee was First Lady Eleanor Roosevelt, who had recently been appointed to the board.[23] G. L. Washington planned a special air show for Mrs. Roosevelt and the other Rosenwald trustees, which included formation flights, acrobatics by the secondary students and instructors, and a parachute jump by Tuskegee student Alexander Anderson.[24] Mrs. Roosevelt described her visit to Kennedy Field in her daily newspaper column, "My Day": "Finally we went out to the aviation field, where a Civil Aeronautics unit for the teaching of colored pilots is in full swing. They have advanced training here, and some of the students went up and did acrobatic flying for us. These boys are good pilots. I had the fun of going up in one of the tiny training planes with the head instructor, and seeing this interesting countryside from the air."[25]

Chief Anderson clearly remembered Mrs. Roosevelt's visit to Kennedy Field and her comment to him that "everybody [had] told her that we [blacks] couldn't fly." Nevertheless, she observed "you are flying all right here. Everybody that's here is flying. You must be able to fly. As a matter of fact, I'm going to find out for sure. I'm going up with you."[26] Anderson recalled that Mrs. Roosevelt's aides were not pleased with her decision to make the flight but found it

First Lady Eleanor Roosevelt visited Tuskegee Institute's airfield in March 1941. She is pictured here next to G. L. Washington (holding hat) and C. Alfred Anderson (third from left). (Courtesy Tuskegee University Archives, Anderson Papers)

impossible to restrain the headstrong first lady, who was determined
to demonstrate her confidence in the flying ability of blacks: "That
caused a lot of opposition among her escorts, you know. As a matter
of fact, they were thinking of calling the President to stop her, but
she was a woman who, when she decided to do something, she was
going to do it. She got in the plane with me and went for a flight. We
had a delightful flight. She enjoyed it very much. We made a tour of
the campus and the surrounding area. We came back, and she said,
'Well, you can fly all right.'"[27]

The air show and the first lady's flight apparently convinced the

Mrs. Roosevelt shown preparing for her flight with C. Alfred Anderson, Tuskegee Institute's chief instructor pilot.(Courtesy Tuskegee
University Archives, Anderson Papers)

Rosenwald Fund trustees that Tuskegee Institute had the expertise to establish and operate an army primary flying school.[28] A week later the Tuskegee Board of Trustees held their spring meeting, discussed the problem of financing the flying field, and learned that the fund would consider favorably a request for a loan of $175,000. The board authorized Patterson to secure the loan provided that "the contract issued by the United States War Department to Tuskegee Institute for the training of cadets be made the sole security against such loan and the payments . . . will be made from the net operating income received from the training program under this contract."[29] The Rosenwald Fund immediately accepted the terms stipulated by the Tuskegee trustees and agreed to lend the institute $175,000 for "the purchase of a suitable site, development of an airfield, construction and installation of facilities and purchase of equipment necessary to meet the requirements of the Air Corps of the United States Army for an approved Civilian . . . Army Flight Training School."[30]

While Patterson sought financing, Washington concentrated on finding an acceptable site for the field and preparing construction plans for the facility. In March, after their trip to New York, Washington brought a black engineer and contractor, Archie A. Alexander, to Tuskegee to discuss the details of constructing the field and associated buildings.[31] Alexander, a native of Des Moines, Iowa, received his engineering degree from the University of Iowa in 1912 and established the contracting firm of Alexander and Repass with a white classmate, M. A. Repass; by the mid-1920s he was widely recognized as one of the most successful black businessmen in the country.[32] Washington and Alexander visited two other primary schools that had just begun to train cadets—Southern Aviation School, at Camden, South Carolina, and Souther Field, near Americus, Georgia.[33] From data gathered from these visits—and the earlier inspection of Darr's school at Albany, Georgia—Washington and Alexander drew up preliminary sketches of the field and buildings and estimated the construction costs.[34]

March 1941 also marked the beginning of the search for a suitable location for Tuskegee's primary flying school. The site on Franklin Road that Washington and officials of the Alabama Aviation Commission had selected for an institute flying field a year earlier was apparently inadequate for primary flight training, as there is no evidence that it was considered at all. Instead, three other sites were considered before a fourth and final location was chosen and construction begun. Patterson and Washington initially considered enlarging Kennedy Field, but officials at Maxwell advised that such a plan would at best provide only the minimum facilities required and that site preparation would be expensive.[35] Next Washington chose a site near the community of Hardaway, some twenty-five miles southwest of campus. Although authorities from Maxwell Field approved the Hardaway site, it was abandoned when soil conditions proved unsuitable for construction of a landing field. When Washington proposed to locate the field on a plot of land northeast of Kennedy Field, on the Vaughn properties, representatives from Maxwell again had no objections. But by early April, the plan to construct a flying field on the Vaughn land was abandoned in favor of some 650 acres of land three miles northeast of campus, owned by S. M. Eich.[36] Officials from Maxwell Field found the Eich lands acceptable, and it was on this site that Tuskegee Institute developed its primary flying field, the airport that would ultimately be known as Moton Field.[37]

Thus by April 1941 Tuskegee Institute had made considerable progress in preparing for its primary flying school. Less than two months after General Weaver and Major Smith had asked Tuskegee to establish such a facility, Washington and Patterson had secured financing, drawn up preliminary plans, and selected a site. Unfortunately, their efforts were not matched by the planners at Air Corps headquarters. The details of the training plan informally presented to Washington in February were subsequently modified several times, and a contract was finally approved between Tuskegee Institute and the War Department in early June.

Shortly after Washington's February meeting at Maxwell Field, the SEACTC had advised headquarters in Washington, D.C., of Tuskegee Institute's willingness to develop a primary flying school and train black cadets under contract with the War Department.[38] The Operations and Training Division at the Air Staff responded several weeks later with a training plan for the school, which contained several important changes from the initial proposal presented by Major Smith. First, the quota had been drastically reduced. Instead of one class of thirty entering training every five weeks to produce 150 black pilots annually, the Air Staff's plan called for only forty-five pilots per annum, with one class of thirty entering training every fifteen weeks and no overlap between classes. Thus instead of planning for a primary school designed to accommodate up to sixty cadets at a time and 300 per year as originally proposed, Tuskegee suddenly had to shift gears and plan for a maximum of thirty students at a time and ninety per year. Second, the initial two primary classes would only number fifteen, not thirty, in order to make room for the men who had entered Tuskegee's secondary course in January, before the plan to contract with Tuskegee for a primary contract school was adopted. Finally, a five-week preliminary course of instruction—preflight—had been added as a prerequisite to primary training.[39]

Preflight training was a new phase in the Air Corps' flight training program, which incoming aviation cadets had to complete before they could actually began their flying training.[40] With the advent of Air Corps expansion in 1939, the pilot training curriculum had been shortened from fifty-two weeks to thirty, leaving precious little time for initial administrative processing, classification testing, and rudimentary military training. By late 1940 the problem became acute, and the preflight schools were proposed as an expedient to provide preliminary processing, classification, and initial military indoctrination prior to a new aviation cadet's arrival at primary school. In late February 1941 General Arnold ordered the establishment of three such schools, one at each of the regional training centers—the Southeast, Gulf Coast, and West Coast Air Corps training centers.[41] The Air Corps had hoped to open the preflight schools immediately, but a number of difficulties delayed the entry of the first class until September. Nevertheless, the plan called for all newly appointed aviation cadets to begin their training at one of the three preflight schools—all cadets, that is, but those with black skin.

The decision to establish preflight schools had to be taken into account by the staff officers charged with developing a "separate but equal" training plan for black aviation cadets. They included it in the training plan that they developed; apparently this was to avert any charges of discriminatory treatment based on the claim that equal training was not being provided. The Air Corps' determination to maintain strict segregation in the training of black and white pilots, however, precluded any consideration of sending black aviation cadets to one of the three regional preflight schools. Instead, they would take that training separately at Tuskegee and then enter flight training at the Tuskegee Institute Primary Flying School.[42]

The addition of preflight school meant that fifteen weeks of training would be required to prepare the aviation cadet for the second phase of flight school—basic training. Thus the starting date for the first class of black aviation cadets was moved forward to mid-June, further reducing the time available to complete the primary school facilities and train the civilian flight instructors. By 14 March Patterson had apparently learned that the plan the Air Corps submitted to the War Department called for the first class to enter training a month earlier than originally proposed. Apprehensive lest valuable time be lost waiting for the plan to be approved, the Tuskegee president sent an urgent wire to Chief of Staff George Marshall: "Proposal for [Primary] training [of] Army cadet flyers approved by Air Corps and presented to War Department for approval. . . . To date have not been advised of action. Necessary we proceed at once if facilities are to be available June 15 as planned. Will appreciate your help on this."[43] General Brett responded to Patterson's wire by directing General Weaver at Maxwell Field to "expedite the conclusion of contractual negotiations with the Tuskegee Institute" and advising the Tuskegee president to contact Weaver immediately, hardly the response Patterson wanted since he was already working closely with officials at the SEACTC.[44]

The War Department did not approve the 11 March plan until the end of the month. Apparently there was some concern there that the criteria for selecting black aviation cadets might be relaxed—the War Department approved the plan "subject to the candidates meeting the same requirements as all other pilot candidates."[45] The Air Staff did not communicate this to the SEACTC until the second week in April, shortly after G. L. Washington had submitted Tuskegee's construction plans to Maxwell Field, plans based on the original proposal outlined by Major Smith in February.[46] Smith advised Washington of the changes, and the Tuskegee administrator scaled down construction plans accordingly.[47]

By the end of April, however, the quotas for black aviation cadets were reduced once again, this time to ten per class. This drastic reduction in student quotas—from classes of thirty every five weeks to classes of ten every fifteen weeks—significantly restricted the income potential for the school, convincing President Patterson that it would be unwise to go ahead with the original plans to construct housing, dining, medical, and instructional facilities at the primary flying field. Thus he met in Washington, D.C., with General Davenport Johnson, chief of the Air Staff's Training and Operations Division, to ask for an exception to the provision in the standard primary school contract, which required the civilian contractor to provide such facilities adjacent to the field. Instead, he offered to house, feed, and instruct the aviation cadets on the Tuskegee Institute campus and transport them to the field for flight training. Johnson agreed to Patterson's request and authorized the SEACTC "to waive such portions of the prescribed specifications for civil contract schools which, according to your judgment, may result in unwarranted expense at Tuskegee." General Johnson noted that Patterson had indicated that he would "provide a separate dormitory at Tuskegee Institute, together with separate messing facilities and the necessary hospital facilities, and class room space for ground school instruction. He is also willing to provide the necessary motor transportation for the movement of students to and from the flying field. If these facilities are satisfactory it appears probable that some arrangement can be made that will restrict the construction at the flying field to that essential for the conduct of flying training."[48]

When Major Smith, at SEACTC, learned of the plan to further reduce the class size to ten, he asked G. L. Washington if Tuskegee Institute was still willing to establish a primary contract school. Washington assured Smith that Tuskegee was ready to proceed if the contract was modified to permit the use of facilities at the institute.[49] On 7 May 1941 Washington submitted the final construction plans for Tuskegee's primary flying school to the SEACTC headquarters,[50] and President Patterson advised Major Smith that "Tuskegee Institute is willing to accept the revised contract for the training of Primary Cadet Fliers. Because of the reduced quota, it is understood that only the essential facilities for housing the ships, doing the necessary shop work, provisions for officers and a student waiting or study room will be made available at the field. The housing, boarding and scheduling of classes for all Cadets will occur on the campus of Tuskegee Institute under conditions to be approved by you."[51]

The following week Washington sought to expedite final approval of the proposed contract by advising authorities at Maxwell Field that the institute's purchase option "for the site at Tuskegee on which we propose to construct facilities for the [Primary] School will expire on June 1. This, you recognize, is an appeal for assistance as far as you can render in effecting a consummation of the contract."[52] The SEACTC offered little encouragement:

> This office has done all within its power to expedite the consummation of the subject contract. Unfortunately, the action by higher authority can not be predicted. It is believed, however, that the contracts, when signed, will be within the 1942 fiscal year, which begins July 1.
>
> It is recommended that, if you have not been able to sign the contract prior to May 30, 1941, immediate action be taken to extend your options through July.[53]

By mid-May 1941 the War Department approved Tuskegee Institute as a primary flying school operator, and the Air Staff authorized the Materiel Division at Wright Field, Ohio, to conclude a contract with Tuskegee, with the first class entering training on 19 July 1941. In addition to the normal ten-week primary curriculum specified in the standard contract, the Air Staff asked the Materiel Division to include a provision for preflight training: "This contract should, however, include provisions for the reception, uniforming, processing, and additional ground school instruction to be given to all students reporting to this Civil Flying School. . . . The pre-flight course of instruction will be given during the 5 weeks preceding the commencement of the 10 weeks flight and ground school instruction period as now provided for in existing contracts." Another change to the standard primary flying school contract requested by the Air Staff involved the student quota. It was to be amended to stipulate that "the number of military students matriculated in any one class [is] not to exceed 10; not more than 3 classes to be under instruction at any one time, of which not more than 2 classes shall receive concurrent flight and ground school instruction, and not more than one class shall receive ground school instruction at any one time."[54]

The Training and Operations Division noted in its instructions regarding the Tuskegee contract that the War Department "desires to complete this contract at the earliest possible date."[55] Four months had elapsed since the plan to establish and train a segregated air unit was announced, a delay that made it impossible to begin preflight training in June as called for in the 11 March plan. Realizing that further delays might be politically dangerous, the Air Staff concluded its instructions to the Materiel Division by directing that the Tuskegee contract "should have precedence over all other contracts with the Civil Flying Schools."[56] Despite this high priority, approval took more than three weeks. Finally, on 7 June 1941, the undersecretary of war approved the contract that established a primary flying school at Tuskegee Institute.[57]

In the end, the school was a much smaller operation than officials at Maxwell Field had originally proposed. Moreover, the thirty men who had been recruited for the CPT secondary course before the plan to establish a primary school materialized were seemingly dropped from consideration by the Air Corps. In the 11 March plan the Air Corps had acknowledged its obligations to the secondary students who had entered training with the legitimate expectation that if they graduated they would be appointed aviation cadets and enter bona fide Air Corps flight training at the basic phase of instruction. But when the quota was reduced to ten per class, the plan to accept the secondary students into basic training was shelved. Only one of these men, George S. Roberts, managed to become a member of the first primary class; not until the following year, after the nation entered the war, were those who had begun training in January 1941 entered directly into basic training.[58]

The final approval of the contract on 7 June provided Tuskegee Institute with definite information on the starting date and the size of the first class. Only six weeks remained before the first class was scheduled to begin preflight training, and eleven weeks before primary flight training was to begin. In the interim, G. L. Washington not only had to prepare adequate housing, mess, instructional, recreational, and medical facilities on campus; he also had to construct an airport and develop the institute's primary school staff, including flight instructors, ground school instructors, and aircraft mechanics.

In early June, just before final approval of the primary contract, Washington had forwarded contractor Archie A. Alexander's construction proposal to President Patterson. The cost figures Alexander submitted, which exceeded Washington's latest estimate of $150,000, provided for construction of the flying field and associated aircraft maintenance buildings, together with remodeling and renovating dormitories and classrooms on campus to accommodate the aviation cadets. Washington explained to Patterson that he believed Alexander's calculations were valid "inasmuch as his figures are based upon estimates and mine upon pure guesswork."[59] Patterson agreed that Alexander's proposal was reasonable and he authorized Washington to proceed with construction.[60]

When he learned that the contract had been approved, Washington immediately contacted Major Smith at Maxwell Field to inquire about the details of activating the primary school. Smith advised him that the school's Air Corps supervisor, Capt. Noel F. Parrish, would soon be on hand to "give the necessary information and facilitate the various assistances needed."[61] Parrish had begun his military career as an enlisted cavalryman in the early 1930s and subsequently received an appointment as an aviation cadet. By 1939 he had earned a regular army commission and was assigned as the assistant supervisor to one of the initial nine civilian primary schools, Harold S. Darr's Chicago School of Aeronautics at Curtiss Field in Glenview, Illinois.[62] His assignment to the school the Civil Aeronautics Authority had designated for the training of black pilots under the provisions of Public Law 18 brought Parrish in contact with Cornelius Coffey and Willa Brown of the Coffey School of Aeronautics.[63] Thus he was the only Air Corps officer who had both experience at a civilian primary school and contact with black aviators. When the navy took over Curtiss Field at Glenview and the Chicago School of Aeronautics shifted its operations to Albany, Georgia, Parrish was ordered to Tuskegee. He later observed that he was perhaps the most logical choice for the first supervisor of the Tuskegee primary flying school: "Since I was the only person who knew anything about this whole affair and the only Air Force officer who had any direct contact with the blacks, it made some sense, I suppose, when the Chicago school was closed—when the Navy took it over, our regular flying school there was closed—just to move me down and send me over to start as the Air Force supervisor at the black primary flying school at Tuskegee."[64] Parrish arrived at Maxwell Field in mid-June and made several visits to Tuskegee to advise Washington on preparations for the first class of cadets who would begin preflight training on 19 July 1941.[65]

Tuskegee's role in the preflight training phase was limited to providing the aviation cadets with quarters, meals, classrooms, physical training facilities, and medical facilities. Part of Alexander's contract included renovation of two campus buildings for the use of the cadets—the Boys' Bath House for barracks, and two rooms in Phelps Hall for classrooms.[66] Washington arranged for meals through the institute's cafeteria and "[c]ontacted various heads of recreational and other campus facilities for firm understandings regarding cadet use of their facilities."[67] Responsibility for actually conducting the preflight training fell to the army and specifically to Capt. Benjamin Oliver Davis, Jr., one of two black line officers in the United States Army.

By 1941 Davis was a well-known figure to most black Americans. The son of Brig. Gen. B. O. Davis, Sr., who had recently become the first black to achieve general officer rank in the United States Army, the younger Davis had made headlines in the black press when he graduated from the United States Military Academy in 1936, the fourth black to do so and the first of the twentieth century.[68] Even though he was "silenced"—none of his fellow cadets would speak to him except in the line of duty—Davis ranked thirty-fifth out of a class of 276.[69] A boyhood flight had sparked his interest in aviation, and when he graduated from the academy he applied for flying duty in the Air Corps. Despite a recommendation from West Point's super–intendent, Maj. Gen. William D. Connor, Davis's application was rejected, and he found himself assigned to the infantry and stationed at Fort Benning, Georgia. By 1938 the young officer was ordered to Tuskegee Institute for ROTC instructor duty, where he remained until late 1940, when he was transferred to Fort Riley, Kansas, to serve as his father's aide.[70] During his tenure at Tuskegee, Davis no doubt communicated his desire to fly to President Patterson, because shortly after the Air Corps finalized its plans to establish a segregated squadron, the Tuskegee president urged his contacts in the Air Corps to return Davis to Tuskegee for flight training in grade.[71] As a regular officer, there was no need for him to complete the preflight course; instead he was appointed commandant of cadets and taught the military indoctrination courses to the first class of aviation cadets.[72]

The first class to enter training at Tuskegee was designated class 42–C, consisting of Davis and twelve aviation cadets.[73] Thus it exceeded the quota of ten that had been established in the contract of 7 June 1941.[74] G. L. Washington realized that when the first class began preflight training it would mark the entry of black Americans into military aviation, and he resolved to celebrate the occasion with

Benjamin O. Davis, Jr., a graduate of West Point and a member of the first class of black pilot trainees to earn their wings, led black flying units in combat during World War II. After the war he continued his military career and retired as a Lieutenant General in the United States Air Force. (Courtesy United States Air Force Historical Research Center, 222.01, January–December 1943, vol. 6)

an inaugural ceremony on the Tuskegee Institute campus on Saturday, 19 July, the day preflight was scheduled to begin. As always, he attended to every detail and by Thursday prepared a detailed plan for the event and submitted it to President Patterson for approval.[75]

In deference to those in the black community opposed in principle to segregated units, Washington omitted any reference to the Ninety-ninth Pursuit Squadron in the official designation of the event, calling it an "Inaugural Ceremony Initiating the Training of Negroes as Military Aviators for the United States Army Air Corps."[76] Because of the close cooperation between Tuskegee Institute and the Southeast Air Corps Training Center at Maxwell Field, Washington invited the SEACTC commander, Maj. Gen. Walter R. Weaver, to represent the War Department and deliver the keynote address.[77] The commander of the black Twenty-fourth Infantry Regiment, posted at nearby Fort Benning, Georgia, was also an invited guest.[78] In his thorough manner, Washington also wired Chief of Staff George C. Marshall and Chief of the Air Corps H. H. Arnold to notify them of the ceremony and solicit "a brief word of greeting to be read as a feature of the inaugural ceremony."[79] In addition to Marshall and Arnold, Washington asked Grove Webster, the CAA official in charge of the Civilian Pilot Training Program, to send felicitations, noting that "By virtue of your close connection with this accomplishment we think it fitting that we receive a word of greeting from you which may be read on this occasion."[80]

The ceremony was scheduled for late afternoon on the Tuskegee campus next to the Booker T. Washington Monument, the institute's traditional site for important events and ceremonies. Washington served as master of ceremonies and platform guests included Maj. James A. Ellison, the white officer designated as the first commander of the Tuskegee Army Air Field (TAAF), Captain Parrish, Captain Davis, and Hilyard R. Robinson, the architect for TAAF.[81] In the presence of the press and army film crews the ceremony began with an invocation by Rev. C. W. Kelly, chaplain of the Tuskegee Veterans Hospital.

Next Washington read telegrams of greetings from Chief of Staff Marshall, Grove Webster of the CAA, and the chief of the Air Corps, General Arnold. Both Marshall and Webster emphasized the segregated nature of the initiative. General Marshall offered his "congratulations upon the inauguration of the Negro Air Corps flying school at Tuskegee Institute. We are confident that its graduates will reflect honor on the nation and distinction upon their race."[82] Grove Webster of the CAA declared that "as director of the Civilian Pilot Training Program in which Tuskegee Institute has participated with such splendid progress, I am pleased to send you greetings and best wishes for the success of the first all-Negro Army Air Corps squadron."[83] Only General Arnold's telegram omitted direct reference to racial segregation: "Democracy finds no finer expression than in making it possible for all Americans to have [the] opportunity for service in the sacred cause of national defense. In this war-torn world every man must give the best that is in him."[84]

In his opening remarks President Patterson noted that there was "no more fitting place for this ceremony to occur than here in the shadow of the statue of the immortal Founder of this institution. The action we take today is another step in the fulfillment of the ideals of participation and expression for which he labored." Speaking on behalf of "the Negro people of America," the Tuskegee president offered "abiding thanks to President Roosevelt, Secretary Stimson, General Marshall, General Weaver, and to all officers and officials of the Air Corps who have worked for this program with interest and enthusiasm." He acknowledged the support of the Rosenwald Fund, noting that Mrs. Roosevelt served on its board: "But for their confidence and desire that Negro youth should have this opportunity, the splendid field on which this training will begin would not now be the proud possession of Tuskegee Institute." Finally, President Patterson confidently predicted success for the young men who would earn their wings at Tuskegee:

> The splendid showing which has been made under Director Washington with the CAA programs demonstrates positively that Negro youth can fly. Our record in the armed forces of this nation is meritorious and of long standing. We go forward, therefore, with confidence and with consecration of purpose to the end that we shall contribute to the aerial defense of this nation and that the Negro people may add to that evidence now mounting in abundance which justifies the full extension of all of the privileges inherent in the concept of American Democracy.[85]

General Weaver followed President Patterson with the keynote address, and the tone of his remarks differed markedly from those of the Tuskegee president's. He began by emphasizing that the Tuskegee Primary Flying School was unique among the forty-two schools of the SEACTC because it was "the first flying school started by the United States Army for Cadets of the Negro race." Thus it was imperative for the participants to "do their utmost because not only are the people of your country, but also the people of your own race are turned in your direction." He laid the responsibility for demonstrating that black Americans were worthy of the opportunity to fly in the Air Corps squarely on the shoulders of the young men who would train at Tuskegee:

> The success of this venture depends upon you. You have the responsibility of laying the foundation; therefore, I cannot impress upon you too much how important it is for you to make a wonderful record. I believe you will. . . . You have an excellent educational background. You have splendid officers. You will have fine mechanics. These are the tools necessary to do your work. You have the best. . . .
> If you knew all the planning that has gone into making this opportunity possible, you would more fully realize the significance and weight of your responsibility.[86]

Following General Weaver's admonitions, the ceremony concluded with brief remarks by Dr. Eugene Dibble, manager of the Tuskegee Veterans Hospital, and E. H. Holland, representing the local American Legion post.[87] After the ceremony, the first class of black aviation cadets began their preflight training. With flight training scheduled to begin on 23 August 1941, only five weeks remained to complete construction of the flying field and develop a corps of instructors and ground support personnel to staff Tuskegee's contract flying school.

Responsibility for recruiting the flying school staff fell to G. L. Washington. When he first discussed the establishment of a contract school with Major Smith in February, Washington had raised the issue of flight instructors, and Smith had assured him that Tuskegee's two most experienced CPT instructor pilots, Chief Anderson and Lewis Jackson, could qualify as primary flight instructors.[88] So Washington had at least two black instructors, but would, of course, need others to take their place in the CPT Program, as well as mechanics qualified to service the primary trainers. For a time it looked as if Tuskegee alumnus John C. Robinson might finally become a member of the institute's aviation staff.

Shortly after his meeting with Major Smith, Washington was in touch with Robinson, who enthusiastically suggested that his NYA-sponsored aviation mechanics training program be transferred from Chicago to Tuskegee. He explained his program's connection with the Chicago School of Aeronautics, suggesting a similar arrangement with Tuskegee's flying school:

> Last year I worked out a program between the Civilian Army School at Glenview and the N.Y.A.; where the N.Y.A. Resident boys would work 100 hours per month with the Civilian Army School. All expenses [were] paid by the N.Y.A., without any expense to the Army School. I scheduled these boys so that at least ten would be on duty at all times. Their duty was helping maintain and service ships used by the Civilian Army Training Program.
>
> They worked under the supervision of three experienced mechanics. These boys had had a minimum of six months' training in my school here and their work was so satisfactory that they were all hired by the Army Program and are being transferred to Georgia; they leave next month. Incidental[ly,] these were white boys that [were] trained under my supervision.
>
> I am sure the same set up can be worked out in connection with your program there. I have fifteen of our boys who have had at least 12 or more months of training, all high school graduates and would like to come down under any condition[s] and work. . . .
>
> I also have two good mechanics who are supervisors here in Aviation and who would like to come down and work. These two men are graduates of Curtiss Wright and have good, very good mechanical background and experience. [With] these two men . . . I could very easily take care of all the Mechanical School instructions on the defense program . . . [and] service and maintain all of the flight equipment that will be put there. In other words with these two mechanics and the N.Y.A. boys I could maintain and service a minimum of 25 planes [and also provide] mechanical instructions.[89]

Robinson suggested that President Patterson work through NYA advisor Mary McLeod Bethune to have the Illinois NYA director declare his equipment surplus so that it could be requested by the Alabama NYA and transferred to Tuskegee, assuring Washington that "If this deal can be turned this would give us a complete Mechanical School at once. There is plenty of Aviation Equipment laying [*sic*] around in Illinois, that is not being used at all, and I sincerely believe through Mrs. Bethune you can get the equipment I am using here."[90]

Over the next several months Robinson continued to urge Bethune, Washington, and Patterson to use their influence to have his Chicago NYA operation transferred to Tuskegee; by May, Alabama NYA officials had agreed to the plan in principle.[91] In the meantime Washington was developing another initiative with the Alabama Department of Education for aeromechanics training funded by the Vocational Education of Defense Workers program and administered by the United States Office of Education. By mid-May he had received approval to hire several instructors, including Robinson, for a ten-week training course for fifteen students scheduled to begin in June.[92] At month's end, however, Robinson abruptly changed his mind and once again declined to accept a position at his alma mater.

In early May Robinson had flown to Tuskegee for a conference with Washington, but returned suddenly to Chicago after receiving a telegram advising him that he had been relieved of his duties. He subsequently notified Washington that he had "definitely straightened out the difficulty" and had accepted a lucrative position with the Chicago Board of Education. He expressed his regrets at the sudden turn of events that precluded his coming to Tuskegee:

> I am very sorry that we could not get together on something definite when I was at Tuskegee, because I definitely set my mind on trying to work out something in connection with Tuskegee if I didn't have to sacrifice too much of the experience and standing that I have in the aeronautical world. . . . It seems as though it just isn't in the books for me to be connected with Tuskegee in a technical way because every time I am approached to make direct negotiations I cannot get together with [you] or something seems to step in up here in Chicago. . . . I regret this situation very much, but under the present conditions that I am working I would be definitely working against myself financially to come down now. Keep in mind that I want to do everything in my power to assist you in any way that I can from this end up here.[93]

Washington responded immediately with a letter that recounted the details of "our efforts to avail ourselves of your services," noting that he had received inquiries from alumni who believed that "we have not been sincere in seeking your services." He observed that since the summer of 1940, when the institute first began flying operations, "we have definitely extended to you your choice of flight instructor, for which you would have had to qualify, ground instructor and a position of mechanic." Washington noted that these offers were still open when Robinson had telephoned in early 1941 to express his interest in "becoming the 99th Squadron Commander . . . , managing our Tuskegee program or heading Aviation Mechanics instruction at Tuskegee Institute." Washington had responded that the position of squadron commander "would be entirely within the province of the Army Air Corps, that what we needed in our program, rather than a manager was men like yourself to . . . make it the best program and that we hoped some day to have Aviation mechanics, at which time you would be given the opportunity to head up this work." He reminded Robinson that after the Department of Education agreed to fund aviation mechanics instruction at Tuskegee, he had been offered the position of head instructor at $250 per month, the "top salary in all of our aviation activities . . . [and] the amount you stated would be necessary for you to give up your work in Chicago." Now it appeared that there was an "opportunity for income in Chicago . . . which Tuskegee could not begin to match." These facts, Washington asserted, should be explained to those "who can not understand why you are not connected with the Tuskegee program." These circumstances notwithstanding, Tuskegee was still interested in Robinson's services as a flight instructor at the primary school. Washington closed by acknowledging the difficulties Robinson faced in coming to Tuskegee and insisted that Tuskegee had made a sincere effort to accommodate him:

> In conclusion, I want to be fair enough to appreciate what has been your problem. I know that you are well entrenched in Chicago and Illinois in your aviation mechanics work and have built up a great deal of good will in that section. To leave this and come down to work in our program even at equal salary would mean in a way, starting over again because our whole program is one of pioneering. If I am correct in this diagnosis, I feel that we should be given full credit for the diligent and continuous efforts to offer you a choice of positions in our organization from the very start, and for the most recent offer to head up Aviation Mechanics.[94]

Washington forwarded copies of the letter to President Patterson and Claude Barnett, who concluded from Washington's assessment of the situation that "Colonel Robinson doesn't seem to be in the least justified in feeling that he was neglected."[95]

Once again Tuskegee Institute and black America's best-known aviator were unable to come to terms. From a personnel perspective, Robinson's refusal was of little consequence to the Tuskegee program, but from a public relations standpoint it was regrettable. The attacks of the National Airmen's Association and the *Chicago Defender* following the announcement of the establishment of a segregated squadron had damaged Tuskegee Institute's prestige and Robinson's active involvement would have helped restore the school's status, a fact which newspaperman Barnett had recognized when he reminded Patterson of the "tremendous . . . publicity value in Colonel Robinson."[96]

Despite Robinson's refusal to affiliate himself with Tuskegee, the resourceful Washington managed to deal with the problem of finding instructors and ground support personnel for the new primary school. As he explained to President Patterson in late June, the most difficult problem was to find a way to transfer CPT instructors to the primary school without curtailing CPT operations:

> We are confronted with a number of problems on flight personnel this summer. The C.P.T. programs [run] from June 15th to September 15th. The Army program starts flying August 23rd. The army flyers must cease full load . . . about July 20th in order to undergo the army instructor's course in preparation to begin army flying August 23rd. Inasmuch as we are shifting two [CPT instructors] to army [primary] . . . from our present personnel, the matter of maintaining C.P.T. instruction in all courses and keeping the required number of instructors per C.A.A. contract has been a puzzle.[97]

By mid-1941 Tuskegee had a staff of seven CPT flight instructors, three blacks (Lewis A. Jackson, George W. Allen, and Charles Alfred Anderson, the chief CPT instructor pilot) and four whites (Joseph T. Camilleri, Dominick J. Guido, Frank Rosenberg, and Forrest Shelton).[98] Washington needed two instructors for the first primary class and a third instructor later in the year, so he drew up a plan to shift Anderson and Shelton from the CPT program to the primary school immediately and bring Jackson over to primary later in the year. As Tuskegee's most experienced flier, Anderson was named chief pilot at the primary school with George W. Allen succeeding him as chief

CPT instructor pilot. Shelton, a Tuskegee native and one of the three young men responsible for establishing Kennedy Field, had been hired as a flying instructor when Tuskegee took over CPT operations from Alabama Air Service. Washington confided to President Patterson that he had selected Shelton as the second primary instructor not only for his record as "one of our best, most cooperative and best liked instructors," but because his race would provide "an advantage in dealing with the [army] supervisor. We have effectively used [Shelton] in this wise with C.A.A. Inspectors in getting over points and concessions."[99] The vacancies on the CPT staff created by the transfer of Anderson, Shelton, and Jackson would be filled by three apprentice instructors, Robert A. Dawson and two alumni of Tuskegee's first CPT class, Charles R. Foxx and Milton P. Crenshaw.[100]

Developing the nonflying primary school staff presented less of a problem for Washington. For his chief of maintenance, he selected Austin P. Humbles, one of the men hired as an instructor under Tuskegee's new aeromechanics training program, administered by the Alabama Department of Education.[101] For the ground school staff Washington once again drew on the resources of nearby all-white Alabama Polytechnic Institute, hiring Warren G. Darty, a recent graduate, as his chief ground school instructor.[102]

Recruiting personnel in time to begin flight training in late August was not the only problem Washington faced in the summer of 1941. Equally important was completion of the construction of the flying field. Work on the facility had begun at the beginning of June, several days before Tuskegee's primary school contract had been finalized. The short interval until the initiation of training—roughly eleven weeks—allowed no time for the preparation of proper construction plans. Consequently when Alexander, the contractor, commenced construction "only a [b]rief descriptive specification and layout design was furnished him." The contract called for "construction of [a] combination hangar-ship-office building at [the] airfield, clearing and grading [the] landing area, provisions for sewer, water and other utilities, and furnishing all classrooms, barracks, shop, hangar and offices with equipment."[103]

The pressure of time and the lack of detailed construction plans placed a special burden on Washington, requiring him to monitor the progress of the job carefully and coordinate the details of the project with Alexander. Washington later recalled that he "kept close on the heels of the contractor, not only to prod, but to render every assistance possible."[104] When heavy summer rains caused delays and put Alexander behind schedule, Washington placed the resources of the institute at his disposal:

> Unusually heavy rainfall that summer impeded [Alexander's] progress considerably. In order to overcome some of the time losses, we made arrangements for him to cut all the members of the 100-foot span trusses in the carpentry shop of Mechanical Industries, and to do other processing of the lumber, so that his work on the trusses at the airdrome would be confined to assembling the trusses. And as soon as the hangar was closed in we made available as foremen Mr. Wagener, for welding and machine work, Mr. Dunham, for masonry, Mr. Owsley, for sheet metal work, and Mr. Sorrell, for painting. These men were master craftsmen and knew where to find good mechanics the contractor needed. Being summer, some student labor was available for employment by the contractor.[105]

Alexander also received valuable assistance from a local white subcontractor, J. H. LaMar, who specialized in heavy construction work. LaMar completed the "grading of [the] flying field, roadwork, well digging, sewer and storm water lines, etc."[106]

As the deadline for commencing flight training approached, it became increasingly apparent that the flying field would not be completed by 23 August. Because of the heavy rains, portions of the landing area were still unusable, and the hangar and refueling facility were not completed.[107] The construction delays must have been particularly frustrating, as other matters relating to the start of flying operations had fallen neatly into place. By 21 August, two days ahead of schedule, all was in readiness for the start of primary flight instruction—the instructors and mechanics had been selected and trained, the full complement of army primary trainers had arrived, and the cadets had completed preflight training.[108]

Rather than delay the start of training while awaiting the completion of the primary field, flying operations were inaugurated on 21 August at Kennedy Field. By 25 August the rains had subsided sufficiently to allow completion of the landing strip at the primary school and the transfer of flying operations from Kennedy Field. The primary school hangar was not completed until late September, however, so the facilities at Kennedy Field were used for storage and maintenance in the interim.[109] Moreover, for several months the landing field at the primary school was plagued with drainage problems, and after heavy rains it was often necessary temporarily to shift training to Kennedy Field. According to the official history of Moton Field, however, the problem was ultimately resolved:

> The condition of the air field in the early stages was not entirely satisfactory, due to the nature of the soil. In wet weather, the field was often unsuitable for flying, particularly solo. On several occasions, after continuous heavy rains, the field could not be used at all for two or three days. After several months, however, with the field settling itself, and with constant rolling and grading, the drainage improved considerably, and almost no time was lost due to unsatisfactory field conditions.[110]

By mid-November 1941 Washington certified that Alexander had fulfilled the terms of his contract for constructing Tuskegee Institute's primary flying school, and the facility met the approval of army inspectors.[111]

Primary flight training began on 21 August with the same complement of students that had entered preflight training on 19 July. None of the cadets had been eliminated, or "washed out," during this initial phase of military indoctrination training, but the real test was yet to come. The principle function of primary training was to determine which students possessed an aptitude for flying and which did not, and to eliminate the latter before they went to the more advanced phases of flight training. If the normal "wash out" rate of forty percent proved accurate in the case of Tuskegee's first class of flying cadets, only seven or eight of the thirteen could be expected to complete the full ten weeks of primary training and continue on to basic flight training.[112]

The training program was identical to that offered at other primary contract schools, consisting of sixty-five flying hours spread over ten weeks and divided into four subphases—presolo, intermediate, accuracy, and acrobatic.[113] The presolo phase introduced the student to the fundamentals of flying a light aircraft and including forced landings, stalls, and spins. In the intermediate phase, after developing additional proficiency in aircraft handling, the cadet completed his solo flight. In the accuracy phase the emphasis was on acquiring

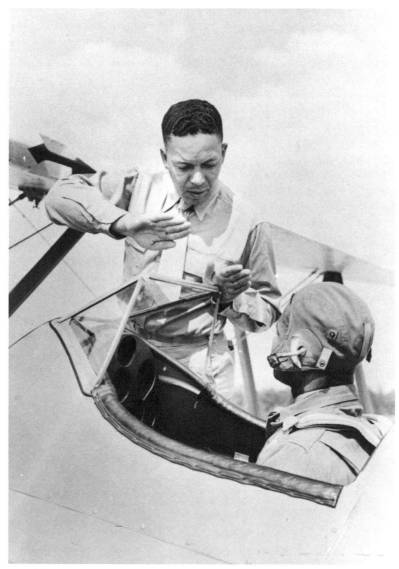

C. Alfred "Chief" Anderson instructs a cadet in primary training at Tuskegee Institute's Moton Field. (Courtesy United States Air Force Historical Research Center, 222.01, June–December 1943, vol. 6)

proficiency in a wide variety of landing approaches and landings. Finally, in the acrobatic phase, the fledgling pilots were required to demonstrate proficiency in such standard maneuvers as loops, slow rolls, and Immelmann turns.[114]

By early November 1941, as the ten weeks of primary training drew to a close, only six of the original thirteen remained. Captain Davis, who had earlier taken flying lessons at his own expense from Chief Anderson, completed the course handily,[115] along with five of the aviation cadets—Lemuel R. Custis, Charles H. DeBow, Jr., Frederick H. Moore, Jr., George S. Roberts, and Mac Ross.[116] With six of thirteen graduating from the primary, the elimination rate came to fifty-six percent, somewhat higher than the standard of forty percent. But with the small numbers it was not necessarily significant, since one less elimination would have put the wash-out rate at forty-six percent, only slightly above the standard. Thus less than ten months after the Air Corps first announced plans to establish a segregated fighter squadron, the first group of black student pilots had passed a significant hurdle—they were ready to leave Tuskegee's contract school and begin training under Air Corps instructors at nearby Tuskegee Army Air Field.

12

Making the Dream
a Reality

The Air Corps' First Black Pilots

By November 1941 Tuskegee Institute had fulfilled its commitment to establish and operate a primary flying school. The first class of black cadets had completed the initial phase of flight training and subsequent classes were under instruction. Yet Tuskegee's contract school was only one component of the overall program that ultimately carved out a small niche for black Americans in the hitherto all-white Army Air Corps. Crucial to the plan for such a program was construction of a separate military flying field near Tuskegee, and the recruitment and training of ground support personnel for the Ninety-ninth Pursuit Squadron. Equally important was the continued progress of the black aviation cadets in flight training. Completion of primary training was indeed a significant accomplishment, but until the first class finished advanced training in March 1942 and received the silver wings of Air Corps pilots, the promise of a black air unit remained unfulfilled. Only then would skeptics in the Air Corps be forced to accept the reality of blacks serving as pilots. And until the first black cadet was rated as an Air Corps pilot, critics in the black community would continue to charge that the War Department had not made good on its promise to establish a segregated squadron. Indeed, as Judge Hastie pointed out in April 1941, "No single enterprise is being watched as closely by Negroes as is the Air Corps training program."[1]

Tuskegee Institute also watched the enterprise closely. Although the school bore no direct responsibility for the construction of the military air field, the recruitment and training of the ground crews, or the training of the pilots after the primary phase, President Patterson and G. L. Washington believed they had a moral obligation to monitor developments in all these areas. Patterson never hesitated to contact officials in Washington, D.C., directly when he believed circumstances or problems threatened progress toward the establishment of the segregated unit. G. L. Washington later explained that he did not consider it always necessary to work through the regional officials at the SEACTC because the War Department had "directed Maxwell Field to seek the cooperation of Tuskegee Institute. [Consequently] Tuskegee Institute was in a position to deal directly with the War Department at Washington, D.C., when this seemed necessary, throughout the life of the military operation at Tuskegee."[2]

By December 1941 the future of black Americans in the air force still seemed to hang in the balance: the military field—Tuskegee Army Air Field (TAAF)—was only partially finished, no cadets had earned their wings, and most of the maintenance crews were at Maxwell Field awaiting completion of barracks and other facilities at Tuskegee. Therefore, when the nation suddenly entered the war after the surprise attack on Pearl Harbor, there were fears that the program might be sacrificed to the war crisis. For a time it seemed as though opponents of the Tuskegee initiative intended to use the war as a convenient excuse for curtailing or postponing the program. Nevertheless, despite several anxious weeks after Pearl Harbor, the War Department not only proceeded with the project but expanded it. By early March, with the graduation of the first pilot training class, black Americans had indeed gained a tenuous foothold in the nation's air arm, some two years after the enactment of the Air Corps Expansion Act (Public Law 18) and its proviso for training black aviators.

The matter of constructing a military flying field in the vicinity of Tuskegee Institute had been a subject of discussion between Air Corps officials and the Tuskegee administration from their first contacts in late 1940. In early December Patterson, Washington, and

Maj. L. S. Smith scouted the environs of Tuskegee for a likely site for an army flying field. They identified three potential sites, one at Tuskegee, a second at Milstead, a small community in Macon County some twelve miles west of town, and a third at Fort Davis, also in Macon County and twelve miles south of Tuskegee. Smith urged Patterson to determine the sentiments of the local authorities regarding the field location; the astute Tuskegee president no doubt needed little urging in this regard, for the administration of the institute had always sought to maintain amicable relations with the surrounding white community. Tuskegee's white town officials balked at the idea of constructing a military facility for a black squadron adjacent to the town, but they were receptive to the Milstead and Fort Davis sites, which would be close enough to benefit the town economically and yet far enough away to ensure a measure of control over the men stationed there.[3]

By early January, after authorities at Maxwell Field had concluded that the Fort Davis site offered the best prospects for developing an operating base for the new black squadron, Patterson obtained the support of the Tuskegee town council. In a telegram to the Alabama congressional delegation, they endorsed the proposal to establish an army flying field in Macon County, declaring Fort Davis as "An ideal site . . . half[way] between Tuskegee and Union Springs on paved highway and main line Rail Road," and noting that eastern Alabama had received little in the way of economic benefits from the defense build-up.[4]

In late February 1941, over a month after the plan to establish a segregated squadron at Tuskegee was announced, the Air Corps convened an Air Site Board of five officers, chaired by Maj. Mark M. Boatner, Jr. of the Army Corps of Engineers and including Maj. James A. Ellison, a white officer who was slated to become the first installation commander of TAAF. On 20 February the board visited the proposed site at Fort Davis and recommended approval, concluding that there were a number of advantages in locating a segregated air unit in the vicinity of Tuskegee Institute:

> The close proximity of Tuskegee Institute makes this site ideal for the training of Negroes, since that Institute furnishes many precepts and examples in conduct and attitude. It is a center of Negro learning and culture, and it has temporary accommodations for Negro personnel.

Further it is an Institute whose leaders exert great influence in the affairs of the Negro race.

The County of Macon in which Tuskegee Institute is situated is predominantly Negro, and a Negro flying field is welcomed by the community.

Tuskegee is predominantly white, while Tuskegee Institute is naturally a Negro community. This condition would assist largely in handling the problem of segregation.[5]

Based on the recommendations of the Air Site Board, preliminary plans for the field were prepared and funds were requested. Efforts came to a standstill in early April, however, when the district engineer discovered that the capacity of the soil at the site to support the proposed buildings and runways was marginal. He concluded that the unfavorable soil conditions would add $600,000–$700,000 to the cost of the project and delay completion by at least six months.[6] When General Weaver, commander of the SEACTC at Maxwell Field, learned of the district engineer's findings, he requested the necessary additional funds, or the authority to select another site. On 19 April the Air Staff authorized General Weaver to select an alternate site and representatives from Maxwell began anew to search for an acceptable location. Among the sites they examined were several adjacent to the eastern and southeastern edges of the town of Tuskegee.[7]

When the white citizens of Tuskegee learned that the Fort Davis site had been abandoned and that Air Corps officials were inspecting land next to the town, they became alarmed and lodged a strong protest to their representatives in Washington, D.C. Almost one hundred Tuskegee citizens signed a petition urging Alabama senators John H. Bankhead and Lister Hill to take "prompt action in protesting to the senators and the Government against the location of a colored aviation camp on the East, or Southeast boundary of Tuskegee."[8] Tuskegee's white citizens feared that the town's traditional pattern of segregation—which consigned blacks to the west side, in the vicinity of the institute and the Veterans Hospital, and whites to the eastern section—would be threatened if a new black facility was established next to the white enclave on the east: "The location of this colored airport would destroy the usefulness of this part of Tuskegee, as well as the Eastend of Tuskegee, for a white residence section. At this time the east end of Tuskegee offers the only outlet of expansion for white citizens of Tuskegee. We most earnestly urge that this camp be located West of Tuskegee Institute, or not within three miles of the Eastend of Tuskegee."[9]

When he received the protest from the Tuskegee whites, Senator Hill telephoned the acting chief of the Air Corps, Maj. Gen. George H. Brett, to inquire about the location of the field. Their conversation, transcribed and preserved in the official headquarters records, provides a rare glimpse into the attitudes of two public officials toward the Tuskegee project:

> Hill: Say old fellow, I understand that your committee—or your engineers have picked out a site down at Tuskegee, Alabama, that lies very close to the town and very near to the white residential section of the town. Senator Bankhead and I have got quite a petition here this morning headed by . . . the Mayor of the town and all the leaders and officials down there protesting the selection of that site. . . .
>
> Brett: Well, I'll turn it over to—what is this man's name—the Special Assistant Secretary of War for—[Note: Brett is referring to the civilian aide to the secretary of war, William H. Hastie.]
>
> Hill: What are you doing—ribbing me a little bit[?]
>
> Brett: You are putting me between the devil and the deep blue sea[.]
>
> Hill: You wouldn't pass the buck—would you[?]
>
> Brett: Well, I don't know—now I get—gosh—I get pounded on that thing[—]
>
> Hill: Oh hell—I know what your trouble is—God almighty I know what it is—don't you think I get some inside of your troubles[?]
>
> Brett: What you do—you send that up to me—I don't know—I thought that had all been coordinated—I don't know why the site was ever picked unless the city council and the Mayor were thoroughly consulted at the time that it was picked—you see what I mean—in other words, they wouldn't pick a site[—]
>
> Hill: I know—unless the local authorities were in accord with it[.]
>
> Brett: Unless the local authorities were in accord with it—now of course, maybe the people at the time weren't taken into the confidence of the local authorities but send it up to me and I'll look it over and see what I can do with it[.]
>
> Hill: All right. General—After you have checked into it and have had somebody find out just what the situation is will you let me know[?]
>
> Brett: Yes—I'll do that[.]
>
> Hill: All right—and I know you won't do anything until we have had an opportunity to talk to you about it[.]
>
> Brett: Yes.[10]

General Brett investigated the matter immediately and learned that the authorities at Maxwell had considered several sites next to Tuskegee and had rejected them in favor of a site near Chehaw, about seven miles northwest of the town. Unable to reach Senator Hill, Brett left word with his secretary on the status of site location for the Tuskegee field:

> Brett: Why you tell the Senator that . . . I don't know what the complaint is about because our—the dope I have on it is—wait a moment—let me look at my map—that the first location surveyed was . . . about 10 miles south of Tuskegee . . . [and] they found soil conditions down there unfavorable—the second location is . . . just north of Cheehaw [sic] which is probably 6 miles north of Tuskegee—now I don't quite understand what the complaint might be . . .
> Hill's secretary: Well, I shouldn't think that was too close—would you?
> Brett: Why now . . . they may have surveyed other localities around Tuskegee but the number 2 location . . . is up north of Cheehaw[.]
> Hill's secretary: Yes I know where that is. That sounds all right to me General—but I'll tell the Senator about it—[11]

The general, however, had some additional information regarding the Tuskegee project that he asked the secretary to pass on to the senator. Brett had at hand a memorandum that Judge Hastie had written to Robert A. Lovett, the assistant secretary of war for air, protesting the fact that the construction plans for the field provided for separate barracks and dining halls for both officers and enlisted men. Hastie noted the Air Corps plan for segregated flying training for black cadets had "already evoked widespread criticism . . . [since] other Arms and Services do not have such separate schools." Nevertheless, he continued, many had

> withheld protests and criticism which they have felt to be justified, because of their desire to see Negroes obtain an opportunity to become Army flyers. However, I can assure you that if in addition to the segregated training school, the Army insists upon the proposed separation of white and colored personnel attached to the same unit, such a nation-wide storm of protest and resentment will arise as to destroy all of the good-will and support of the Negro public with reference to the Army program. . . . I cannot overemphasize the catastrophic effect of the arrangement now proposed upon morale.[12]

Brett related the contents of Hastie's memorandum to the senator's secretary and their brief conversation clearly shows the general's attitude toward social contacts between blacks and whites, one of the few instances in which a high-ranking Air Corps officer's candid comments on racial matters were preserved in the official record:

> Brett: All right—now you might tell the Senator that I just read a long letter from our Special Assistant Secretary of War Mr. Hastie who is intensely interested in the colored people, who claims that he notes that they are building separate barricks [*sic*] for white and separate barricks for colored people and separate barricks for white officers and he objects most strenuously—says that the whites must live with the blacks—
>
> Hill's secretary: Good night—
>
> Brett: Now you might tell Senator Hill . . . that this little thing he is talking about is only . . . one small phase of the whole problem. Oh yes, they are demanding . . . non-segregation[.]
>
> Hill's secretary: Well, I don't think that is going to work in the South[.]
>
> Brett: Well . . . I don't know why it should work in the North—I happen to be from Cleveland and I'm sure I don't want to live with a nigger—I don't know why there is any difference.
>
> Hill's secretary: I don't know why it should work anywhere General[.]
>
> Brett: Well . . . you might tell him that's a sidelight on the whole question.[13]

Hastie's information regarding segregated facilities at TAAF was indeed accurate. In mid-March the Building and Grounds Division in the Air Staff had forwarded proposed construction plans to the adjutant general's office. The plan observed that the projected base population of 596 included some twenty-six whites (eleven officers and fifteen enlisted men) and included the following notation concerning messing facilities: "White instructors will mess in their barracks. Colored cadets will mess with colored officers."[14] Assistant Secretary Lovett responded to Judge Hastie's inquiry by noting that the establishment of separate facilities at the Tuskegee field was "identical with the general procedure followed in all other Army units where white and colored officers and enlisted men serve in the same outfit." Thus, Lovett concluded, it was apparent "that the branch of the Army charged with construction of the facilities is conforming to regularly established practice and that there is no departure in the case of the Air Corps."[15] But Hastie pointed out to Lovett that the Air Corps followed established procedures when they enforced racial segregation and ignored them when they did not:

> Your memorandum says in substance that segregation in the barracks and mess halls as planned at this new installation, is in accord with the general practice of the Army. Some time ago I submitted a recommendation in opposition to the establishment of this separate training base for Negroes on the ground that it was contrary to the established policy in training for other Arms and Services. In such circumstances, the present reliance upon such policy is not persuasive. The Army Air Corps has seen fit to establish this anomalous special school for Negroes and now proposes to jeopardize the morale of the school and to invite widespread public indignation by racial separation of personnel within the unit. As I see it, the question is not what is or has been done in other circumstances, but whether this is the correct thing to do under the peculiar circumstances of the present case. I feel strongly as I indicated before, that the present proposal is a serious mistake.[16]

The issue of separate facilities for the white personnel assigned to TAAF remained unresolved throughout the tenure of the first commander, Maj. James A. Ellison. When Col. Frederick V. H. Kimble replaced Ellison as post commander in January 1942, this white officer reaffirmed the policy of segregated quarters and dining facilities and broadened it to include toilet facilities.[17] Only after Noel Parrish assumed command of TAAF in December 1942 were the rigid segregation policies relaxed. Although Parrish did not totally eliminate segregation, he reduced it where he could and made an effort to enforce the War Department directives against discrimination, which began to appear in 1943.[18]

The alarm of the Tuskegee whites over the prospects of a military airfield for a black flying unit on the town's eastern border and Judge Hastie's protest over segregated facilities had little long-term effect on the progress toward building TAAF. By the end of May 1941 the adjutant general approved the Chehaw site and directed the quartermaster general to obtain title to the land.[19] Acquisition of the land took several weeks, and by the end of June condemnation proceedings on some 1,650 acres of land—involving eight property owners—were complete, with the government scheduled to take possession at the end of July.[20]

Before construction could begin, however, a general contractor had to be selected. From the outset, President Patterson and other proponents of the Tuskegee project had hoped that the War Department would award the contract to a black company. Associated Negro Press director and Tuskegee board member Claude Barnett had realized immediately that the establishment of a black squadron at Tuskegee could provide important new opportunities for black contractors. In February 1941, following the War Department's announcement of plans to form a segregated flying unit, Barnett had suggested to President Patterson that a black firm should receive the contract for the military field: "Archie Alexander . . . is in Washington, working very hard to secure the bid for the Tuskegee airport from the War Department. I am sure you know him well, but believe it would be a fine stroke if a Negro was able to get it."[21] By March, while G. L. Washington worked with Alexander on plans for Tuskegee Institute's airfield, Patterson wired the district engineer in Mobile that he fully supported Alexander's efforts to secure the contract for the military field.[22]

Despite Patterson's endorsement, the contract did not go to Alexander, but to another black contracting company that also had the strong support of the Tuskegee president, McKissack and McKissack of Nashville, Tennessee. Patterson had firsthand experience with the work of this firm, which had recently completed the construction of the institute's Infantile Paralysis Hospital.[23] Moreover, he apparently had a personal relationship with one of its principle members, Calvin McKissack, who on at least one occasion received the Tuskegee president as an overnight guest in his Nashville home.[24] In late April Patterson had lent his full support to the firm's bid for the contract when he provided McKissack with a letter of introduction to the local authorities:

> Tuskegee Institute, as you know is extremely interested in the development of the Aviation Pursuit Squadron in this community. We are anxious that this contract for the building of the physical facilities be given to the firm of McKissack and McKissack, licensed contractors in the state of Alabama.
> We will greatly appreciate your cooperation.[25]

Ever cautious in racial matters, the War Department—which had never awarded a major contract to a black company—had requested

McKissack and McKissack to obtain letters of support from local officials and businesses indicating that they would cooperate, or at least not object, if the project was awarded to a black general contractor. Despite Patterson's strong endorsement, however, the McKissacks encountered resistance from some local whites. Fearful lest an opportunity that he believed should rightfully go to a Negro firm slip away, the Tuskegee president wrote Assistant Secretary of War Robert P. Patterson in early May to outline the reasons he believed the War Department should award the contract to McKissack and McKissack, the recalcitrance of the local whites notwithstanding.

President Patterson began by explaining to the assistant secretary that the firm had been asked to obtain "letters indicating the attitude of local people and others who would be in position to offer encouragement and cooperation in the development of this project." Some letters, the Tuskegee president continued, were "freely given while others are being withheld, and . . . the question of race is entering into the picture." Despite this unfortunate circumstance, Patterson assured the assistant secretary that "if the contract is given to the firm of McKissack and McKissack, it is my opinion they will experience no difficulty because of their race."[26] A policy of using qualified black professionals in federal projects undertaken for the benefit of black Americans, he pointed out, had been established some years earlier when the federal government located a Veterans Hospital next to the institute:

> This question was gone into in 1927 in connection with the personnel of the all Negro Veterans Hospital located just two miles from the Institute, and the position was firmly established that in the case of those projects and undertakings concerning the Negro race that members of said race of professional ability should, as a matter of fairness and common decency, be given a chance to render the professional services in connection therewith. I believe the process which was necessary in order to establish this right was sufficiently thorough to set a precedent which will hold. I am therefore urging favorable consideration of the firm of McKissack & McKissack in connection with the 99th Pursuit Squadron Base.[27]

Moreover, he concluded, the exclusion of black firms from "the opportunities enjoyed by white contractors throughout the nation in . . . developments of plants concerned with National Defense" provided ample justification for "giving this opportunity to members of the Negro race."[28]

Apparently President Patterson's arguments were persuasive. On 19 May 1941 Assistant Secretary Patterson sent a memorandum to President Roosevelt advising him of the decision to use McKissack and McKissack for the military field at Tuskegee: "We are making a contract with a firm of negro contractors for the construction of the Air Corps school at Tuskegee. The school is for the 99th Pursuit Squadron, a new unit made up of negro pilots and other negro personnel." Patterson took care to point out to Roosevelt that to his knowledge, "this is the first time that the War Department has made an important contract for construction with a negro construction company," perhaps hoping that the president might find the information useful in his efforts to head off a march on Washington scheduled for 1 July, organized by black labor leader A. Philip Randolph to protest racial discrimination in national defense programs.[29] Patterson was also sensitive to the first lady's interest in racial matters and took care to send her a copy of the memorandum as well.[30] The following month, as the threatened march on the capital loomed near, Patterson outlined for Mrs. Roosevelt the steps the War Department had taken to expand opportunities for black Americans to participate in the national defense program and avert the march. He reminded her of the decision to use a black contractor for the $1.5 million project at Tuskegee and, in words that echoed those of President Patterson, declared that it provided "real recognition of the place of the Negro business and professional men in the national defense program."[31]

President Patterson and G. L. Washington did not limit their involvement in the military side of the Tuskegee project to the selection of a site and the awarding of the construction contract. They also closely monitored the recruitment and training of the enlisted ground crews and the nonflying officers. In January 1941, the week after the Air Corps announced that it intended to establish a segregated squadron, Washington submitted the names and qualifications of almost one hundred men interested in nonflying duty with the Ninety-ninth Pursuit Squadron to Major Smith at Maxwell.[32] In early March 1941 Washington provided information on an additional thirty-five men interested in ground duty with the Ninety-ninth.[33] Most of the men whose names Washington supplied to the Air Corps were graduates or students at major black institutions, including Tuskegee, Wilberforce University, Hampton Institute, and Virginia State College.

The question of where the ground crews and nonflying officers would be trained had been settled at a December 1940 conference between General Weaver, Major Smith, President Patterson, and Director Washington. As the four were discussing the details of the plan to establish a segregated squadron, General Weaver suddenly proposed that the Air Staff's recommendation to train ground personnel at the Air Corps Technical School—located at Chanute Field, Illinois—be abandoned and a separate technical school for blacks be established at Tuskegee Institute. The proposition caught Patterson and Washington off guard, and before they could respond, Major Smith interceded and argued forcefully against the general's proposal. As director of training, Smith was closely involved with the preparation of the training plan for the segregated pursuit squadron, and he no doubt realized the added difficulties the establishment of a separate technical training center would impose on both the SEACTC and Tuskegee Institute. He argued that the success of the Tuskegee project depended in large part on the effective and expeditious training of the ground crews, and there was no substitute for the high-quality training available at Chanute. The general, perhaps unwilling to continue the discussion in the presence of Patterson and Washington, let the matter drop, and the plan that went forward to the Air Staff recommended that the black ground crews receive their training at the Air Corps Technical School. Both Patterson and Washington were relieved that Weaver had not insisted on conducting the technical training at Tuskegee Institute, which would have required them to house and train several hundred men on short notice.[34]

Major Smith's stand convinced Washington that he was sincerely interested in developing a quality program for blacks, rather than simply establishing it as a sop to deflect criticism against the Air Corps. Washington's respect for Smith comes through clearly in his account of the incident:

Major Smith challenged the General in the presence of visitors . . . on the matter of sending enlisted men to Chanute Field, rather than setting up facilities at Tuskegee Institute for the training. It is true that the Air

Force [*sic*] could have built temporary structures on our campus, fully equipped the facility for training, and furnished the military personnel to conduct the training. But at best, it could not in so short a time duplicate staff and facilities at the best school the Air Corps had for the training in question. Certainly the men would have been later entering training, and thus later arriving at the Tuskegee Army Airfield,— probably delaying the beginning of operations there.

This stand was characteristic of Major Luke S. Smith. He was always insistent on the best training and the best officers for the project. . . .

I think tribute should be paid Major Smith. And it was fortunate that General Weaver delegated full authority to him in regard to setting up Primary, Basic, and Advanced training at Tuskegee.[35]

On 21 March 1941 the Ninety-ninth Pursuit Squadron was activated when the first group of black recruits arrived at Chanute Field near Rantoul, Illinois, to begin training as mechanics and ground crew specialists at the Air Corps Technical School. The first arrivals, who had enlisted at Maxwell Field, were soon joined by recruits from other fields across the nation, and by May some 250 black recruits were in training at Chanute.[36] The nonflying officers also trained there—on 9 June six blacks began training as communications, weather, armament, and engineering officers.[37]

In early June, as the number of black recruits at Chanute began to swell, President Patterson asked Chief of Staff George C. Marshall's permission to visit the Air Corps Technical School to meet with the men and observe the progress of the training. Marshall quickly agreed and even offered to fly the Tuskegee president to Chanute in a military aircraft. Patterson accepted the invitation and asked the general to permit G. L. Washington to accompany him on the trip. Marshall, cognizant of the difficulties Patterson faced because of his support for the segregated unit at Tuskegee, directed General Arnold to arrange for air transportation, explaining that he had "suggested the plane—somewhat to meet the opposition to [Patterson] developed in Chicago and Harlem."[38] The effect of the cooperation was offset several days later, however, when Marshall directed that there be no press release regarding the trip to Chanute.[39] Gen. Davenport Johnson, the Air Staff officer in charge of arrangements for the trip, told Patterson that the Air Corps had "been embarrassed due to so many calls for use of Army airplanes to ferry personnel not connected with the military establishment, and would therefore ap–

preciate your cooperation in releasing no publicity regarding this flight."[40] Patterson assured Johnson that he would "respect your wishes in the matter of publicity in connection with this trip. No release whatever in regard to it will be made."[41]

Patterson and Washington left from Maxwell Field on the morning of 19 June. When they landed at Chanute, they were greeted by installation commander Col. R. E. O'Neill and several members of his staff. After lunching at the officers' club, the pair met with the black trainees, toured their quarters, and had an evening meal at their mess hall. Colonel O'Neill had arranged for his guests to stay overnight in visiting officers' quarters, and he placed a chauffeured staff car at their disposal. Patterson was duly impressed with the efforts of his hosts, noting that "no detail was left unattended to make our stay pleasant and profitable."[42] Indeed, the only unpleasant incident of the trip came at the hands of the local civilian manager of a theater in Rantoul, who denied Patterson admission to a movie until the Tuskegee president declared himself a guest at Chanute Field and threatened to report the establishment to the War Department.[43] The following morning Patterson and Washington toured the training facilities of the school. After lunching again with Colonel O'Neill at the officers' club, the two Tuskegee administrators boarded their aircraft and returned to Maxwell Field. Patterson commented to Marshall afterwards that from beginning to end he and Washington were accorded every courtesy by both the flight crew and the staff at Chanute Field.[44]

As for the men of the Ninety-ninth, Patterson told Marshall that he was favorably impressed with "the treatment and progress these men are making and receiving." Colonel O'Neill commented that the presence of black trainees on the post had caused no problems, and Patterson agreed: "it was evident from casual observation that race relations were of the best." "It was gratifying," he continued, "to observe the keen interest and enthusiasm demonstrated by post officials . . . and instructors with respect to the general welfare, adjustment and training of these men of the 99th Pursuit Squadron." The men had made satisfactory progress in their training thus far and were expected to finish in October. Based on the performance of the men to date, Colonel O'Neill said that he expected additional classes of black recruits to enter training at Chanute, a policy which Patterson enthusiastically supported.[45]

President Patterson offered several recommendations to General Marshall. He acknowledged that once the initial cadre of ground personnel completed their training at Chanute, the Air Corps might opt for special training courses at TAAF. Nevertheless, he urged that the army provide for "at least a limited number of Negroes in shops at Chanute Field."[46] He tactfully suggested an easing of the rigid segregation that had been imposed on the men since their arrival on post:

> I would recommend that some thought be given to the present separate arrangements which exist for the housing and feeding of the 99th Pursuit Squadron as compared with all other members of the unit [i.e., the other students at the Air Corps Technical School]. The present arrangement is working very we[ll], but a word of caution seems in order that this may not crystallize into a fixed policy of segregation. I am impressed with the present leadership at the post and I believe that whatever arrangement they desire to carry through can be accomplished with success.[47]

Finally, he urged that the Air Corps use the black press to publicize the progress of the Ninety-ninth and to announce opportunities for service in the air arm that were open to black Americans. Such measures were warranted, Patterson explained, because the institute had received a number of inquiries regarding the Air Corps from black Americans who were uninformed on the new racial policies in the air arm. Thus he concluded "that preparations for the establishment of various opportunities . . . for training and service [in the Air Corps], and [the] correct procedure[s] for enlistment in the 99th Pursuit Squadron are not clear in the minds of Negro citizens."[48]

Marshall never responded directly to Patterson's report, and there is no indication that any of the latter's recommendations were acted upon.[49] Nevertheless, Patterson's comments regarding the continuation of training for black technical specialists and his carefully worded suggestion that segregation not be "crystallized" into policy belie the Uncle Tom portrait painted by many of his critics. Certainly, his approach was far from the confrontational tactics of NAACP secretary Walter White or even Civilian Aide William H. Hastie, who until his resignation in 1943 kept his criticism within official War Department channels. Instead, the Tuskegee president preferred to follow the example set by his predecessors, Booker T. Washington and R. R. Moton, gently prodding and using his influence to attempt to work behind the scenes and promote changes that were often similar to those advocated by the more outspoken critics of the American caste system.

By the time Patterson and Washington returned from Chanute Field, work was about to begin on the new military field. A black architect, Hilyard R. Robinson of Washington, D.C., had been selected by the Army Corps of Engineers as the project architect and engineer. Patterson put the facilities of the institute at Robinson's disposal, and the architect reciprocated by expressing his desire to "make the fullest practicable use of Tuskegee graduates, students, etc., as may be determined upon for the mutual benefit of all concerned."[50] After the district court ruled in favor of the federal government at the end of June, the property owners were given until the end of July to surrender the land, with 10 July set as the date that preliminary site preparation could begin on portions of the land not being used by the owners.[51] Work commenced on 12 July when logging crews began clearing timber from the construction site, and within ten days grading operations were under way.[52]

In mid-August, with the first class of cadets scheduled to complete primary training in early November, Major Ellison reported to his headquarters at Maxwell that the army engineer in charge of the project had just made a drastic revision in the expected completion date of the field. Instead of the original date of 8 November, the project engineer estimated that the field would not be completed until the end of January 1942.[53] Authorities at SEACTC headquarters were shocked to learn that the field was so far behind schedule, and they immediately notified the division engineer that construction should be accelerated so that basic flight training could begin in November as planned: "The project at Tuskegee is considered by the War Department as No. 1 priority due to political pressure that [is] . . . being brought to bear upon the White House and War Department to provide pilot training for colored applicants. For this reason training must be initiated on schedule regardless of cost."[54] The division engineer in turn requested authorization from Washington, D.C., to officially upgrade the priority of the project as well as an additional $85,000, the amount needed to cover overtime pay and other expenses associated with an accelerated construction schedule.[55] The Air Staff, however, refused to support the SEACTC's request for top priority and additional funds. Brig. Gen. George E. Stratemeyer told General Weaver that in light of "the relatively small number of students concerned and the lack of money for construction purposes, it is impracticable to comply with this request." He directed that training plans for the black cadets be revised to conform to the 31 January 1942 completion date and suggested that facilities at the Dothan Advanced Flying School, eighty-five miles south of Tuskegee, be used in the interim.[56]

By mid-September Washington realized that progress on the military field was lagging. He wrote Major Ellison to ask if Tuskegee Institute should plan to continue to provide quarters for the first primary class after its graduation in November.[57] A week later he wrote Maj. L. S. Smith, SEACTC director of training, on the same matter, and also to inquire about "the possibility that Tuskegee Institute will be called upon to permit basic flight instruction on the primary field."[58] Smith responded: "It is impossible at the present to give you an adequate reply . . . pending the results of a study now being made by Major James A. Ellison on the possibility of opening the Basic School on November 8."[59]

At the end of September, after the slow progress of the construction at Tuskegee came to the attention of General Marshall, the inspector general's office sent an officer to Alabama to investigate the matter and apprise the chief of staff on the status of the project. On 25 September Lt. Col. S. N. Karrick wired his findings to Marshall through the inspector general, reporting that the project lagged well behind schedule because the change in sites had delayed the start of construction by four months. Work had only recently begun on nine of the forty-five buildings called for in the plan; grading operations were fifty percent complete; and only one of the four runways planned for the field was expected to be complete by 15 November. Karrick explained that "essential completion is anticipated by December 15th," and that "final completion of the project as now authorized with roads, utilities etc. is anticipated January 31st."[60] Finally, he agreed with the division engineer that facilities capable of supporting limited operations could be provided by 15 November, "if the work is accelerated by expensive overtime work, additional equipment, and the raising of the priority rating from the present A–1–e to a priority rating that will insure prompt delivery of heating, plumbing, electric and hydraulic equipment." Karrick, however, concurred with the Air Staff's assessment, concluding that "the completion of this project at the advanced date does not seem of sufficient importance to warrant the additional expense and the revision of priority of supply with the consequent detrimental effect on other projects."[61]

Marshall referred Karrick's findings on the situation at Tuskegee to General Arnold, who replied that he had the situation at Tuskegee well in hand.[62] In a 30 September memorandum, Arnold noted that his chief of operations and training, together with the commander of the SEACTC, had already conducted "a very thorough and complete survey of the facilities at Tuskegee" and found that the field would indeed not be ready by 8 November, when the first class was scheduled to begin basic flight training. Consequently, the Air Corps planned to enter the first class of black cadets into basic training "at Dothan, Alabama, or some other similar station within the Southeast Air Corps Training Center, this in order that they be properly housed and suitably trained pending the completion of the Basic and Advanced School at Tuskegee." Arnold concluded with a resounding declaration that the Air Corps would spare no effort to ensure that the training of the black cadets went smoothly, despite the delays in construction: "In view of the fact that this is the first class of Negro Aviation Cadets to ever receive flying training in the Military Establishment, every phase of their training, administration, and other factors are being very carefully scrutinized by this office. Due to the nation-wide publicity accorded this training, it can be readily appreciated why such care is necessary. It is not planned to inaugurate Basic and Advanced Training at Tuskegee until the field is practically complete in every detail and is entirely satisfactory."[63]

A scant two weeks after Arnold issued these assurances that training would not begin at Tuskegee until the construction of the field was complete, the Air Staff authorized the SEACTC to abandon its plans to conduct basic training for the first class at another field. On 9 October General Weaver advised the Air Staff that revised training plans within the SEACTC made it "necessary to initiate training at Tuskegee" rather than at some alternate location.[64] Since only the runway would be complete by November, Weaver requested $3,000 to construct a "temporary tent camp and temporary sanitary facilities to accommodate troops."[65] Arnold's assurances

This aerial photograph of Tuskegee Army Air Field, taken in late December 1941, shows only one runway complete. (Courtesy United States Air Force Historical Research Center, 222.01, January 1939–7 December 1941, vol. 3)

notwithstanding, on 13 October the Air Staff complied with Weaver's request and allotted the commander at Tuskegee $3,000 "for the provision of simple tent camp facilities."[66]

Thus on 8 November 1941, when the first class of black aviation cadets began basic flying training at TAAF, it was hardly an installation "complete in every detail" and "entirely satisfactory" as General Arnold had promised at the end of September. Only one of the four runways planned for the field had been completed and it had not yet been paved. Four tents at the end of the runway served as a parachute room, a communications center, a cadet ready room, and a supply and maintenance hangar. Two miles from the runway the main "Tent City" was erected, providing classrooms, administrative offices, and quarters for the enlisted men and aviation cadets.[67] Perhaps one of the most trying jobs during these early days was the difficult task of providing meals to the cadets and enlisted men, described here by the official historian:

> There was a general Mess Hall which consisted of merely four walls, a roof and sand floor. Food was prepared under field conditions, with gasoline pump stoves. An out-door tin stove was used to boil water for sanitary cooking and drinking purposes. Water for the station was hauled in on a water trailer from the town of Tuskegee. Under such perilous circumstances, it is remarkable that the mess personnel was capable of adequately serving approximately eighty-five persons daily, this number comprising the enlisted and Aviation Cadet personnel.[68]

Although crude, these arrangements "were not particularly uncomfortable during the mild weather of November 1941, but when the weather became more severe in December and the winter rains set in they were a real trial to the pioneer class."[69] When the rains came, they transformed the partially completed field—still undergoing grading and excavation and lacking an adequate drainage system—into a "bottomless morass." Indeed, the mud became such a problem that the men often joked that soon the commander would be forced to furnish all personnel "long poles with flags at the top so that when they disappeared the rescue squad would know where to dig."[70]

Another problem relating to the TAAF construction program also surfaced in the latter months of 1941, one potentially more serious than simple construction delays. In late September President

Patterson learned that the project was running well over cost, primarily because clearing, grading, and drainage costs at the Chehaw site were significantly higher than cost estimates based on the original site at Fort Davis. In order to stay within the budget, War Department planners were considering a curtailment in the project.[71] When he learned of this development, Patterson wired Assistant Secretary of War Robert P. Patterson to express his concern over the fact that the field "will not be completed as originally planned." Acknowledging that the shortage of funds was a widespread problem throughout the national defense program, he pointed out that the Tuskegee project was unique; it was the "only Negro aviation project in [the] nation."[72]

By mid-October Patterson learned the full extent of the budget shortfall when he received a letter from the contractor, Calvin McKissack, outlining his estimates of the additional funds "necessary to carry the work to a final completion and to provide adequate facilities for the training schedule of the Squadron" and future expansion of the installation. In all, McKissack estimated an additional $1.2 million would be required, if additional funds were allocated immediately so that work could continue without interruption. Moreover, he predicted that costs would be even higher if the work was discontinued when funds were depleted and then restarted at a later date. McKissack urged Patterson to "use your good offices and influence to see if sufficient funds can be appropriated to make full completion of this project possible."[73]

Patterson responded immediately to McKissack's plea. He sent a copy of the contractor's letter to Assistant Secretary Patterson and solicited his assistance in securing additional funds. He emphasized the importance of timely action, pointing out that "once the work is terminated it can be initiated again only at great cost." The Tuskegee president assured the assistant secretary that any efforts to "bring this project to full completion I am sure will be appreciated by all of the Negro people of America and will be a credit to the United States Government."[74]

Brig. Gen. A. H. Carter, a member of Assistant Secretary Patterson's staff, responded to this inquiry. He assured President Patterson that "not only is the project being developed in accordance with the original plans but substantial additions have been made to the original estimate" of $1.8 million, bringing the total funds allocated to the Tuskegee field to almost $2.4 million.[75] Current plans for the Tuskegee field, Carter continued, "will result in a satisfactory project from an operating point of view and it will be completed on the same basis as other similar stations." Carter noted that the plans for drainage suggested by McKissack and McKissack had been approved; he dismissed the contractor's request for additional buildings because "all of the buildings in the original plan plus the additional buildings subsequently approved are under way." General Carter's closing comment, intended to reassure President Patterson that the project would be completed properly, highlighted the close connection between Tuskegee Institute and the military program at TAAF in the minds of officials at the War Department: "I am satisfied that this project, when completed, will be a credit to *Tuskegee Institute* and also to the Air Corps."[76]

The scare over construction curtailment subsided for several weeks, but reappeared in late November. As before, it was the contractors who sounded the alarm, but they changed their tactics somewhat. This time the white subcontractors (C. G. Kershaw Contracting Company and Daugette–Millican Company) took the lead role by wiring Alabama congressmen Henry B. Steagall and Joe Starnes, as well as Senators Bankhead and Hill, to apprise them of the problem and to solicit their aid in obtaining additional funds for the completion of this important defense project in their home state:

> We were advised orally today by local U.S. Engineers that a large portion of Tuskegee airport construction will be curtailed early in December for lack of funds occasioned by change of project site before work was begun which necessitated increase of drainage by $200,000 and excavation by 3 to 4 million cubic yard[s] costing $354,000 extra. To suspend work now and resume in the spring means considerable extra cost. . . . We understand from government officials that this project has functioned efficiently and economically and we are substanuating [*sic*] by mail the above reasons for allotment exhaustion. This is the only Negro defense project in the entire defense program. Alabama as well as the thirteen million Negros [*sic*] of the U.S. will justly look with pride upon it if completed. . . . McKissack and McKissack as prime contractors join us in requesting your help and cooperation in securing [an] additional $500,000 [in] emergency funds immediately for completion of project.[77]

For his part, President Patterson wrote a strongly worded, two-page letter to Assistant Secretary of War Patterson. He outlined the difficulties at Tuskegee and his dissatisfaction over the War Department's failure to appreciate that there were important differences between the Tuskegee project and other Air Corps projects.[78] In addition, the Tuskegee president forwarded copies of his letter to a number of influential individuals and asked for their assistance in securing funds for the project. Besides Generals Marshall and Arnold in the War Department, Patterson sent copies to Eleanor Roosevelt, Vice President Henry A. Wallace, Sen. Lister Hill, and Cong. Henry B. Steagall. He also solicited the support of influential blacks including Mary McLeod Bethune, Claude A. Barnett, and Truman K. Gibson, William Hastie's assistant.[79]

Patterson's letter noted that work at TAAF would stop on 7 December because "all monies appropriated for this project will have been expended by that time." He laid the blame for the funding shortfall squarely on the shoulders of the War Department, charging that the "lack of funds is due entirely to a failure of army engineers to estimate costs adequately." He pointed out that only one of the four runways had been paved. This was only "one example of many essentials which are being left off" and it threatened to "seriously curtail" the flying training program. Acknowledging the additional funds that had been provided for the project, he nevertheless cautioned that "the amount made available was highly inadequate to meet the needs" and urged that the $0.5 million requested by the contractors be provided as an emergency appropriation to prevent an interruption in construction. While acknowledging similar funding problems with other projects, he pointed out that opportunities for black Americans to become pilots depended totally on the progress of construction at Tuskegee: "I realize that this is not an unusual case and that there are probably several other fields being dealt with in exactly the same manner. The chief difference, however, lies in the fact that this is the only field which is available to Negro fliers, whereas already established fields are providing a splendid opportunity for the training of white youth and for that matter any other nationality except Negroes."[80]

Finally, Patterson vented his frustration over the incessant delays and problems that had plagued the entire project since its inception a year earlier. He told the War Department that he and others who had collaborated in the plan to establish a segregated unit, in the face of demands for integration, were distressed at the slow progress of construction at Tuskegee. If the War Department failed to follow through on its promise to provide segregated opportunities, Patterson warned, then the more radical segments of the black community would be encouraged to seize the initiative:

Those of us who have worked diligently for the success of this project are seriously disturbed. . . . In the first place this project has been maligned as a "Jim-Crow" set-up, yet there were members of the Negro group who, in an effort to serve the best interests of the nation in this period of emergency and at the same time wanted to see progress by evolution rather than revolution, stood out for the development of this as a separate project. I need not say that failure to complete this project, at least to a degree acceptable for wise use, is not only a source of embarrassment to those who have supported the above position, but it also constitutes just the sort of argument which will be used in the future to contribute to the evidence indicating the undemocratic behavior of the federal government in the treatment of its Negro citizens.[81]

Equally embarrassing to Patterson and the other conservative black leaders who had supported the Air Corps' proposal for a segregated program was the agonizingly slow progress of the flying training program:

There are yet only five cadets at the Squadron Base taking actual flight training, although the program has dragged along for months. Tuskegee Institute through the interest of the Julius Rosenwald Fund has borrowed nearly $200,000 in order to develop a primary field adequate to take care of at least fifty flying cadets, yet only ten continue to dribble in at each five-week interval. This goes on in spite of the fact that the Air Corps is constantly announcing its plan to double its standing strength and making every effort to encourage enrollment in this branch of the service. I am sure you will agree then that this is not just another Air Corps project whose completion may indifferently await the regular appropriation of funds, but that this being the only opportunity available to Negroes, we think deserves special consideration in order that this opportunity which has been promised them may be forthcoming in full measure.[82]

The Air Staff's reaction to President Patterson's letter to the assistant secretary of war clearly demonstrates their failure to appreciate or understand the delicate situation in which Patterson's support for the Tuskegee project had placed him. In an internal memorandum to the secretary of the Air Staff, the Training Division commented on the Tuskegee president's charges, noting that he had "been actively interested in this program from the outset." The Training Division declared that "considerable effort" had been expended to satisfy Patterson, and cited as an example his trip to Chanute Field in an army aircraft. The memorandum accused the Tuskegee president of initiating an ill-advised campaign aimed at "greatly enlarging the negro training project and the elimination of the War Department policy for segregation." This comment clearly shows that at least some members of the totally white Air Staff could not distinguish between the demands of groups such as the NAACP and the NAAA, who were calling for full integration of the air arm, and the conservative approach of moderate black leaders such as Patterson, who were willing to accept segregation in exchange for a modicum of progress. Thus the Air Staff concluded that Patterson's attempts to monitor the progress of the Tuskegee project were "contributing in no way to the present program."[83]

Assistant Secretary Patterson, however, apparently understood the divisions in the black community and realized the validity of President Patterson's observation; if the War Department did not follow through on its commitment to establish a segregated program at Tuskegee, it would play into the hands of those advocating total integration. Claude Barnett probably made this point clear when he spoke to Assistant Secretary Patterson in early December. Over dinner the assistant secretary assured Barnett that the War Department would follow through on the program and develop a suitable facility at Tuskegee. Barnett quickly wired Patterson the good news: "Everything moving along. Field will be completed. May lose some features originally contemplated. You will receive letter being prepared by General Carter. I sat at dinner with Secretary Patterson who regretted not seeing you when you were here."[84] A week later, when he thanked Barnett for his "prompt and efficient" handling of the matter, Patterson reported that "We seem to be getting results. The work is going on according to schedule, and if this new angle can be averted, the cadets should all be in the barracks by the middle of next month."[85]

The "new angle" that Patterson hoped could be averted was an alarming rumor that began circulating shortly after the attack on Pearl Harbor. On 16 December 1941 Patterson sent an urgent telegram to the former commander of the SEACTC, Gen. Walter

Weaver, who had recently been transferred to Washington, D.C., as the acting chief of the Air Corps. The Tuskegee president explained that he had just been informed of a "plan to use Tuskegee 99th Pursuit base in [the] training of British fliers." Although he acknowledged the importance of cooperating with the Allies, Patterson protested any move to sacrifice the miniscule program at Tuskegee just to train a few more British pilots. Indeed, he cautioned, "Negro people will be greatly distressed . . . unless [the] present quota of colored aviation cadets is materially increased and selections are made from those having had previous CAA experience . . . permitting an increased number of those enlisting to successfully complete training."[86]

The rumor of a British takeover, coming fast on the heels of construction delays and work stoppages, so alarmed Patterson that he supplemented his traditional behind-the-scenes approach with an active press campaign. He told Barnett that "we ought to head them off with a little publicity," that a "reasonable, unanimous expression of alarm and protest by the Negro press throughout the nation would tend to stem or head off" any War Department plans detrimental to the Tuskegee initiative.[87] Patterson directed one of his assistants to prepare an anonymous press release and forward it to Barnett. The release recounted rumors that "Negro flyers . . . will be pushed entirely off the field." "Why is it," the report asked, "that British flyers must be trained at Tuskegee, the only field available to Negroes? What has happened to the dozens of other fields spotted all over the country? They are available to every other race and nationality except the Negro." The release speculated that the War Department "considers the Base now under construction . . . too good for Negroes," in light of the recent increases in appropriations: "Could it be that the Army feels $3,000,000 is too much to spend on Negroes?" Black Americans, the release stated, were willing to share the facilities at Tuskegee but should not allow their tenuous foothold in the nation's air arm to be sacrificed in the process: "No one objects to sharing the field with the British or any other flyers, particularly during this emergency and until the 99th Pursuit Squadron actually has its full strength. Under no circumstances, however, should we sit idly by while the cadets of the 99th are pushed off the field entirely."[88] Finally, the press release warned that any tampering with the Tuskegee project constituted a potential threat to national unity, as it would seriously undermine the morale of black Americans, some ten percent of the population:

> We may wonder how the Japanese were able to make a surprise attack on Pearl Harbor, but the Army is seeing to it that no one need doubt as to why the morale of Negro soldiers is hitting a new low. . . . Certainly at this crucial time, a betrayal of Negro flyers, and what is vastly more serious, a betrayal of the confidence of the Negro people, would be particularly disastrous. "Where there's smoke there's fire"—Let us hope someone will discover the fire behind this rumor and put it out before it grows into a major conflagration. "KEEP 'EM FLYING"!!![89]

The army, however, dismissed rumors that they planned to convert the Tuskegee field into a British pilot training base. In early January 1942 General Arnold and Brig. Gen. George E. Stratemeyer, Weaver's assistant, assured President Patterson that the program at Tuskegee would proceed according to plans. Arnold told Patterson that he was not aware of "any plan to transfer the present training activities from Tuskegee, in order to permit the training of British personnel at that place."[90] Stratemeyer likewise discounted the rumor, declaring that the War Department did not plan "to disrupt the training at Tuskegee."[91] The source of the rumor was never disclosed, but it may have originated from a concurrent plan to establish a British training base in Tuscaloosa, another Alabama town with an Indian name that might easily be confused with Tuskegee by those unfamiliar with the geography of the state.[92]

Even though the rumor regarding British training was spurious, it nevertheless provided Patterson with a forum to urge that opportunities for blacks to enlist as aviation cadets be expanded and that preference be given to applicants who had completed CPT secondary training. Both G. L. Washington and Capt. Noel Parrish, the Primary School commander, were convinced that procedures for selecting black aviation cadets should be modified, based on the results of the training since its inception in late August.

In late November Washington urged Patterson to appeal to the chief of the Air Corps for an increase in the number of black aviation cadets appointed to training at Tuskegee. Washington had discussed the matter with Captain Parrish and they had "agreed that a letter to the *Chief of the Air Corps in Washington, D.C.* would be the thing . . . based up[on] lack of enough entering to feed the scheduled number to the 99th Squadron school."[93] Patterson concurred. He asked General Arnold for a fifty percent increase in the size of each entering class, from ten to fifteen, noting that only six of thirteen in the original class had graduated from primary and that four of eleven were expected to graduate from the second class.[94]

Parrish followed up with his own letter to the chief of the Air Corps, submitted through SEACTC headquarters at Maxwell Field.[95] Unlike Patterson, Parrish did not call for an increased quota, but instead recommended the adoption of an improved method for selecting black cadets, an initiative which he believed was "the most important single improvement which could now be accomplished in the training of Negro flight students." He questioned the usefulness of appointing black cadets on the normal "first come, first served" basis and recommended instead that "during the initial period of Negro pilot training the selection of candidates be based upon some indication or demonstration of superior ability rather than upon priority of application." Parrish considered such a policy advisable because

> developing a Negro flying unit is unusual in many ways and it differs in one important respect from the problem of training and organizing any other type of unit. The senior pilots of any other unit would be men of superior ability and experience. For a Negro unit such men are not available. The senior pilots will have very little advantage in experience. Great responsibilities will devolve upon them as a result of their having been trained a few months earlier than the majority of the pilots in their unit. It is obvious that these men should be of superior abilities and potentialities.[96]

Parrish acknowledged that there were many ways to screen blacks awaiting appointment as aviation cadets; however, he suggested that giving priority to applicants who had successfully completed CPT secondary was the simplest and easiest to implement. He predicted that "by using only C.P.T. Advanced graduates for the next few classes, the standard of proficiency could be raised considerably and the percentage of graduates could be greatly increased." Parrish noted that there was a considerable backlog of black secondary graduates on the aviation cadet waiting list and urged that they be selected for training immediately, rather than continuing to send men to Tuskegee who were completely unfamiliar with aviation: "Of the last class of ten to report to this flying school, five have never been in an airplane in their lives. Of the five who have flown, three had their first flight as students in the C.P.T. course. This situation is typical."[97]

A primary reason black aviation cadets were unexposed to aviation, Parrish explained, was because "opportunities for Negroes to fly and to become familiar with aviation in general are somewhat restricted." He argued that "the C.P.T. program may be much more useful in Negro flying training than it has proved to be in the training of white students, simply because it has offered the only opportunity for any large number of Negroes to become familiar with aviation in any way." By selecting cadets for the next few classes from the pool of CPT secondary graduates, Parrish concluded, individuals of superior ability would receive their training first, and thus provide them with the requisite rank and experience to assume leadership positions in black flying units.[98]

Parrish's recommendations, prepared shortly before Japan attacked Pearl Harbor, took on new meaning for G. L. Washington after the nation entered the war. On 11 December 1941 Washington advised CPT Program officials of Parrish's proposal and requested that Tuskegee Institute's CPT quotas be expanded. He predicted that with the nation at war "Quotas are very likely to step up suddenly on army pilot training." If, at the same time, the War Department approved Parrish's request, Washington warned that there might not be "sufficient secondary graduates to meet their needs."[99]

Approximately a week later, amid rumors of the British takeover, Washington told President Patterson that the time had come for resolute action. Unless decisive steps were taken to protect the program at Tuskegee, Washington warned, its opponents in the War Department would use the national emergency to curtail or discontinue the Tuskegee initiative. Indeed, Washington warned, "the real issue is not a matter of English Cadets coming to the 99th." Instead, he learned from an anonymous source that the War Department had other plans for TAAF now that the nation was at war: "It is, as I was informed, a matter of turning the base into an Instructor Training School and redistribution of Negro ground technicians to other units in the country. Flying Instructors are coming in there in large numbers. More planes have been ordered than is needed for cadet training. All this seems to be the Major's [Ellison's?] planning." This information, Washington cautioned, was received "in confidence and is second hand but comes fairly straight."[100]

Washington urged Patterson to contact President Roosevelt directly and obtain a public statement endorsing the Tuskegee program: "If the President declares himself favorably[,] speedy action will follow and any indifferent air corps officials will be settled down to a realization that the job is to be done."[101] Washington's reasons for urging such action were threefold. First, he suspected that there were elements in the Air Corps and the War Department that remained opposed to the idea of admitting blacks to the air arm:

> Over the months of contact with air corps officials and officers I have come to believe that there is division on their appraisal of Negro youth's ability to measure up. Some apparently have no doubts. Others call for demonstration and experiment. Some would not be convinced even in the face of successful demonstration. It is logical to think that this cross-section of opinion obtains with officials of the War Department. The White House route would reduce opinion obstacles in the way of expansion of training and giving Negro youth a fair opportunity.[102]

Second, Washington believed that the performance of black students in both CAA and military training programs showed that black Americans could fly at a level equivalent to white students: "Negro youth has demonstrated under the Civil Pilot Training program that he can fly, do the ground work subjects, and that he is no different in this respect from any American youth. The experience at the primary field under a seasoned Captain of the Air Corps is that training has proceeded as normally expected in any such unit. Those cadets that have gone to Advanced Flying School are all still in training with no accidents or damage to planes. The demonstration is more than over." Finally, Washington contended that in light of the war crisis the program for training Negro pilots should not be curtailed but rather expanded. For some months prior to the Pearl Harbor attack, he reported, Air Corps officials had been in a quandary over "what to do with Negro military pilots after [they completed] training and [were] ready for service." With the nation at war, such issues seemed to Washington irrelevant: "Today this should give no concern with the many outlying posts to be defended by the air corps. I somehow believe that this single factor is an important consideration standing in the way of expansion. To the contrary it would seem even [greater] reason for expansion in order that certain posts may be entirely protected by Negro personnel." Moreover, the demand for qualified pilot trainees made policies that restricted the participation of qualified applicants simply because of their race almost unpatriotic: "The United States is at war. Military pilot training has been increased throughout the country. The army is finding it more and more difficult to keep its [voluntary] recruitment of air corps trainees in line with the expanded program. For this program Negro youth constitute an untapped source."[103]

Ironically, while Patterson and Washington expressed their concerns over construction delays, threats of work stoppage, rumors that TAAF would be given over to the British, and smaller aviation cadet quotas, an initiative was under way within the War Department to expand the black presence in the nation's air arm by establishing a second black pursuit squadron. The parallels between the establishment of the second squadron and the establishment of the first squadron a year earlier are striking. The Ninety-ninth had been formed in response to an intense pressure campaign that reached a fever pitch in late 1940, over a month after the War Department had decided in principle to establish a segregated unit. Yet black Americans were kept in the dark for several months while detailed plans were being prepared and approved, a delay which convinced many that the War Department intended to keep the Air Corps all-white. By fall 1941 the focus of the pressure campaign had shifted from admitting blacks to the Air Corps to expanding black participation beyond the limits of a single air unit. The waiting list of black aviation cadet applicants, all approved as qualified by cadet examining boards, numbered over two hundred. Only a handful, however, were selected to enter training at Tuskegee every five weeks. Once again suspicions of Air Corps duplicity surfaced, and the black press and various black organizations suspected that the War Department was intentionally restricting the level of black participation in the nation's air arm, even though the number of qualified applicants far exceeded the requirements of a single black squadron.[104]

Until November 1941, Air Staff officers resisted efforts to modify the selection process or enlarge quotas, just as they had opposed the establishment of the first segregated squadron a year earlier. They insisted that the Tuskegee project was an "experimental" program and therefore could not be enlarged until the results were evaluated. In October 1941 for example, when Judge Hastie urged that opportunities for blacks to enlist as aviation cadets be expanded, Col. St. Clair Street, executive officer to Assistant Secretary of War for Air Robert A. Lovett, responded that expansion was not warranted since the black pilot training program was only in the experimental phase: "At the present time, in view of the experimental nature of the training, it is believed that the only course of action open to the Air Corps is the continuation of the present policy; that is, limiting training in both pilot and non-pilot status to those candidates who can be absorbed into vacancies in existing negro aviation units."[105]

By the beginning of November 1941, however, the Air Staff was directed to prepare "a plan to organize and activate another Pursuit Squadron (colored)."[106] The plan that General Arnold received from the Air Staff three weeks later recommended that the second black pursuit squadron be based at TAAF. The costs of additional facilities at the installation to accommodate the unit were estimated at roughly $750,000, with construction of these facilities to take eight months.[107] On 8 December 1941 Arnold submitted his recommendations to General Marshall regarding the "Constitution and Activation of the 100th Pursuit Squadron (Colored)."[108] He briefly reviewed the progress to date regarding the formation of the Ninety-ninth Pursuit Squadron and the establishment of TAAF and commented that "If the training of negro pilots is to continue, it will be necessary to organize additional Air Corps units (colored)." Moreover, he noted, "There has been some criticism with reference to the long delay in ordering aviation cadets (colored) for flying training, due to the limited requirements and facilities at the Air Corps Basic and Advanced Flying School at Tuskegee." Thus Arnold recommended the establishment of a second black pursuit squadron "to provide units to which Aviation Cadets (Colored) may be assigned upon graduation from the flying school and to increase the number of flying cadets per class." Enlisted personnel were to receive their technical training at Chanute Field, following the same procedures as for the Ninety-ninth. The new squadron could not, however, be organized until "approximately 8 months after funds have been made available for the additional construction for the expansion of the Tuskegee Air Base."[109]

Although the plan was approved in late December 1941, the War

As of June 1942, four types of aircraft were used for flight training at Tuskegee. From top to bottom: the PT-17 (for primary training), the BT-13 (for basic training), the AT-6 (for advanced training), and the P-40 (for transition training). (Courtesy United States Air Force Historical Research Center, 222.01, December 1941–January 1943, vol. 5)

Department released no information on the new squadron until 1942. Indeed, the decision apparently remained a closely guarded secret until mid-January, for it was not discussed in a conference between Assistant Secretary Patterson, Judge Hastie, Claude Barnett, and President Patterson at the War Department earlier in the month. On the seventeenth, however, Assistant Secretary Patterson advised the Tuskegee president "that the Air Corps will soon announce the activation of an additional Negro squadron. This will be the 100th Pursuit Squadron which will be based at Tuskegee."[110] In a brief note of thanks to Judge Hastie for arranging the conference at the War Department, President Patterson commented that he had just learned of the new squadron, declaring it "good news and entirely in line with our needs here, as well as for the opportunity the Air Corps should extend to the Negro group."[111] Hastie's response highlights the differing philosophies of the two men: "As you know, I have never agreed that the Tuskegee project is a sound approach to the integration of the Negro into the Army Air Forces. However, so long as I am in the Department I shall continue my efforts to the end that the Negro may obtain the maximum possible benefit out of the enterprises which the Department undertakes."[112]

Despite Hastie's critical attitude, Patterson, Barnett, Washington, and other blacks who supported the Tuskegee initiative had reason to be optimistic. Not only was a second squadron to be organized, but the War Department had also agreed that blacks on the aviation cadet waiting list who had completed CPT secondary would be ordered to Tuskegee for training on a priority basis.[113] Then in early February 1942 the Air Staff took steps to raise the quota of students entering primary training at Tuskegee from ten to twenty per class to obtain "sufficient pilots to man a second pursuit squadron being activated at the Tuskegee Air Base."[114] After the dark days of November and December 1941 when it seemed as though the entire Tuskegee project was about to collapse, these developments must have been encouraging to black Americans such as President Patterson who advocated gradual change and segregated opportunity.

Finally, on 7 March 1942 the first class of flying students graduated from advanced training at TAAF to become the nation's first black military pilots.[115] Only two black Americans had ever served as military aviators—Eugene Bullard, the "Black Swallow of Death" who flew with the French in World War I, and John C. Robinson, who flew in the service of Haile Selassie in Ethiopia's war with Italy. But never had a black American served as a pilot in the armed forces of his own nation until the first class at Tuskegee completed training. Of the original thirteen who had entered flight training in August, only five remained. Four were aviation cadets, who received commissions as second lieutenants in the Air Corps Reserve—Lemuel R. Custis, Charles H. DeBow, Jr., George S. Roberts, and Mac Ross. The fifth was Capt. Benjamin O. Davis, Jr., who was transferred on graduation from the infantry to the Air Corps, thus becoming the first black American to hold a regular commission in the air arm.[116]

A milestone had been passed. More than anything else—more than establishing squadrons, constructing runways, and training ground crews—when Tuskegee's first class received their wings it symbolized the crossing of a threshold. It meant that black Americans had finally gained a foothold in the nation's air arm, after more than three years of struggle. The keynote speaker for the occasion, Brig. Gen. George E. Stratemeyer, sensed the importance of the event: "I am sure that everyone present, as well as the vast unseen audience of your well wishers, senses that this graduation is an historic moment, filled with portent of great good."[117] Thus by March 1942 it was apparent that black Americans had earned a small, albeit segregated, presence in the air forces of the nation, a presence which was growing. The Air Corps had been forced to accept black Americans into its ranks a year before the nation entered World War II, a step that might have been labeled as treasonous once the country mobilized for war. So by forcing the Air Corps to admit blacks during peacetime, the foundation had been laid for the participation of black Americans in aerial combat in the skies over North Africa and Europe.

The first black Americans to earn the wings of Air Corps pilots graduated from Tuskegee in March 1942. This photograph shows the five members of the first class with an instructor shortly before graduation. From left to right: George S. Roberts, Benjamin O. Davis, Jr., Charles H. DeBow, Jr., R. M. Long (instructor), Mac Ross, and Lemuel R. Custis. (Courtesy United States Air Force Historical Research Center, 222.01, January 1939–7 December 1941, vol. 3)

13

Conclusion

By 1942 Tuskegee, Alabama, had emerged as the center of black military flight training. Tuskegee's central role in the training of black pilots during World War II is reflected in the collective title that the men who trained there have assumed—the Tuskegee Airmen.[1] Three factors coalesced by late 1940 to push Tuskegee into the forefront as the site of the only training facility at which black Americans might earn the wings of an army pilot—the campaign for Air Corps participation, the establishment of a viable aviation program at Tuskegee Institute, and the Air Corps' search for an expedient means of complying with the War Department's demand that it train and accept black pilots.

The campaign for black participation in the Air Corps, which appeared after 1938, was itself the product of three separate but related elements—the black public's military orientation, blacks' growing air-mindedness, and the emergence of civil rights as a national issue. For black Americans, the opportunity for military service had always been an important concern.[2] It provided not only social status and economic stability, but perhaps most important, the presence of black Americans in the nation's armed forces provided tangible evidence of citizenship. In the late 1930s, as the likelihood of an armed forces build-up loomed large, the issue of military service became a central concern for many black Americans, especially after the *Pittsburgh Courier* launched its campaign in 1938.

By 1938 the black public had also become air-minded. This was a reflection of the attitude of the American public at large, and it was also a consequence of the growing numbers of African Americans involved in aviation and the wide coverage of their activities by the black press. Although only a few black Americans were involved in aviation prior to 1939, reports of their exploits appeared frequently in black newspapers and magazines, especially during the 1930s, when one black pilot flew for Ethiopia and several others undertook long-distance publicity flights. By 1939, as the United States began its prewar build-up of civilian and military aviation, several hundred black Americans were actively involved in aviation. The example of these aerial pioneers stimulated the black public's interest in aviation and highlighted the limited opportunities open to blacks who aspired to a career in aviation. Without the efforts and examples of America's black air pioneers, the opportunities that opened up in the CPT Program and the Air Corps after 1939 might not have been forthcoming. Therefore, the activities of these early black fliers are an important, and often overlooked, prologue to the campaign for Air Corps participation.

The final element that precipitated the Air Corps participation campaign was the emergence of civil rights as a national issue. During Franklin D. Roosevelt's second term, the hopes and aspirations of black Americans rose as the federal government exhibited an unprecedented responsiveness to their economic and social welfare. This shift was due to the pressure applied by civil rights organizations such as the NAACP, the growing importance of the black vote, a new concern for racial justice among political organizations and labor unions, and the presence of individuals within the Roosevelt administration who were interested in promoting civil rights initiatives. Thus by 1938 the rising expectations of black Americans gave a new urgency to demands that racial exclusion in military aviation be abandoned.

The second factor that contributed to the establishment of Tuskegee as the single site for the training of black military pilots during World War II was the establishment of a viable aviation program at Tuskegee Institute. From the outset of the campaign for Air Corps participation in 1939, Tuskegee Institute was considered by many black activists and lobbyists as a likely site for the training of black army pilots. This was due in part to the association of the institute earlier in the decade with the aeronautical exploits of two of its alumni, Albert E. Forsythe and John C. Robinson. Equally important were the efforts of another influential alumnus, Claude A. Barnett. From his vantage point in Chicago at the head of the Associated Negro Press, he realized the public relations potential of aviation. Consequently, he encouraged and prodded his alma mater's aeronautical efforts.

The individual who was largely responsible for establishing and developing an aviation program at Tuskegee Institute, however, was G. L. Washington. Although not an aviator himself, Washington nevertheless believed that aviation could offer black Americans new possibilities for economic advancement. He seized for Tuskegee Institute the opportunity offered by the CPT Program and used his considerable talents as an organizer and administrator to establish and develop a credible aviation program at the institute. By the summer of 1940, less than a year after Tuskegee entered the CPT Program, the school was the only black institution in the nation offering advanced CPT courses.

Washington was aided in his efforts to establish a CPT Program at Tuskegee by several Alabama whites, an example of racial cooperation that was engendered by the institute's willingness to follow southern rules of racial etiquette and not openly challenge segregation. At the outset, Leslie Walker, a native of Macon County, lent a sympathetic ear at CAA headquarters; he facilitated the all-important waiver that admitted the institute to the CPT Program even though the nearest suitable field was forty miles away in Montgomery. White pilot J. W. Allen's willingness to cooperate and conduct the initial flight training was also a crucial element in the establishment of the school's aviation program. The two API professors, Robert G. Pitts and B. M. Cornell, likewise made important contributions to the fledgling Tuskegee initiative when they agreed to teach the first ground school. Finally, the API president, L. N. Duncan, provided critical support to the institute's expanding program when he agreed to allow the temporary use of the Auburn flying field in 1940 for the first CPT secondary course.

The contributions of Washington, Tuskegee alumni, and sympathetic whites notwithstanding, the institute's sudden surge to the forefront of black aviation during the first six months of 1940 would not have been possible, without the benefit of the CPT Program, the first tangible achievement of the Air Corps participation campaign. Thanks to the efforts of interested black aviators and lobbyists, and the support of several key legislators, the Civilian Pilot Training Act, passed by Congress in 1939, contained a proviso that prohibited racial discrimination. The advent of federally funded pilot training gave Tuskegee Institute the wherewithal to establish its aviation program after several false starts earlier in the decade. As a result, the institute was an early beneficiary of the campaign for admitting blacks to the Air Corps.

During 1940, as black activists and the black press intensified their assault on the Air Corps' policy of racial exclusion, the third and final factor that led to Tuskegee's ascendancy as the single site for training black military pilots emerged—the Air Corps' search for an expedient means of complying with the War Department's demand that it accept and train blacks. For the Air Corps leaders the question of admitting black Americans was, until the fall of 1940, simply an unwelcome annoyance that they hoped would resolve itself. When it did not, they sought to render it impotent by interpreting P.L. 18 in such a way that they did not actually have to train blacks as military pilots. Indeed, circumstantial evidence strongly suggests that this was, in fact, the intent of the congressional committee that prepared the final wording of the legislation, even though the *original* intent of the measure was to force the Air Corps to admit blacks and train them as pilots. After black leaders and the black press exposed this subterfuge, and as more and more blacks demonstrated an ability to fly in the CPT Program, pressure grew. The Air Corps could no longer postpone the inevitable.

After the Selective Training and Service Act was passed in September 1940 and because a presidential election in which the black vote was deemed crucial loomed in November, the Roosevelt administration began to react to demands for racial equality in the armed forces. Once the service and ground components of the army responded to these realities by accepting greater numbers of blacks, the War Department could no longer permit the Air Corps to exclude blacks; the General Staff directed General Arnold and his staff to develop plans for a segregated air unit. Finally, the Air Corps had to act. Although Chicago was initially considered the most appropriate site for such a unit, the Tuskegee plan soon emerged as the most palatable solution to the leadership of the Air Corps—a segregated program endorsed and supported by the administration of one of the nation's most respected black institutions.

For several reasons the Air Staff settled on Tuskegee, Alabama, as the preferred site for complying with the requirement to admit blacks. An important factor, of course, was Tuskegee's newly established role as the center of black aviation. The indefatigable efforts of G. L. Washington and the outstanding performance of the Tuskegee CPT students had earned the respect of CAA officials. The Air Staff apparently consulted with the CAA—in 1940 the only federal agency with any experience in training black pilots—as they began to formulate plans for a segregated unit, and learned of the progress that had been made at the famous school of Booker T. Washington. Tuskegee's reputation as a conservative institution that accepted segregation as an inescapable fact of life and encouraged its students to focus their energies on becoming productive workers no doubt reassured the Air Corps leadership. This, perhaps, was the underlying reason that Tuskegee won out over Chicago. The confrontational tactics of NAAA spokesmen such as Enoch Waters at the meetings held in Chicago early in 1940 may well have convinced the Air Staff that maintaining segregation—a paramount concern to the Air Corps leadership—would be at risk in a northern, urban locale.

Another unspoken factor that almost certainly weighed in favor of Tuskegee was the relationship the school had with the Roosevelt administration. President Roosevelt had visited the campus during his second term, a powerful endorsement of its philosophy, as was the first lady's visit in 1940, which culminated in her well-publicized flight with Chief Anderson. Those in control at the Air Corps probably realized, perhaps unconsciously, that placing the black squadron at Tuskegee would be looked on favorably by the White House and might offset some of the criticism coming from the black press and the NAACP.

Geography was also on the side of Tuskegee. It was located in a rural area in a region that offered favorable year-round flying weather. Moreover, it was only forty miles from Maxwell Field, headquarters of the Southeast Air Corps Training Center, one of the three regional training centers responsible for overseeing the Air Corps' rapidly expanding flight training facilities. Thus when the Air

Staff found itself saddled with the requirement to admit blacks, the establishment of a segregated flying unit, in cooperation with Tuskegee Institute, quickly emerged as the most expeditious and logical solution to the problem. Indeed, given the mounting pressure campaign that emerged by the fall of 1940, the Air Staff no doubt concluded that if it failed to establish a token black unit, it might lose all control of the situation and find the door to integration swinging open.

Black America's concern over the role of its young men in the nation's air arm did not diminish, however, once the decision to establish a segregated squadron and base it at Tuskegee was announced in January 1941. Instead, the focus shifted as a host of new problems and issues surfaced. The overriding concern, one which divided black Americans and plagued the Tuskegee project throughout the war, was the segregation issue. Those who supported Tuskegee's cooperation with the Air Corps agreed with President Patterson and G. L. Washington—holding out for integration was futile and played into the hands of those who wanted to keep the Air Corps all-white. Opponents of segregation, like the NAACP and the NAAA, argued that accepting segregation in the Air Corps, where none had existed before (because no blacks had served in the army's air arm), perpetuated a practice that ran counter to democratic principles.

For young black men eager to fly and fight, however, the segregation issue was often eclipsed by their intense desire for an opportunity to prove themselves. Most would have agreed with Bigger Thomas, the tragic protagonist in Richard Wright's 1940 novel *Native Son*, who declared confidently "I *could* fly a plane if I had a chance."[3] Many realized that any opening available to them was contingent on finding a "window of opportunity" based on age—men who had reached their twenty-seventh birthday were ineligible for flight training—and marital status—only unmarried men were eligible. They might understand, in the abstract, the principled stand of those who opposed segregated training and segregated units. But their intense desire to fly predisposed them to take advantage of any chance they were offered. Thus they jumped at the opportunity to enter CPT, to take advanced CPT courses, and to become aviation cadets, even if it meant they were "jim crowed." The lure of flight, the elite status of military pilots, and the desire to serve their country inevitably worked against the NAACP's efforts to hold out for integration.

For men like F. D. Patterson, G. L. Washington, Walter White, and William H. Hastie, all too old to aspire to glory and adventure in the skies over Europe and the Pacific, the matter of how blacks would serve in the Air Corps was just one aspect of much broader problems that they faced. For Patterson and Washington, their obligation was to use the name and resources of Tuskegee Institute to provide—in Patterson's phrase—segregated opportunities following the accommodationist philosophy of Booker T. Washington. White and Hastie saw the mounting pressure for black participation in the air arm as an opportunity to undermine the traditional army policy of segregation. No precedent of segregated training or units existed in the Air Corps because blacks had been excluded from participating; thus the proponents of integration hoped to use the policy of exclusion to advantage and compel the admission of blacks to the Air Corps on an integrated basis.

When officials at Tuskegee cooperated with the Air Corps in establishing a segregated program, they undercut the strategy of the NAACP and others who hoped to forestall the extension of segregated units to the air arm. In cooperating with the Air Corps, however, Patterson and Washington assumed an obligation to ensure that the segregated program succeeded. When delays occurred and obstacles were encountered, they found themselves caught in the crossfire. The attacks of the integrationists were matched by white officers in the Air Corps, who resented Patterson's appeals to Assistant Secretary of War Patterson, General Marshall, and Eleanor Roosevelt.

Patterson was not, however, opposed to integration per se. His admonition to the commander at Chanute that the segregation of the black trainees not be "crystallized" into policy and his refusal to acquiesce when a northern theater manager refused him admission showed that he was not oblivious to the evils of the American caste system. He was, nevertheless, an evolutionist and was willing to accept segregation in exchange for even marginal progress and broader opportunities.

Late in 1941, when construction delays and curtailments threatened the future of the Tuskegee program, Patterson felt compelled to contact officials at the War Department and within the Roosevelt administration to urge that the project be completed as planned. Failure to do so, he warned, would strengthen the position of the integrationists and would undermine black leaders like Patterson who supported evolutionary progress toward racial equality. Indeed, he believed that the NAACP was "trying to cram their ideology down the throats of government officials"; he warned his friend Claude Barnett that "unless those of us who are able to think sanely on the whole question of integration of Negroes in the defense program step in, we aren't going to get much out of it."[4]

Despite the controversies that surfaced after the decision to admit blacks to the Air Corps on a segregated basis was announced, a military flight training program for blacks was indeed established at Tuskegee. Admittedly, it was a token, segregated initiative, fraught with difficulties, but in retrospect it was a significant achievement that ultimately played an important role in the desegregation of the independent air force established after the war.

Given the attitude of the Air Corps leadership toward the admission of blacks, even on a segregated basis, it is highly unlikely that black Americans would have been trained as pilots if the Air Corps participation campaign had not been launched. Moreover, the campaign was timely; it was initiated at the outset of the prewar expansion of the Air Corps and thus produced results by the time the United States entered the war. Once the nation was at war, it might have been too late for black Americans to exert pressure on the federal government for new opportunities in the Air Corps, for such tactics might well have been declared divisive and detrimental to the war effort.[5]

Despite the criticisms of the NAACP, the NAAA, and some black editors, the role of Tuskegee Institute in the opening of the Air Corps to black Americans was a crucial one. The question of whether greater gains could have been made had not Tuskegee Institute been available and willing to cooperate with the Air Corps in 1940 is problematic. The point is that in 1939 Tuskegee's interest in aviation, dormant since 1936, had been reawakened by the CPT Program. G. L. Washington performed yeoman's service when he established Tuskegee's aviation program under very difficult conditions and made it a credible operation that earned the enthusiastic support of CAA officials and brought the school to the attention of the Air Corps. After the War Department directed the establishment of a segregated squadron, the Air Corps turned to Tuskegee Institute because it offered a number of advantages unmatched by any other black institution—an advanced CPT Program, a favorable locale, close proximity to a regional flying training center, and a reputation for working within the segregated system.

Thus when the Air Corps was finally forced to make some concession to demands that it admit blacks, Tuskegee provided a convenient means of resolving the problem in a manner acceptable to the leadership of the Air Corps. Given the racial attitudes of Arnold and his staff, it seems likely that if the Tuskegee option had not been available, the project would have been delayed even longer than it was and might not have been far enough advanced by the time the nation entered the war to guarantee its survival. Moreover, without the active involvement of Tuskegee Institute, there might have been no individual like President Patterson to monitor the progress of the project and serve as its unofficial ombudsman when problems arose. Tuskegee's reputation as the nation's leading black educational institution gave Patterson the wherewithal to lobby federal officials up to and including the president and the first lady.

Although no evidence was found that points to any intentional effort to sabotage the program, the problems were myriad and Patterson's active involvement may have forestalled the squadron's premature death due to malignant neglect. Perhaps the most inexplicable difficulty encountered during the early stages of the program was the series of reductions in the class sizes. The progressive reductions in the initial class quotas from thirty students to ten may have been based on legitimate—though unstated—considerations. They might also, however, have been subtle attempts to reduce Tuskegee's profit margin for the primary contract to an unacceptable level in hopes that the school would renege on its commitment to participate in the training. Had the institute backed out of the primary training contract, it would have provided the Air Corps with an excuse to abandon, or at least postpone, the training of black pilots. Patterson, however, never wavered. In this and subsequent difficulties he steadfastly supported the program and used his influence and the institute's resources to foster the establishment of a black flying unit.

Equally important during these early days of the Air Corps' training program at Tuskegee were the circumstances that led to the involvement of two men who would subsequently play important roles in the development of black participation during the war, Noel Parrish and B. O. Davis, Jr. During his tenure as the first army supervisor of Tuskegee's primary school, Parrish exhibited an attitude toward the training program at Tuskegee uncharacteristic of most white Air Corps officers; he not only cared about effective training but also came to understand and empathize with the difficult situation that the black cadets in his charge faced. The effects of his enlightened attitudes toward black participation in the air arm continued throughout the war—he remained at Tuskegee after the first class graduated, serving as director of training at TAAF for most of 1942 and as commander of TAAF from 1943 until the war ended.

Parrish's involvement with the Tuskegee program, which had important immediate and long-term consequences, might never have happened without the establishment of the demonstration unit at the Coffey School of Aeronautics in Chicago. Indeed, like the lobbying efforts that resulted in amendments to P.L. 18 and the CPT act, this was an important contribution of the black aviation community in Chicago. Even though the demonstration unit proved to be a dead-end as far as bona fide Air Corps training was concerned, it nevertheless served to bring Parrish to Tuskegee, a development which had unforeseen but important ramifications for the future of black Americans in the air arm.

B. O. Davis, Jr., who completed his flight training with the class that graduated in March 1942, was also to play an important role in the future of black Americans in the Air Force. As a member of the first class he set the example and provided the leadership for the first black pilots in the Air Corps. As the first black officer to lead black units into combat, he helped to dispel the notion that blacks lacked a capacity for leadership and that black troops would not take orders from officers of their own race. Davis's performance as an air commander provided postwar Air Force decision makers with clear evidence that race was not a barrier to effective leadership.

Thus by 1942 the framework for black participation in aerial combat during World War II was in place, the result of almost three years of campaigning for black participation in the Air Corps. Since 1939 air-minded blacks and civil rights activists—building on the black public's traditional interest in military service and encouraged by the emergence of civil rights as a national issue—had clamored for the right of black Americans to serve their nation as military pilots. True, America's black leaders had not spoken with one voice. The campaign for Air Corps participation did not escape the long-standing dispute over whether to work within the segregated system or refuse compromise and demand full integration. Nevertheless, the Air Corps leadership, under pressure from the War Department, was finally forced to open its ranks to blacks and train at least a few as pilots, its most prestigious specialty. Tuskegee Institute's concerted efforts, along with the sincere interest of a few empathetic white officers—particularly Parrish—made the program a reality. After 1942 the continued coverage by the black press and the constant pressure from Judge Hastie inside the War Department and the NAACP from without, guaranteed that the black public's interest in the air arm remained high. Thus for the Air Corps there was no turning back, although there was considerable reluctance to go any further than absolutely necessary. Ultimately, however, the Ninety-ninth Pursuit Squadron deployed to the Mediterranean theater of operations and flew its first combat mission on 2 June 1943.[6]

The first step in a long journey toward racial equality in the air force was taken in the early months of 1942. By then it was obvious that black Americans had gained a foothold in the nation's air arm. After 1942 there was no turning back. The Air Corps had been forced to open its ranks to black Americans, but insisted that segregation prevail. Undaunted, Davis and his four comrades who earned their wings in the first class, along with the hundreds who followed them, showed a skeptical nation that they could fly and fight. Without the gains that had been made by 1942, these achievements would not have been possible. And even greater gains were made after the war: in 1949, following the establishment of an independent air arm, the United States Air Force became the first branch to fully implement Pres. Harry S. Truman's executive order directing the desegregation of the armed forces.[7]

Notes

Abbreviations

AAF	Army Air Forces
API	Alabama Polytechnic Institute, Auburn, Alabama
CAA	Civil Aeronautics Authority
CHS	Chicago Historical Society, Chicago, Illinios
CPT Program	Civilian Pilot Training Program
FDRL	Franklin D. Roosevelt Library, Hyde Park, New York
GC	General Correspondence
IGAC	Interracial Goodwill Aviation Committee
LC	Local Correspondence
Lib. Cong.	Manuscript Division, Library of Congress, Washington, D.C.
NA	National Archives, Washington, D.C.
NAAA	National Airmen's Association of America
NAACP	National Association for the Advancement of Colored People
NYA	National Youth Administration
RFC	Reconstruction Finance Corporation
RG	Record Group
SEACTC	Southeast Air Corps Training Center, Maxwell Field, Alabama
TAAF	Tuskegee Army Air Field, Alabama
TUA	Tuskegee University Archives, Tuskegee, Alabama
USAF	United States Air Force
USAF-HRC	United States Air Force Historical Research Center, Maxwell Air Force Base, Alabama
YUL	Manuscripts and Archives, Yale University Library, New Haven, Connecticut

1. Aviation and Tuskegee Institute: The Early Years

1. *Tuskegee Messenger*, June 1934, p. 11; *Chicago Defender* (nat. ed.), 9 May 1936, p. 12. Robinson had originally left Chicago flying his own two-place biplane in the company of two fellow black aviators from Chicago, Cornelius R. Coffey and Grover Nash. Coffey accompanied Robinson and Nash flew alongside in his single-seat monoplane. En route Robinson's plane was damaged beyond repair. He borrowed Nash's craft and flew on alone, with his companions completing the trip to Tuskegee by bus. For details of the trip, see *Chicago Defender* (nat. ed.), 26 May 1934, p. 1; 2 June 1934, p. 2; 2 May 1936; and 9 May 1936, p. 12.

2. For examples of coverage of Robinson's visit in the black press, see *Pittsburgh Courier* (city ed., suburban), 2 June 1934, p. 5, and *Chicago Defender* (nat. ed.), 16 June 1934, p. 3; for coverage in a white paper, see *Montgomery Advertiser*, 25 May 1934, p. 18.

3. *Chicago Defender* (nat. ed.), 2 June 1934, p. 2. Because Robinson's airport in Robbins, Illinois—a black community south of Chicago—had recently been destroyed in a freak storm, he may have come to Tuskegee hoping to obtain a teaching position, after he convinced the administration to construct an airport; see *Chicago Defender* (nat. ed.), 2 May 1936.

4. G. L. Washington to John C. Robinson, 7 July 1936, G. L. Washington Folder, LC 1936, Frederick Douglass Patterson Papers, TUA.

5. See, for example, *Chicago Defender* (nat. ed.), 23 May 1936, p. 1, and 30 May 1936, p. 7; *New York Times*, 18 May 1936, p. 11, and 19 May 1936, p. 6; New York *Amsterdam News*, 23 February 1936, in John W. Kitchens and Jonell Chislom Jones, eds., *Microfilm Edition of The Tuskegee Institute News Clippings File* (Tuskegee, Ala.: Tuskegee Institute, 1978), reel 51, frame 900 (hereafter cited as *Tuskegee Clippings File*, with reel numbers and frame numbers separated by a colon, i.e., 51: 900); and *St. Louis Argus*, 29 May 1936, in *Tuskegee Clippings File* 51: 901. Robinson's service with Ethiopia is discussed fully below; the air force that he commanded contained less than twenty outmoded aircraft.

6. There is no standard history of Tuskegee Institute. The following historical sketch is based on Louis Harlan's two-volume biography of Booker T. Washington, *Booker T. Washington: The Making of a Black Leader, 1856–1901* (New York: Oxford Univ. Press, 1972) and *Booker T. Washington: The Wizard of Tuskegee, 1901–1915* (New York: Oxford Univ. Press, 1983); Allen W. Jones, "The Role of Tuskegee Institute in the Education of Black Farmers," *Journal of Negro History* 60 (April 1975): 252–61; Carl S. Matthews, "The Decline of the Tuskegee Machine, 1915–1925: The Abdication of Political Power," *South Atlantic Quarterly* 75 (Autumn 1976): 460–69; Pete Daniel, "Black Power in the 1920s: The Case of Tuskegee Veterans Hospital," *Journal of Southern History* 36 (August 1970): 368–88; Manning Marable, "Tuskegee Institute in

the 1920's," *Negro History Bulletin* 40 (November–December 1977): 764–68; William H. Hughes and Frederick D. Patterson, eds., *Robert Russa Moton of Hampton and Tuskegee* (Chapel Hill: Univ. of North Carolina Press, 1956); Addie Louise Joyner Butler, *The Distinctive Black College: Talladega, Tuskegee and Morehouse* (Mutuchen, N.J.: Scarecrow Press, 1977); and Robert J. Norrell, *Reaping the Whirlwind: The Civil Rights Movement in Tuskegee* (New York: Alfred A. Knopf, 1985).

7. Harlan, *The Making of a Black Leader,* p. 140.

8. Ibid., pp. 217–20.

9. Matthews, "Decline of the Tuskegee Machine," passim.

10. Harlan, *The Making of a Black Leader,* pp. 279–80.

11. *Tuskegee Messenger,* 16–30 July 1927, p. 3.

12. Ibid.; Hughes and Patterson, *Robert Russa Moton,* pp. 88–89.

13. Daniel, "Black Power in the 1920s," pp. 369–72.

14. There is no authoritative biography of Moton; the best available source is Hughes and Patterson, *Robert Russa Moton.*

15. The title of Tuskegee's chief executive officer was changed by the Board of Trustees from principal to president in 1933 (*Minutes of the Meeting of the Board of Trustees of the Tuskegee Normal and Industrial Institute, October 27, 1933,* p. 8).

16. Autobiographical statement by Frederick D. Patterson for Carnegie Study of the Negro in America, December 1939, Bro–By Folder, GC 1939, Patterson Papers, TUA.

17. Editorial written for the New York *Amsterdam News,* January 1941, Box 33, Folder 59, Addresses, Speeches, and Statements, Patterson Papers, TUA.

18. Anderson and Forsythe were the first blacks to complete a *round-trip* transcontinental flight. The first one-way transcontinental flight by blacks was completed in October 1932 by Thomas Allen and James Herman Banning (Lonnie G. Bunch, III, "In Search of a Dream: The Flight of [James] Herman Banning and Thomas Allen," *Journal of American Culture* 7 [Spring/ Summer 1984]: 100–103). Anderson and Forsythe are frequently credited with the first transcontinental flight of any kind; see, for example, Harry A. Ploski and Warren Marr, comps. and eds., *The Negro Almanac: A Reference Work on the Afro-American,* 3d ed. (New York: Bellwether, 1976), s.v. "Black Firsts," p. 1044. Very little research has been published on either the Banning–Allen team or the Forsythe–Anderson partnership, and it is based almost exclusively on newspaper accounts. See Joseph J. Corn, *The Winged Gospel: America's Romance With Aviation, 1900–1950* (New York: Oxford Univ. Press, 1983), pp. 59–60, for a brief discussion of Forsythe and Anderson. Two extremely useful sources on the Forsythe–Anderson flights are the Interracial Goodwill Aviation Committee (IGAC) Folder, GC Box 191, in the Robert

Russa Moton Papers, TUA, and Oral History Interview of C. Alfred "Chief" Anderson by James C. Hasdorff, 8–9 June 1981, typed transcript (120 pp.), K239.0512–1272, USAF Collection, USAF–HRC.

19. In 1929 William J. Powell and James Herman Banning inadvertently landed in Mexico through a navigational error and became the first American blacks to fly across international borders, albeit unintentionally (William J. Powell, *Black Wings* [Los Angeles: Ivan Deach, Jr., 1934], pp. 115–35; Des Moines *Iowa Bystander*, 16 November 1929, p. 1).

20. *Official Program: Christening of the Booker T. Washington* and *Our Race Soars Upward: The Pan American Goodwill Flight Outlined*, pamphlets in IGAC Folder, GC Box 191, Moton Papers, TUA. The Anderson–Forsythe flight was not the first Pan–American Goodwill Flight; in 1926–27 the U.S. Army completed an air tour of South America also known as the Pan-American Goodwill Flight.

21. *Newsweek*, 29 September 1934, p. 35; Anderson interview, p. 27.

22. Anderson interview, p. 28.

23. Ibid., p. 26.

24. Ibid., pp. 1–6, 14, 25.

25. Julia Goens to Moton, 2 September 1934, IGAC Folder, GC Box 191, Moton Papers, TUA.

26. Goens to G. Lake Imes, 5 September 1934, IGAC Folder, GC Box 191, Moton Papers, TUA.

27. Telegrams, Goens to Imes, 11 September 1934, and Imes to Goens, n.d., IGAC Folder, GC Box 191, Moton Papers, TUA.

28. Telegrams, Moton to Daniel W. Armstrong, n.d., and Armstrong to Moton, 13 September 1934, IGAC Folder, GC Box 191, Moton Papers, TUA.

29. Telegram, James Edmund Boyack to Imes, 13 September 1934, Folder 1567, GC Box 186, Moton Papers, TUA.

30. Telegrams, Boyack to Imes, 15 September 1934; Robert G. Lyon, Clarence Chamberlain, Roscoe Turner, J. Erroll Boyd, Emile Burgin, and Clyde Pangborn, all to Moton, 15 September 1934; in IGAC Folder, GC Box 191, Moton Papers, TUA.

31. *Montgomery Advertiser*, 14 September 1934, p. 16; *Official Program: Christening of the Booker T. Washington*, IGAC Folder, GC Box 191, Moton Papers, TUA.

32. Speech, MS, n.d., IGAC Folder, GC Box 191, Moton Papers, TUA; *Tuskegee Messenger* 10 (October 1934): 1, 8.

33. Publicity brochure and MS list, n.d., IGAC Folder, GC Box 191, Moton Papers, TUA.

34. *Tuskegee Messenger* 10 (October 1934): 1, 8; Rev. H. V. Richardson to Moton, 6 October 1934, R. R. Moton Folder, LC Box 64, Moton Papers, TUA;

and MS telegram, Richardson to Goens, n.d., IGAC Folder, GC Box 191, Moton Papers, TUA.

35. Imes to Albert E. Forsythe, 28 September 1934, IGAC Folder, GC Box 191, Moton Papers, TUA. For information on the Guggenheim Fund, see Richard P. Hallion, *Legacy of Flight: The Guggenheim Contribution to American Aviation* (Seattle: Univ. of Washington Press, 1977).

36. Imes to Forsythe, 28 September 1934, IGAC Folder, GC Box 191, Moton Papers, TUA.

37. Boyack to Imes, 25 September 1934, IGAC Folder, GC Box 191, Moton Papers, TUA.

38. Imes to Forsythe, 28 September 1934, IGAC Folder, GC Box 191, Moton Papers, TUA.

39. Forsythe to Moton, 1 October 1934, IGAC Folder, GC Box 191, Moton Papers, TUA.

40. Forsythe to Imes, 2 October 1934, IGAC Folder, GC Box 191, Moton Papers, TUA.

41. Ibid.

42. Ibid.

43. Imes to Boyack, 4 October 1934, IGAC Folder, GC Box 191, Moton Papers, TUA.

44. Ibid.

45. Ibid.

46. Forsythe to Imes, 21 October 1934, IGAC Folder, GC Box 191, Moton Papers, TUA.

47. Ibid.

48. Imes to Forsythe, 19 October 1934, IGAC Folder, GC Box 191, Moton Papers, TUA. Despite Moton's reluctance to support the project in accordance with Forsythe's proposals, Tuskegee was nevertheless listed as a sponsor in the predeparture press releases; see *New York Times*, 8 November 1934, p. 20, and *Pittsburgh Courier*, 10 November 1934, p. 1.

49. Forsythe to Imes, 24 October 1934; telegrams, Goens to Moton, 1 November 1934, and Moton to Goens, 6 November 1934, in IGAC Folder, GC Box 191, Moton Papers, TUA.

50. Except as noted, the following sketch of the flight is based on reports in the *New York Times*, the *Pittsburgh Courier* (nat. ed.), the December 1934 *Tuskegee Messenger*, and the Anderson interview, pp. 35–45. Reports of the flight in the *New York Times* can be found in the following issues: 8 November 1934, p. 20; 10 November 1934, p. 17; 11 November 1934, p. 29; 13 November 1934, p. 10; 9 December 1934, p. 25; 12 December 1934, p. 16; 14 December 1934, p. 11; and 15 December 1934, p. 6. Reports in the *Pittsburgh*

Courier can be found in the following issues: 10 November 1934, p. 1; 17 November 1934, p. 1; 24 November 1934, p. 2; and 22 December 1934, p. 1.

51. Forsythe to Imes, 2 December 1934, IGAC Folder, GC Box 191, Moton Papers, TUA.

52. Paul M. Pearson, Governor of the Virgin Islands, to the Advisory Council, 12 December 1934, Negroes 1934 Folder, Secretary's Correspondence, Records of the Office of the Secretary of Agriculture, Record Group 16, National Archives, Washington, D.C. (photocopy in Black Wings Collection, National Air and Space Museum, Washington, D.C.). For a copy of the San Juan reception program and a full-page story from the San Juan (Puerto Rico) *El Mundo*, 10 December 1934, see IGAC Folder, GC Box 191, Moton Papers, TUA.

53. Anderson interview, pp. 43–45.

54. Forsythe continued his aeronautical proselytizing; see, for example, his article in the October 1935 *Tuskegee Messenger* (vol. 11, p.7) entitled, "The Outlook for the Negro in Aviation."

55. Anderson interview, pp. 47–48.

56. Forsythe to Imes, 24 October 1934, IGAC Folder, GC Box 191, Moton Papers, TUA.

57. William Randolph Scott, "A Study of Afro-American and Ethiopian Relations, 1896–1941" (Ph.D. dissertation, Princeton University, 1971), pp. 29–30.

58. John Hope Franklin, *From Slavery to Freedom: A History of Black Americans*, 5th ed. (New York: Alfred A. Knopf, 1980), pp. 422–23.

59. Malaku E. Bayen to Claude Barnett, 3 January 1935, Barnett to Bayen, 8 January 1934 [*sic,* 1935], Box 170, Folder 9, Claude A. Barnett Papers, CHS.

60. William Randolph Scott, "Colonel John C. Robinson: The Condor of Ethiopia," *Pan-African Journal* 5 (Spring 1972): 61–62; Bayen to Barnett, 6 May 1935, and Robinson to Barnett, 3 June and 28 November 1935, Box 170, Folder 9, Barnett Papers, CHS.

61. Scott, "Colonel John C. Robinson," pp. 62–63. Robinson was also serving as a correspondent for the Associated Negro Press, writing under the name of Wilson James; see Barnett to E. G. Roberts, 17 July 1935, Box 170, Folder 9, Barnett Papers, CHS.

62. Barnett to Robinson, 19 October 1935, Box 170, Folder 9, Barnett Papers, CHS.

63. See Scott, "A Study of Afro-American and Ethiopian Relations," pp. 217–48; see also chapter 3.

64. Joe Louis, *Joe Louis: My Life* (New York: Harcourt, Brace, Javonovich, 1978), p. 48.

65. *Campus Digest: The Voice of the Tuskegee Student,* 28 September 1935.

66. Ibid., 12 October 1935, p. 1, and 26 October 1935, p. 1.

67. Scott, "Colonel John C. Robinson," pp. 62–63.

68. On at least one courier mission Robinson was attacked by Italian fighters; he managed to evade his attackers and delivered his messages safely at Addis Ababa (Robinson to Barnett, 21 November 1935, Box 170, Folder 9, Barnett Papers, CHS).

69. Scott, "Colonel John C. Robinson," p. 64; Barnett to Robinson, 22 April 1936, Box 170, Folder 9, Barnett Papers, CHS.

70. See Andrew Buni, *Robert L. Vann of the Pittsburgh Courier: Politics and Black Journalism* (Pittsburgh: Univ. of Pittsburgh Press, 1974), pp. 246–48, for a sketch of Joel Rogers's activities as the *Courier's* war correspondent.

71. Robinson to Barnett, 28 November 1935, and Barnett to Robinson, 31 December 1935, 22 January 1936, and 16 April 1936, in Box 170, Folder 9, Barnett Papers, CHS.

72. Barnett to Patterson, 7 May 1936 (with enclosed Associated Negro Press release on William J. Powell) and 9 May 1936, in Barnett Correspondence 1936–1942, LC 1937, Patterson Papers, TUA.

73. Barnett to Patterson, 12 May 1936, and Patterson to Barnett, 13 May 1936, in Barnett Correspondence 1936–1942, LC 1937, Patterson Papers, TUA.

74. Barnett to Patterson, 14 May 1936, and wires, Boyack to Barnett, 13 May 1936, Patterson to Barnett, 14 May 1936, and Barnett to Patterson, 14 May 1936, in Barnett Correspondence 1936–1942, LC 1937, Patterson Papers, TUA; Barnett to Boyack, 13 May 1936, Box 170, Folder 9, Barnett Papers, CHS.

75. Barnett to Patterson, 18 and 20 May 1936, Barnett Correspondence 1936–1942, LC 1937, Patterson Papers, TUA. The basis for the claim that the proposed aviation school at Tuskegee would be "the first aviation school under Negro auspices" is unclear because Barnett was fully aware of black aviator William J. Powell's school in Los Angeles. Perhaps he did not consider Powell's operation "under Negro auspices" because it had New Deal sponsorship (see chapter 3).

76. Barnett to Patterson, 18 and 20 May 1936, Barnett Correspondence 1936–1942, LC 1937, Patterson Papers, TUA. For the *New York Times* story, see 18 May 1936, p. 11.

77. Barnett to Patterson, 18 May 1936, Barnett Correspondence 1936–1942, LC 1937, Patterson Papers, TUA.

78. Barnett to Patterson, 20 May 1936, Barnett Correspondence 1936–1942, LC 1937, Patterson Papers, TUA.

79. *New York Times,* 19 May 1936, p. 6.

80. Barnett to Patterson, 20 May 1936, Barnett Correspondence 1936–1942, LC 1937, Patterson Papers, TUA.

81. *New York Times*, 24 May 1936, p. 3.

82. Barnett to Boyack, 28 May 1936, Box 170, Folder 9, Barnett Papers, CHS. Although the *Chicago Defender*, 30 May 1936, estimated the crowds at the airport and the Grand Hotel to be 5,000 and 20,000 respectively, Barnett's more conservative estimates of 3,000 and 8,000, which he conveyed privately to Boyack in the correspondence cited above, are probably more accurate.

83. *Chicago Defender*, 30 May 1936, pp. 1–2; see also Scott, "Colonel John C. Robinson," pp. 64–65.

84. Paul Carter, *The Twenties in America*, 2d ed. (Arlington Heights, Ill.: AHM Publishing, 1975), p. 67 (quoting an essay in *Time* on the occasion of the fortieth anniversary of Lindbergh's flight).

85. Various black aviators came close but all, for various reasons, fell short: the difficulties of Forsythe and Anderson have been examined earlier in this chapter; the careers of other possible black air heroes, such as Hubert Julian, Bessie Coleman, and James Herman Banning are discussed in chapter 3.

86. Barnett to Patterson, 3 June 1936, and telegram, Barnett to Patterson, 4 June 1936, Barnett Correspondence 1936–1942, LC 1937, Patterson Papers, TUA; see also Box 171, Folder 1, Barnett Papers, CHS, for additional correspondence relating to Barnett's orchestration of Robinson's activities.

87. *Tuskegee Messenger*, July–August 1936, p. 1.

88. Washington to Robinson, 7 July 1936, G. L. Washington Folder, LC 1936, Patterson Papers, TUA.

89. Robinson to Barnett, 1 July 1936, Box 171, Folder 1, Barnett Papers, CHS.

90. Washington to Robinson, 7 July 1936, G. L. Washington Folder, LC 1936, Patterson Papers, TUA. Robinson's 1 July 1936 letter and the unsigned agreement were not found among the Patterson Papers; their general contents can, however, be inferred from Washington's 7 July 1936 letter, a three-page response to Robinson.

91. Washington to Robinson, 7 July 1936, G. L. Washington Folder, LC 1936, Patterson Papers, TUA.

92. Ibid.

93. Ibid.

94. Linda O. McMurry, *George Washington Carver: Scientist and Symbol* (New York: Oxford Univ. Press, 1981), pp. 43–69.

95. Washington to Robinson, 7 July 1936, G. L. Washington Folder, LC 1936, Patterson Papers, TUA.

96. Ibid.

97. *Baltimore Afro-American*, 29 August 1936, in *Tuskegee Clippings File* 52: 366; Robinson to Washington, 4 November 1936, Washington to Robinson, 20 November 1936, and Washington to Patterson, 20 November 1936, G. L. Washington Folder, LC 1936, Patterson Papers, TUA.

2. Black Americans and the Military

1. Harvard Sitkoff, *A New Deal for Blacks. The Emergence of Civil Rights as National Issue: The Depression Decade* (New York: Oxford Univ. Press, 1978).

2. Public Law 18, 3 April 1939 (H.R. 3791); Civilian Pilot Training Act, 27 June 1939 (H.R. 5619).

3. Ulysses Lee, *United States Army in World War II. Special Studies: The Employment of Negro Troops* (Washington, D.C.: Office of the Chief of Military History, United States Army, 1966), p. 55 (hereafter cited as *Employment of Negro Troops*).

4. See, for example, Alan M. Osur, *Blacks in the Army Air Forces During World War II* (Washington, D.C.: Office of Air Force History, 1977); Morris J. MacGregor, Jr., *Integration of the Armed Forces, 1940–1965*, Defense Studies Series (Washington, D.C.: Center of Military History, United States Army, 1981); Richard M. Dalfiume, *Desegregation of the U.S. Armed Forces: Fighting on Two Fronts, 1939–1953* (Columbia: Univ. of Missouri Press, 1969); A. Russell Buchanan, *Black Americans in World War II* (Santa Barbara, Calif.: Clio Books, 1977); Neil A. Wynn, *The Afro-American and the Second World War* (New York: Holmes and Meier Publishers, 1975).

5. U.S. Congress, Senate, *Hearings Before the Committee on Military Affairs, United States Senate, Seventy-sixth Congress, First Session, on H.R. 3791, An Act to Provide More Effectively for the National Defense by Carrying Out the Recommendations of the President in His Message of January 12, 1939, to the Congress*, 76th Cong., 1st sess., 17 January–22 February 1939, pp. 311–16, quotations from p. 312.

6. U.S. Department of Commerce, Bureau of the Census, *Negro Aviators*, Negro Statistical Bulletin no. 3, January 1939; and *Negro Aviators*, Negro Statistical Bulletin no. 3, September 1940, containing retrospective data for the years 1935, 1936, 1937, and 1939.

7. Sitkoff, *New Deal For Blacks*, p. i, ix.

8. Gunnar Myrdal, *An American Dilemma: The Negro Problem and Modern Democracy* (New York: Harper and Brothers, 1944), p. 74.

9. Lee, *Employment of Negro Troops*, pp. 3–4. Although Lee's observations were focused on the interwar years, they apply with equal validity to black Americans of the post–World War II era. With the elimination of the racial quotas that characterized the segregated armed forces, blacks have consis-

tently enlisted in numbers proportionately larger than their representation in the population at large, and their retention rates exceed those of white recruits. Indeed, for those familiar with the history of black participation in the nation's armed forces, the parallels between black attitudes and patterns of service before the Second Reconstruction and after are striking. For commentaries on the current status of blacks in the armed forces, see "Success Story: Blacks in Uniform," *Wilson Quarterly* 8 (Spring 1984): 80–81; and Charles Moskos, "Success Story: Blacks in the Army," *The Atlantic* 257 (May 1986): 64–72.

10. Quotation from Jack D. Foner, *Blacks and the Military in American History: A New Perspective* (New York: Praeger, 1974), p. 3; for the text of the act, see Morris J. MacGregor and Bernard C. Nalty, eds., *Blacks in the Military: Basic Documents* (Wilmington, Del.: Scholarly Resources, 1977), 1: 3.

11. Foner, *Blacks and the Military*, p. 3.

12. Ibid., pp. 6–7; Bernard C. Nalty, *Strength for the Fight: A History of Black Americans in the Military* (New York: Free Press, 1986), pp. 10–11.

13. Foner, *Blacks and the Military*, p. 27; MacGregor and Nalty, *Basic Documents*, 1: 218.

14. MacGregor and Nalty, *Basic Documents*, 1: 187; Foner, *Blacks and the Military*, pp. 20–21; Nalty, *Strength for the Fight*, pp. 19–20.

15. Nalty, *Strength for the Fight*, p. 5; for the text of the act, see MacGregor and Nalty, *Basic Documents*, 1: 3–4.

16. Foner, *Blacks and the Military*, pp. 4–5.

17. Ibid., pp. 6–19; Nalty, *Strength for the Fight*, pp. 10–18.

18. Nalty, *Strength for the Fight*, p. 12; Foner, *Blacks and the Military*, pp. 8–9.

19. Nalty, *Strength for the Fight*, p. 18.

20. Ibid., p. 23.

21. Ibid.; Foner, *Blacks and the Military*, p. 23.

22. C. Peter Ripley, ed., *The Black Abolitionist Papers* (Chapel Hill: Univ. of North Carolina Press, 1985), vol. 1, *The British Isles, 1830–1865*, p. 545.

23. Foner, *Blacks and the Military*, pp. 23–25; Nalty, *Strength for the Fight*, pp. 23–25.

24. Although some authorities maintain that blacks did not participate in the Mexican War, Robert Ewell Greene has shown that African Americans were involved in that conflict, albeit on a very limited basis. See Robert Ewell Greene, *Black Defenders of America, 1775–1973* (Chicago: Johnson Publishing, 1974), pp. 43–49.

25. John Greenleaf Whittier, "The Black Men of the Revolution and War of 1812," *National Era*, 22 July 1847, p. 1. The *National Era*, established in

1847, achieved national prominence in 1851 with the first publication of Harriet Beecher Stowe's *Uncle Tom's Cabin*, which ran serially through 1852.

26. Ibid.

27. William Cooper Nell, *Services of Colored Americans in the Wars of 1776 and 1812* (Boston: Prentiss and Sawyer, 1851; reprint, New York: AMS Press, 1976, 2d ed., Boston: R. F. Wallcut, 1852; various reprints, 1854 [Canada], 1862, and 1894), and *The Colored Patriots of the American Revolution* (Boston: Robert J. Wallcut, 1855; reprint, New York: Arno Press, 1968). Nell cites the influence of Whittier in *The Colored Patriots*, p. 8.

28. Introduction by Benjamin Quarles to the Arno Press reprint of Nell, *The Colored Patriots* (1968).

29. See, for example, ibid., pp. 119–20, 126, 150, 216–20, 230, 237–38.

30. Dred Scott v. Sandford, 19 Howard (U.S.), 414–17, 15 L. Ed. 705 (1884).

31. Leslie H. Fishel, Jr. and Benjamin Quarles, comps., *The Negro American: A Documentary History* (Glenview, Ill.: Scott, Foresman and Co., 1967), pp. 222–23.

32. William Lloyd Garrison, *The Loyalty and Devotion of Colored Americans in the Revolution and War of 1812* (Boston: R. F. Wallcut, 1861); George Henry Moore, *Historical Notes on the Employment of Negroes in the American Army of the Revolution* (New York: Charles T. Evans, 1862). Both works cited have been reprinted in *The Negro Soldier: A Select Compilation* (New York: Negro Universities Press, 1970).

33. Much has been written on the role of black soldiers in the Civil War. The two standard book-length treatments are Benjamin Quarles, *The Negro in the Civil War* (Boston: Little, Brown and Co., 1953), and Dudley Taylor Cornish, *The Sable Arm: Negro Troops in the Union Army, 1861–1865* (New York: Longmans, Green, 1956). For more concise surveys see Bernard C. Nalty, "Civil War and Emancipation," chap. 3 of *Strength for the Fight*; Foner, "Blacks in the Civil War," chap. 3 of *Blacks and the Military*; and Allan Nevins, "Shape of Revolution: Chattels No More," chap. 20 in *The War for the Union: War Becomes Revolution, 1862–1863* (New York: Charles Scribner's Sons, 1960). For a concise account of the engagements involving black troops, see Mark M. Boatner, III, *The Civil War Dictionary* (New York: David McKay Co., 1959), pp. 584–85.

34. Boatner, *The Civil War Dictionary*, p. 584; Nalty, *Strength for the Fight*, p. 43; Foner, *Blacks and the Military*, p. 48; Franklin, *From Slavery to Freedom*, pp. 221, 224.

35. Foner, *Blacks and the Military*, pp. 48–50; Nalty, *Strength for the Fight*, pp. 44–46; Barbara C. Ruby, "General Patrick Cleburne's Proposal to Arm Southern Slaves," *Arkansas Historical Quarterly* 30 (Autumn 1971): 193–212.

36. Quoted by Burghardt Turner in his foreword to Arthur E. Barbeau and

Florette Henri, *The Unknown Soldiers: Black American Troops in World War I* (Philadelphia: Temple Univ. Press, 1974), p. xii.

37. Lee, *Employment of Negro Troops*, p. 4.

38. Ibid., p. 3.

39. Nalty, *Strength for the Fight*, pp. 51–52; Foner, *Blacks and the Military*, pp. 52–54.

40. Foner, *Blacks and the Military*, pp. 52–53; quotation from Lee, *Employment of Negro Troops*, p. 23.

41. The three standard sources on the role of the black regulars on the western frontier are William H. Leckie, *The Buffalo Soldiers: A Narrative of the Negro Cavalry in the West* (Norman: Univ. of Oklahoma Press, 1967); Arlen L. Fowler, *The Black Infantry in the West, 1869–1891* (Westport, Conn.: Greenwood Publishing, 1971); and John M. Carroll, ed., *The Black Military Experience in the American West* (New York: Liverwright, 1971).

42. Foner, *Blacks and the Military*, pp. 56, 69. The cavalrymen were transferred from the adverse conditions of the frontier in 1891, shortly after the final Indian uprising that ended with the tragic massacre at Wounded Knee, as a reward for their record of distinguished service during the Indian Wars.

43. Nalty, *Strength for the Fight*, pp. 51–52.

44. Leckie, *The Buffalo Soldiers*, p. 26. According to Leckie, the origin of the term *buffalo soldiers* is undetermined; he notes, however, that most contemporary observers assumed that it was based on the similarity between the hair of the black soldiers and the coat of the buffalo.

45. Ibid., p. 259; Foner, *Blacks and the Military*, p. 54.

46. "Historical Sketch of the 10th Cavalry, 1866–1891," in MacGregor and Nalty, *Basic Documents*, 3: 74.

47. Marvin Fletcher, *The Black Soldier and Officer in the United States Army, 1891–1917* (Columbia: Univ. of Missouri Press, 1974), p. 32; Foner, *Blacks and the Military*, p. 75.

48. Quoted in Fletcher, *The Black Soldier and Officer*, p. 32.

49. Ibid., pp. 32–47.

50. Foner, *Blacks and the Military*, p. 84

51. Nalty, *Strength for the Fight*, p. 65.

52. Foner, *Blacks and the Military*, pp. 86–87. The only black volunteers to see combat during the war were the members of a black company in the Sixth Massachusetts, which participated in the invasion of Puerto Rico.

53. Ibid., pp. 89–91.

54. Fletcher, *The Black Soldier and Officer*, pp. 158–59.

55. Rayford W. Logan, *The Betrayal of the Negro from Rutherford B. Hayes to Woodrow Wilson* (New York: Collier Books, 1965), p. 335.

56. Richard Carl Brown, "Social Attitudes of American Generals, 1898–1940" (Ph.D. dissertation, University of Wisconsin, 1951), p. 177.

57. Foner, *Blacks and the Military,* pp. 95–98. For full treatment of the Brownsville affair, see John D. Weaver, *The Brownsville Raid* (New York: W. W. Norton, 1970), and Ann J. Lane, *The Brownsville Affair: National Crisis and Black Reaction* (Port Washington, N.Y.: Kennikat Press, 1971).

58. A year after the court of inquiry concluded its investigation, the first issue of *The Crisis,* the official organ of the National Association for the Advancement of Colored People, carried an article protesting the treatment of the Brownsville soldiers. In 1972 the names of the black soldiers were cleared when the army, at the request of Attorney General Richard Kleindienst, changed their records to indicate they had been honorably discharged (see Nalty, *Strength for the Fight,* p. 97).

59. Nalty, *Strength for the Fight,* p. 18.

60. W. E. B. Du Bois, "Close Ranks," *Crisis* 16 (July 1918): 111.

61. W. E. B. Du Bois, editorial, *Crisis* 16 (September 1918): 216.

62. Ibid., p. 217.

63. Ibid.

64. Quoted in Barbeau and Henri, *The Unknown Soldiers,* p. 7.

65. Nalty, *Strength for the Fight,* p. 142.

66. Foner, *Blacks and the Military,* pp. 109–10; see also Barbeau and Henri, *The Unknown Soldiers,* pp. 10–14.

67. Foner, *Blacks and the Military,* p. 112.

68. Nalty, *Strength for the Fight,* p. 112; Franklin, *From Slavery to Freedom,* pp. 325–26.

69. Foner, *Blacks and the Military,* p. 111.

70. Lee, *Employment of Negro Troops,* p. 5. The best study of black soldiers during World War I is Barbeau and Henri, *The Unknown Soldiers.*

71. Foner, *Blacks and the Military,* p. 121.

72. B. Joyce Ross, *J. E. Spingarn and the Rise of the NAACP, 1911–1939* (New York: Athenaeum, 1972), pp. 81–102; Barbeau and Henri, *The Unknown Soldiers,* pp. 56–69.

73. The rationale for this decision has never been fully explained; see, for example, Barbeau and Henri, *The Unknown Soldiers,* pp. 27–28, and Lee, *Employment of Negro Troops,* p. 16.

74. Barbeau and Henri, *The Unknown Soldiers,* pp. 111–36.

75. Ibid., pp. 137–63.

76. The definitive study on the Houston Riot is Robert V. Haynes, *A Night of Violence: The Houston Riot of 1917* (Baton Rouge: Louisiana State Univ. Press, 1976); for a more concise, earlier version of Haynes's research on the subject, see "The Houston Mutiny and Riot of 1917," *Southwestern Historical Review* 76 (April 1973): 418–39. Nalty, *Strength for the Fight*, pp. 101–6, provides a useful summary of the incident.

77. For an analysis and discussion of the study, see Osur, *Blacks in the Army Air Forces*, pp. 1–6, and Alan L. Gropman, *The Air Force Integrates, 1945–1964* (Washington, D.C.: Office of Air Force History, 1978), pp. 1–3.

78. Osur, *Blacks in the Army Air Forces*, p. 5; Gropman, *The Air Force Integrates*, p. 2.

79. Lee, *Employment of Negro Troops*, p. 36.

80. Ibid., pp. 29–30.

81. Ibid., pp. 23–29.

3. The Black Public Becomes Air-minded

1. *Atlanta Daily World*, 8 October 1938, in *Tuskegee Clippings File* 57: 381; Corn, *Winged Gospel*, p. 12.

2. Corn, *Winged Gospel*, p. vii.

3. Charles A. Lindbergh, "Aviation, Geography, and Race," *Reader's Digest* 35 (November 1939): 64.

4. Kenneth Brown Collings, "America Will Never Fly," *American Mercury* 38 (July 1936): 292.

5. See, for example, rebuttals to Collings's article by George S. Schuyler, "Negroes in the Air," *American Mercury* 39 (December 1936): xxviii–xxx, and William J. Powell, "Negroes in the Air," *American Mercury* 41 (May 1937): 127.

6. See Joel Rogers's syndicated feature "Your History" in *Pittsburgh Courier* (nat. ed.), 6 May 1944, p. 7.

7. George Edward Barbour, "Early Black Flyers of Western Pennsylvania, 1906–1945," *Western Pennsylvania Historical Magazine* 69 (April 1986): 95–97.

8. Henry E. Baker, "The Negro in the Field of Invention," *Journal of Negro History* 2 (January 1917): 34; Carter G. Woodson and Charles H. Wesley, *The Negro in Our History*, 12th ed., rev. and enl. (Washington, D.C.: Associated Publishers, 1972), p. 464; U.S. Patent Office, *Official Gazette* 90 (20 February 1900): 1506.

9. *Christian Recorder*, 21 January 1911, in *Tuskegee Clippings File* 242: 671.

10. Marshall, b. ca. 1885, was originally from Macon, Georgia; he died

suddenly of appendicitis in 1916 (*Crisis* 4 [September 1912]: 223; *Southern Workman*, August 1916, in *Tuskegee Clippings File* 242: 680; and U.S. Patent Office, *Official Gazette* 182 [10 September 1912]: 365).

11. Baker, "The Negro in the Field of Invention," p. 34; Woodson and Wesley, *The Negro in Our History*, p. 464; U.S. Patent Office, *Official Gazette* 185 (December 1912): 487. Madison, a graduate of Tuskegee Institute, apparently gave up his dreams of designing an airplane and became a plumbing engineer; in 1931 he patented a radiator bracket for which he received a lucrative U.S. Army contract in 1942 (*Pittsburgh Courier* [nat. ed.], 21 February 1942, p. 19).

12. Baker, "The Negro in the Field of Invention," p. 34; Woodson and Wesley, *The Negro in Our History*, p. 464; U.S. Patent Office, *Official Gazette* 185 (17 December 1912): 692; James de T. Abajian, comp., *Blacks in Selected Newspapers, Censuses, and Other Sources: An Index to Names and Subjects*. (Boston: G. K. Hall and Co., 1977), 3: 352, citing *Oakland Sunshine*, 20 December 1913, and *Western Review*, June 1915.

13. Baker, "The Negro in the Field of Invention," p. 34, gives the name as J. E. Whooter of Missouri; Woodson and Wesley, *The Negro in Our History*, p. 464, reports the name as H. E. Hooter of Missouri. Neither name appears in the "Alphabetical list of patentees," U.S. Patent Office, *Annual Report for the Commissioner of Patents for the Year 1914* (Washington, D.C.: Government Printing Office, 1915), but a John E. McWhorter of St. Louis appears in the "Alphabetical list of inventions" as the inventor of an "aeroplane" (see U.S. Patent Office, *Official Gazette* 208 [3 November 1914]: 84).

14. Hayden, whose name also appears as Lucian Headin and L. Arthur Headen, reportedly went abroad to market his device after the United States government refused to adopt it; he was supposedly commissioned a second lieutenant in the Royal Flying Corps after Great Britain adopted the device. See *New York Age*, 18 January 1912 (with photo of Hayden at controls of an aircraft), in *Tuskegee Clippings File* 242: 672–73, and Des Moines *Iowa State Bystander*, 16 February 1912, p. 2 (reported in both as Lucian Headin); Des Moines *Iowa State Bystander*, 28 February 1913, p. 3 (reported as L. Arthur Headin); and *New York Age*, 27 April 1918, and *Raleigh Independent*, 4 May 1918, in *Tuskegee Clippings File* 242: 689 (reported as L. A. Hayden). Hayden's name (in any of its various spellings) does not appear in the "Alphabetical list of patentees" of the Commissioner of Patents annual reports for the years 1910 through 1918, nor is there any entry for an aeronautical "stabilizer" or "equalizer," the reported name of Hayden's invention, in the "Alphabetical list of inventions" for those years.

15. Barbour, "Early Black Flyers," p. 97.

16. Ibid., pp. 97–98.

17. *Atlanta Independent*, 28 October 1911, p. 11.

18. Barbour, "Early Black Flyers," p. 98; Elizabeth Ross Haynes, *The Black*

Boy of Atlanta[: A Biography of Richard Robert Wright, Sr.] (Boston: House of Edinboro, 1952), p. 81.

19. Des Moines (Iowa) *Bystander,* 22 June 1917, p. 3.

20. Lee, *Employment of Negro Troops,* p. 55.

21. Ibid.; Charles H. Williams, *Negro Soldiers in World War I: The Human Side,* with an introduction by Benjamin Brawley (New York: AMS Press, 1970; reprint of 1923 ed. published under the title *Sidelights on Negro Soldiers*), p. 53; Barbeau and Henri, *The Unknown Soldiers,* p. 63.

22. Bullard was not, as some sources claim, a member of the Lafayette Escadrille, the French unit officially known as Spad Squadron 124 and manned exclusively by some twenty–five American volunteers. He was, instead, part of the Lafayette Flying Corps, a designation that applies to all of the roughly two hundred Americans who flew with the French, most of whom, like Bullard, were scattered in various squadrons throughout the French Air Service (see Philip M. Flammer, *The Vivid Air: The Lafayette Escadrille* [Athens: Univ. of Georgia Press, 1981], p. x and passim).

23. P. J. Carisella and James W. Ryan, *The Black Swallow of Death: The Incredible Story of Eugene Jacques Bullard, the World's First Black Combat Aviator* (Boston: Marlborough House, 1972); James Norman Hall and Charles Bernard Nordhoff, eds., *The Lafayette Flying Corps* (Boston: Houghton Mifflin, 1920), 1: 151; Walt Brown, Jr., ed., *An American for Lafayette: The Diaries of E. C. C. Genet, Lafayette Escadrille,* with an introduction by Dale L. Walker (Charlottesville: Univ. of Virginia Press, 1981), pp. 99–100; Flammer, *The Vivid Air,* p. 3; Nalty, *Strength for the Fight,* p. 124.

24. Bricktop, with James Haskins, *Bricktop* (New York: Atheneum, 1983), pp. 82, 84–88, 118–19, 145, 167, 203. The only black known to have publicly acknowledged Bullard's wartime flying record during the interwar years was George S. Schuyler in his rebuttal to Kenneth Brown Collings, see note 5 this chapter. In September 1941 Bullard wrote F. D. Patterson and Walter White asking for their assistance in recruiting American blacks for Free French flying forces, but there is no indication that either recognized Bullard's unique status as the only American black with a confirmed aerial victory to his credit (Bullard to Walter White, 6 September 1941, Group II, Series A, Box 647, U.S. Army Air Corps, General Folder, National Association for the Advancement of Colored People Papers, Manuscript Division, Library of Congress [hereafter cited as IIA–647, NAACP Papers, Lib. Cong.]; Bullard to F. D. Patterson, 10 September 1941, and n.d.; Patterson to Bullard, 16 September 1941, all in Bro–By Folder, GC 1941, Patterson Papers, TUA).

25. Elois Patterson, *Memoirs of the Late Bessie Coleman, Aviatrix: Pioneer of the Negro People in Aviation* (n.p., 1969); Kathleen L. Brooks-Pazmany, *United States Women in Aviation, 1919–1929* (Washington, D.C.: Smithsonian Institution Press, 1983), p. 8; Marianna W. Davis, ed., *Contributions of Black Women to America* (Columbia, S.C.: Kenday Press, 1982), pp. 497–98; David Young

and Neal Callahan, *Fill the Heavens with Commerce: Chicago Aviation, 1855–1926* (Chicago: Chicago Review Press, 1981), p. 156. Some sources maintain that Coleman was the first black—man or woman—to earn a flying license; if, however, the reports of Lucian Arthur Hayden earning a license in France by 1913 are accurate (see above), then Coleman was the second black and the first black woman licensed to fly.

26. There are two book-length works on Julian, both of which should be used with caution: Hubert Julian, as told to John Bulloch, *Black Eagle* (London: The Adventurers Club, 1965), and John Peer Nugent, *The Black Eagle* (New York: Stein and Day, 1971). More reliable are "Hubert F. Julian: The Black Eagle of Harlem," chap. 7 in Scott, "A Study of Afro-American and Ethiopian Relations," pp. 217–48; Robert A. Hill, Emory J. Tolbert, and Deborah Forczek, eds., *The Marcus Garvey and Universal Negro Improvement Association Papers*, Vol. 4, *1 September 1921–2 September 1922* (Berkeley: Univ. of California Press, 1985), pp. 1059–60n. 1; and Robert G. Weisbord, "Black America and the Italian-Ethiopian Crisis: An Episode in Pan-Negroism," *Historian* 34 (February 1972): 230–41.

27. *Crisis* 22 (October 1921): 273; Richmond (Virginia) *Planet*, 6 August 1921, in *Tuskegee Clippings File* 242: 700. The accuracy of the title "Dr." and Julian's status as a student at McGill University have not been confirmed but are likely apocryphal, based on later biographical sketches. Julian did receive a patent for his safety device; see U.S. Patent Office, *Official Gazette* 286 (24 May 1921): 732. The offers from the Curtiss and Gerni companies are unconfirmed.

28. Franklin, *From Slavery to Freedom*, pp. 354–56.

29. Hill et al., *Marcus Garvey Papers*, 4: 1048–49, 1059–60; Franklin, *From Slavery to Freedom*, p. 355; Theodore Draper, *The Rediscovery of Black Nationalism* (New York: Viking Press, 1970), p. 51.

30. Nugent, *Black Eagle*, pp. 40–47; Julian, *Black Eagle*, pp. 58–65; Hill et al., *Marcus Garvey Papers*, 4: 1060.

31. Nugent, *Black Eagle*, pp. 48–50; Julian, *Black Eagle*, pp. 70–71.

32. Nugent, *Black Eagle*, p. 63.

33. Ibid., pp. 121–22.

34. Ibid., pp. 53–127; Julian, *Black Eagle*, pp. 81–135.

35. Editorial, reprinted in *Crisis* 43 (March 1936): 83.

36. Editorial, *Pittsburgh Courier* (nat. ed.), 25 June 1927, sec. 2, p. 1.

37. Charles A. Lindbergh, *"We": The Famous Flier's Own Story of His Life and Transatlantic Flight, Together with His Views on the Future of Aviation* (New York: Grosset and Dunlop, 1927), pp. 54–60.

38. Herbert R. Northrup, Armand J. Thieblot, Jr., and William N. Chernish, *The Negro in the Air Transport Industry*, The Racial Policies of American

Industry, Report no. 23 (Philadelphia: Univ. of Pennsylvania Press, 1971), p. 28.

39. Foreman subsequently announced that he would open a flying school in Los Angeles, and then he disappeared from public view. *Pittsburgh Courier* (nat. ed.), 12 March 1927, p. 1, and 17 September 1927, p. 5; *Chicago Bee,* 4 June 1927, in *Tuskegee Clippings File* 30: 466; and Powell, *Black Wings,* pp. 53–55.

40. *Opportunity* 5 (July 1927): 216.

41. *Pittsburgh Courier* (nat. ed.), 16 July 1927, p. 5.

42. *New York Times,* 4 August 1927, p. 14.

43. *Pittsburgh Courier* (nat. ed.), 22 October 1927, pp. 1, 5.

44. *Crisis* 43 (June 1936): 189–90.

45. Abajian, *Blacks in Selected Newspapers,* 1: 82, citing the *Los Angeles News Age,* 24 June 1921, p. 1.

46. Nugent, *Black Eagle,* pp. 51–52. The similarity of the name of Julian's organization to that of the NAACP may not have been coincidental; several years earlier Julian was reportedly rebuffed by NAACP executive secretary James Weldon Johnson when he sought the organization's support for his transatlantic flight attempts. Weldon's excuse that the NAACP concentrated on attacking racism and discrimination in court proceedings prompted Julian to suggest that a more appropriate name for the organization might be "[National] Association for the Defense of Colored People" (Julian, *Black Eagle,* pp. 68–69).

47. *Opportunity* 7 (November 1929): 353; *Crisis* 35 (September 1928): 304, 306.

48. Only one extant copy of *Bessie Coleman Aero News* has been located, Vol. 1, no. 1 (May 1930), held by the Moorland–Spingarn Research Center, Howard University, Washington, D.C.

49. Powell, *Black Wings,* pp. 1–95; *Bessie Coleman Aero News* 1, no. 1 (May 1930): 13. Banning earned his license only a year after the establishment of the Bureau of Air Commerce, the federal agency empowered to issue pilots licenses; prior to 1926 the only licenses issued in the United States were at the state level.

50. *Bessie Coleman Aero News* 1, no. 1 (May 1930): 6, 21; Des Moines *Iowa Bystander,* 12 October 1929, p. 1.

51. *Bessie Coleman Aero News* 1, no. 1 (May 1930): 5.

52. Northrup et al., *The Negro in the Air Transport Industry,* p. 28

53. George E. Hopkins, *The Airline Pilots: A Study in Elite Unionization* (Cambridge, Mass.: Harvard Univ. Press, 1971), p. 71.

54. *Chicago Defender* (nat. ed.), 11 April 1936, p. 2.

55. The Challenger Air Pilot Association was known briefly as the Challenger Aero Club (*Chicago Defender*, 30 May 1936, p. 5).

56. *Newsweek*, 29 September 1934, p. 36.

57. Powell, *Black Wings*, p. 87.

58. *East Tennessee News*, 13 June 1935, in *Tuskegee Clippings File* 48: 998.

59. See, for example, *Chicago Defender* (nat. ed.) 28 March 1936, p. 3, and 9 May 1936, p. 12; *Pittsburgh Courier* (nat. ed.), 30 January 1937, p. 10. Bessie Coleman has remained a compelling figure in black history. See, for example, J. Goodrich, "Salute to Bessie Coleman," *Negro Digest* 8 (May 1950): 82–83; Anita King, "Brave Bessie: First Black Pilot," *Essence* 7 (May 1976): 36; "They Take to the Sky: Group of Midwest women follow path blazed by Bessie Coleman," *Ebony*, May 1977, pp. 88–96; and "Drive Launched to Honor Coleman with [U.S. Postal] Stamp," *Chicago Defender*, 5 May 1986, p. 13.

60. *Pittsburgh Courier* (nat. ed.), 16 May 1931, p. 10. At least two readers responded to "Call of the Wings." Ed Sanders of New Orleans proposed the establishment of a National Aviation Fund, chaired by Tuskegee president R. R. Moton, to support flying training among blacks. John W. Greene, Jr., of Boston called Levette's verses "horribly inaccurate," explaining that the real problem was not fear of flying but the high cost of flight instruction and the refusal of the army and navy to accept blacks as aviation cadets (*Pittsburgh Courier* [nat. ed.], 6 June 1931, sec. 2, p. 2).

61. Bunch, "In Search of a Dream," pp. 100–103. For a more critical account see Powell, *Black Wings*, pp. 155–77.

62. *Crisis* 40 (April 1933): 92; Powell, *Black Wings*, p. 176.

63. Mary J. Washington, "A Race Soars Upward," *Opportunity* 12 (October 1934): 300; Des Moines *Iowa Bystander*, 21 July 1933, p. 1, 28 July 1933, p. 1, and 4 August 1933, p. 1.

64. See chap. 1.

65. See chap. 1, and Scott, "Colonel John C. Robinson," pp. 59–69.

66. Enoch P. Waters, *American Diary: A Personal History of the Black Press* (Chicago: Path Press, 1987), pp. 196–97, 202.

67. Ibid., p. 195.

68. Ibid., pp. 196–97.

69. Chauncey E. Spencer, *Who is Chauncey Spencer?* (Detroit: Broadside Press, 1975), p. 140.

70. Ibid., p. 140; Waters, *American Diary*, p. 201. For a reprinted copy of the NAAA certificate of incorporation, dated 16 August 1939, see Spencer, *Who is Chauncey Spencer?* pp. 148–49; the year 1939 on the reprint of the certificate may be a printing error—Waters cites 1937 as the year the state charter was granted and Spencer's narrative seems to confirm Waters's

account. Concurrently with the formation of the NAAA, Coffey, Brown, and Waters established and obtained a state charter for a flying school under the name Coffey School of Aeronautics (Waters, *American Diary,* p. 204).

71. Waters, *American Diary,* p. 204.

72. *Crisis* 43 (June 1936): 189–90.

73. *Pittsburgh Courier* (nat. ed.), 6 June 1931, sec. 2, p. 2.

4. Civil Rights Emerges as a National Issue

1. Lee, *Employment of Negro Troops,* p. 55.

2. Sitkoff, *New Deal for Blacks.* The following discussion of black Americans and civil rights during the depression decade is based primarily on the work of Sitkoff. Other studies of the period that proved especially helpful include John B. Kirby, *Black Americans in the Roosevelt Era: Liberalism and Race,* Twentieth-Century America Series (Knoxville: Univ. of Tennessee Press, 1980); Nancy J. Weiss, *Farewell to the Party of Lincoln: Black Politics in the Age of FDR* (Princeton: Princeton Univ. Press, 1983); and Raymond Wolters, *Negroes and the Great Depression: The Problem of Economic Recovery,* Contributions in American History no. 6 (Westport, Conn.: Greenwood Publishing, 1970).

3. Sitkoff, *New Deal for Blacks,* p. 58.

4. Ibid., p. 59.

5. Ibid., p. 82.

6. Ibid., pp. 82–83.

7. Ibid., p. 10.

8. Ibid., pp. 30–31.

9. Ibid.

10. Franklin, *From Slavery to Freedom,* p. 385.

11. Ibid., p. 362.

12. Kirby, *Black Americans,* p. 98, quoting from the *New York Times,* 5 April 1931.

13. Sitkoff, *New Deal for Blacks,* p. 40, quoting *Chicago Defender,* 21 October 1932.

14. Sitkoff, *New Deal for Blacks,* p. 20.

15. Ibid., p. 41.

16. Raymond Wolters has examined the effect of the AAA and the NRA on black Americans in great detail in *Negroes and the Great Depression,* pp. 3–218.

17. Ibid., pp. 78–79.

18. Ibid., pp. 213–15.

19. Sitkoff, *New Deal for Blacks*, p. 55.

20. Ibid., p. 58.

21. Kirby, *Black Americans*, pp. 98–100.

22. Ibid., p. 155.

23. Weiss, *Farewell*, p. 181.

24. Ibid.

25. Ibid., p. 180, quoting *Time*, 17 August 1936, p. 10.

26. Sitkoff, *New Deal for Blacks*, pp. 139–215.

27. The role of Eleanor Roosevelt as black America's spokesperson in the administration is discussed in Kirby, *Black Americans*, pp. 76–96, and Weiss, *Farewell*, pp. 120–35.

28. Weiss, *Farewell*, pp. 136–56.

29. Sitkoff, *New Deal for Blacks*, pp. 88–89, 95.

30. Myrdal, *An American Dilemma*, p. 74.

31. Lee, *Employment of Negro Troops*, p. 55.

32. Gerald W. Patton, *War and Race: The Black Officer in the American Military, 1915–1941* (Westport, Conn.: Greenwood Press, 1981), p. 128.

33. Nalty, *Strength for the Fight*, pp. 128–29; Lee, *Employment of Negro Troops*, p. 25.

34. Lee, *Employment of Negro Troops*, p. 26

35. Nalty, *Strength for the Fight*, p. 129.

36. At least one black passed as white and joined the Air Corps during the interwar years; Reginald Thair enlisted in January 1931 and served in the Panama Canal Zone for three years (Osur, *Blacks in the Army Air Forces*, p. 150).

37. *Chicago Defender*, 12 May 1934, p. 12.

38. Franklin D. Roosevelt, *The Public Papers, Speeches, and Addresses of President Franklin D. Roosevelt*, edited by Samuel Rosenman (New York: Macmillan, 1938–50), 3: 111.

39. *Craftsmen Aero News* 1 (January 1937): 4–5; Claude A. Barnett to F. D. Patterson, with undated press release attached, 7 May 1936, Barnett Correspondence 1936–1942, LC 1937, Patterson Papers, TUA; *Opportunity* 15 (May 1937): 156.

40. John B. Kirby, ed., *New Deal Agencies and Black America in the 1930s* (Frederick, Md.: University Publications of America, 1983), 18: 251, 297, 319, 614–19 (containing a copy of *Negro Aviators*, 15 August 1936), and 742. A copy of the 15 August 1936 issue is also cataloged in the holdings of the Library of Congress. Issues for 1937, 1939, and 1940 are cited in Daniel W.

Lester, Sandra K. Faull, and Lorraine E. Lester, comps., *Cumulative Title Index to United States Public Documents, 1789–1976* (Arlington, Va.: United States Historical Documents Institute, 1979–82), 11: 194. Other copies of *Negro Aviators* have been found in the Tuskegee University Archives (January 1939 issue in U.S. Department of Commerce Folder, GC 1939, Patterson Papers, TUA) and the National Archives (September 1940 and April 1942 issues in Air Corps—General and Air Corps—Aviation Cadets Folders, Subject File, 1940–1947, Civilian Aide to the Secretary of War, Records of the Secretary of War, RG 107, NA [hereafter cited as Civilian Aide Records, RG 107, NA]).

41. Kirby, *New Deal Agencies*, 18: 325.

42. See for example the Des Moines *Iowa Bystander*, 15 May 1936, p. 2; *Crisis* 43 (June 1936): 189–90; and *Pittsburgh Courier*, 16 October 1937, in *Tuskegee Clippings File* 54: 591.

43. Von Hardesty and Dominick Pisano, *Black Wings: The American Black in Aviation* (Washington, D.C.: National Air and Space Museum, Smithsonian Institution, 1983), pp. 4–5; *Pittsburgh Courier*, 21 May 1938, in *Tuskegee Clippings File* 57: 382.

44. *Chicago Defender*, 28 May 1938, in *Tuskegee Clippings File* 57: 382. Several months after Nash and Cable made their flights, Earnest L. Gayden of Topeka, Kansas, asked President Roosevelt to support an airmail flight by another black pilot, a request which the president's secretary declined; see Gayden to Roosevelt, 31 August 1938, and Stephen Early to Gayden, 2 September 1938, Official File (hereafter cited as OF) 93, FDRL.

45. *Papers of the NAACP*, Part 1, *Meetings of the Board of Directors, Records of Annual Conferences, Major Speeches, and Special Reports, 1909–1950* (Frederick, Md.: University Publications of America, 1981), 9: 939–43.

46. Foner, *Blacks and the Military*, p. 130, quoting Houston.

47. For a full account of Vann's activities as a journalist, businessman, and politician, see Buni, *Robert L. Vann*. Except where noted, chapter 12 ("Twilight: 1938–1940," pp. 299–324) of Buni's study forms the basis for the following account of the *Courier's* armed forces participation campaign.

48. William Manchester, *The Glory and the Dream: A Narrative History of America, 1932–1972* (New York: Bantam Books, 1974), p. 175.

49. Wilkins to White, 26 April 1938, quoted in Buni, *Robert L. Vann*, p. 307, emphasis added.

50. *Pittsburgh Courier*, 19 February 1938, in *Tuskegee Clippings File* 60: 164.

51. Ibid.

52. Vann survey letter, quoted in full in Buni, *Robert L. Vann*, pp. 303–4.

53. Roy Wilkins to Walter White, 26 April 1938, quoted in Buni, *Robert L. Vann*, pp. 307–8.

54. Wesley Frank Craven and James Lea Cate, eds., *The Army Air Forces in World War II* (Chicago: Univ. of Chicago Press, 1948–58; reprint, Washington, D.C.: Office of Air Force History, 1983), vol. 6, *Men and Planes*, 171–72.

55. *Pittsburgh Courier*, 3 December 1938, in *Tuskegee Clippings File* 57: 376.

56. *Atlanta Daily World*, 8 December 1938, in *Tuskegee Clippings File* 60: 175.

57. *Chicago Defender*, 10 December 1938, in *Tuskegee Clippings File* 57: 378.

58. E. Franklin Frazier, *Negro Youth at the Crossways: Their Personality Development in the Middle States* (Washington, D.C.: American Council on Education, 1940), pp. 165–66.

5. The Aviation Legislation of 1939

1. Williamson Murray, *Luftwaffe* (Baltimore, Md.: The Nautical and Aviation Publishing Company of America, 1985), pp. 19–20; Craven and Cate, *Army Air Forces in World War II*, 6: vi–vii, 7–11.

2. Craven and Cate, *Army Air Forces in World War II*, 6: 10; Michael S. Sherry, *The Rise of American Air Power: The Creation of Armageddon* (New Haven: Yale Univ. Press, 1987), p. 82.

3. Quoted in Craven and Cate, *Army Air Forces in World War II*, 6: 10.

4. Quoted in ibid.

5. Roosevelt, *Public Papers and Addresses*, 8: 1–12.

6. Ibid., 8: 51–52.

7. Ibid., 8: 71–72.

8. Public Law 18, *Statutes at Large*, 53: 555–60 (1939); Civilian Pilot Training Act, *Statutes at Large*, 53: 855–56 (1939).

9. No satisfactory full-length study of the Civilian Pilot Training Program has yet been published. The best source currently available is Dominick A. Pisano, "A Brief History of the Civilian Pilot Training Program, 1939–1944," *National Air and Space Museum Research Report, 1986* (Washington, D.C.: Smithsonian Institution Press, 1986), pp. 21–41, which is a synopsis of the author's research for a forthcoming doctoral dissertation. Less reliable is Patricia Strickland, *The Putt-Putt Air Force: The Story of the Civilian Pilot Training Program and the War Training Service (1939–1944)* (Washington, D.C.: U.S. Dept. of Transportation, Federal Aviation Administration, [1971]). See also Robert H. Hinckley and JoAnn J. Wells, *"I'd Rather Be Born Lucky than Rich": The Autobiography of Robert H. Hinckley*, Charles Redd Monographs in Western History, no. 7 (Provo, Utah: Brigham Young Univ. Press, 1977), pp. 79–85, 92–98, and John R. M. Wilson, *Turbulence Aloft: The Civil Aeronautics Administration amid Wars and Rumors of Wars* (Washington, D.C.: U.S. Dept. of Transportation, Federal Aviation Administration, 1979), pp. 96–106.

10. *Complete Presidential Press Conferences of Franklin D. Roosevelt,* with an introduction by Jonathan Daniels (New York: Da Capo Press, 1972), 12: 319–21. A copy of the CAA press release of 26 December 1938 on the Civilian Pilot Training Program is attached to an unsigned Memorandum to the President, 21 December 1938, OF 2955, 1938 Folder, FDRL.

11. Pisano, "Brief History of the Civilian Pilot Training Program," p. 7.

12. The first bill relating to the Civilian Pilot Training Program was H.R. 5093, introduced by Rep. Clarence F. Lea on 16 March 1939. Following hearings on H.R. 5093, Lea introduced another bill, H.R. 5619, on 6 April, which was debated in the House and ultimately approved as the Civilian Pilot Training Act. Similar legislation was also introduced in the Senate (S. 2119) on 8 April by Sen. Pat McCarran, but ultimately H.R. 5619 was passed in lieu of S. 2119, (*Congressional Record,* 76th Cong., 1st sess., pp. 2899 [16 March 1939], 3942 [6 April 1939]; *Congressional Record,* 76th Cong., 1st sess., p. 3977 [8 April 1939]).

13. U.S. Congress, House of Representatives, *Training of Civil Aircraft Pilots: Hearings Before the Committee on Interstate and Foreign Commerce, House of Representatives, on H.R. 50[9]3,* 76th Cong., 1st sess., 20 and 27 March 1939, p. 7. The H.R. number was erroneously printed as 5073 on the cover.

14. Ibid., p. 22.

15. See, for example, ibid., pp. 7, 10, 15, 21–22, 26, 42–53, 60–62.

16. Pisano, "Brief History of the Civilian Pilot Training Program," p. 25.

17. U.S. Congress, Senate, *Hearings before the Committee on Military Affairs, United States Senate on H.R. 3791.*

18. *Congressional Record,* 76th Cong., 1st sess., p. 1222 (7 February 1939); "Legislation Relating to the Army Air Forces Training Program, 1939–1945," rev. ed., Army Air Forces Historical Studies no. 7 (Headquarters, Army Air Forces, April 1946), p. 20, #101–7B, USAF Collection, USAF-HRC.

19. Craven and Cate, *Army Air Forces in World War II,* 6: 172–73.

20. Public Law 18, *Statutes at Large,* 53: 556–57 (1939).

21. Craven and Cate, *Army Air Forces in World War II,* 6: 10.

22. Ibid., 6: 428.

23. Ibid., 6: 454. One of the problems that the student of World War II flight training encounters is the profusion of terms for the various phases of civilian and military training. Adding to the difficulty is the tendency on the part of the participants to use many of the terms loosely and/or interchangeably; they understood the meaning, but the researcher examining the documents decades later often finds the loose terminology confusing. The following is a list of terms as they are used in this study. The list is based on my best judgment as to the most commonly accepted term for each phase of training. Alternate terms are included.

Civilian Pilot Training Program (this became the War Training Service in 1942 and supported various aspects of army and navy training):

Elementary training. The first phase of CPT Program training, which led to a private pilot's license. Sometimes called primary training or the Private Pilot Training Course.

Advanced training. All CPT Program training beyond elementary training; not to be confused with army advanced training, described below. Includes the following courses of instruction: *Secondary; Cross-country; Instrument;* and *Apprentice instructor.*

[Note: CAA officials frequently asserted that students who completed elementary and secondary training had received the equivalent of army primary training. Although army officials were skeptical of this claim, pointing out that they had no control over the training standards, some students who had completed CPT Program elementary and secondary were allowed to bypass army primary training and were entered directly into army basic.]

Army flight training:

Primary training. The first phase of army training. Until mid–1939 administered at the Air Corps Training Center in San Antonio by Air Corps flight instructors; subsequently, Air Corps aviation cadets took their primary training at army-supervised civilian flying schools from civilian flying instructors. Sometimes called elementary training, but not to be confused with CPT Program elementary training.

Basic training. The second phase of army training. Administered by Air Corps instructors at military flying fields.

Advanced training. The third phase of army training; not to be confused with CPT Program advanced training, described above. Administered by Air Corps instructors at military flying fields. On completion of this training aviation cadets received their wings and became commissioned officers or flight officers.

Transition training. The fourth phase of army training, in which the new pilot learned to fly a combat aircraft.

24. Craven and Cate, *Army Air Forces in World War II,* 6: 455; U.S. Congress, Senate, *Hearings before the Committee on Military Affairs, United States Senate on H.R. 3791,* pp. 39–40.

25. Roosevelt, *Public Papers and Addresses,* 8: 73, emphasis added.

26. File memorandum, M. H. McIntyre, Secretary to the President, 13 January 1939, referring letter from William R. Johnson, North Carolina State Board of Charities and Public Welfare, Raleigh, N.C., to Franklin D. Roosevelt, 10 January 1939, to Chairman, CAA, OF 249, January-June 1939 Folder, FDRL.

27. Helene Weissman to Roosevelt, 4 January 1938 [*sic,* 1939], OF 93, 1936–1937 Folder, FDRL.

28. William J. Trent, Jr., to McIntyre, with enclosures, 31 January 1939, and memorandum, R. B. to M. H. M[cIntyre], 3 February 1939, OF 93, 1936–1937 Folder, FDRL.

29. D. Ormande Walker to Roosevelt, 23 January 1939, January–June 1939 Folder, OF 249, FDRL, emphasis added.

30. Arthur Howe to Roosevelt, 18 January 1939, with enclosed letter, Grady P. Anderson to Roosevelt, 17 January 1939, OF 249, January-June 1939 Folder, FDRL.

31. Grove Webster, Chief, Private Flying Development Division, CAA to Trent, 26 January 1939, OF 93, 1936–1937 Folder, FDRL; Robert Hinckley to Col. Edwin M. Watson, 30 January 1939, OF 249, January–June 1939 Folder, FDRL.

32. McIntyre to Howe, 21 February 1939, OF 249, January–June 1939 Folder, FDRL.

33. McIntyre to Walker, 21 February 1939, OF 249, January–June 1939 Folder, FDRL.

34. Brown specifically mentioned the interest of Scott and Cable in his testimony; see U.S. Congress, Senate, *Hearings before the Committee on Military Affairs, United States Senate on H.R. 3791,* p. 316.

35. Ibid., pp. 311–12.

36. *Congressional Record,* 76th Cong., 1st sess., pp. 1737 (22 February), 2367 (7 March 1939).

37. Ibid., pp. 2055 (1 March 1939), 2368 (7 March 1939).

38. Ibid., p. 2369 (7 March 1939).

39. Ibid., pp. 2367–68 (7 March 1939).

40. U.S. Congress, House of Representatives, *Supplemental Military Appropriations Bill for 1940: Hearings Before the Subcommittee on Appropriations, House of Representatives, on the Supplemental Appropriation Bill for 1940 [H.R. 6791],* 76th Cong., 1st sess., 16 May–5 June 1939, "Statement of Hon. H. H. Schwartz a United States Senator from the State of Wyoming," pp. 342–44.

41. *Congressional Record,* 76th Cong., 1st sess., p. 2367 (7 March 1939).

42. Ibid., p. 2368 (7 March 1939).

43. Bridges initially offered his amendment as a substitute for the Schwartz amendment. Senator Miller objected, pointing out that the Bridges amendment also contained provisos prohibiting discrimination on the basis of race, creed, or color that were not included in the Schwartz amendment, which only addressed the issue of schools for black pilot trainees. The presiding officer ruled that the Bridges amendment was not acceptable as a

substitute, but could be offered as an amendment to the Schwartz amendment, and it was on this basis that it was considered by the Senate (*Congressional Record*, 76th Cong., 1st sess., pp. 2368–69 [7 March 1939]). It is not clear from the proceedings whether the tabling of the Bridges amendment as soon as it was offered, its rejection as a substitute, and its subsequent acceptance as an amendment to the amendment were merely parliamentary formalities or matters of greater consequence. Bridges later commented that the maneuvers which resulted in the Schwartz amendment taking precedence over his amendment were simply "to allow a Democrat to put it through purely as a matter of congressional courtesy" (U.S. Congress, Senate, *Military Establishment Appropriation Bill for 1941: Hearings Before the Subcommittee of the Committee on Appropriations, United States Senate . . . on H.R. 9209*, 76th Cong., 3rd sess., 30 April–17 May 1940, p. 368).

44. *Congressional Record*, 76th Cong., 1st sess., pp. 2369–70 (7 March 1939).

45. Ibid., p. 2370.

46. Ibid.

47. For statements, discussions, and misunderstandings regarding the Air Corps plans to use civilian flying schools and the CAA plan for pilot training in colleges and universities, see U.S. Congress, Senate, *Hearings before the Committee on Military Affairs, United States Senate on H.R. 3791*, pp. 39–40, 54–58, 150–56, 159–60.

48. Ibid.

49. *Congressional Record*, 76th Cong., 1st sess., pp. 2367–70 (7 March 1939).

50. Memorandum, Lt. Col. Hume Peabody, Chief, Plans Section, Office of the Chief of the Air Corps, to Chief of the Air Corps, 3 March 1939, quoted in "Legislation Relating to the Army Air Forces Training Program, 1939–1945," pp. 19–20, 101–7B, USAF Collection, USAF-HRC.

51. Harry H. Woodring to Sen. Morris Sheppard and Rep. Andrew H. May, 10 March 1939, 101–7B, USAF Collection, USAF-HRC.

52. *Congressional Record*, 76th Cong., 1st sess., pp. 2449, 2463 (8 March 1939).

53. Ibid., p. 2824 (16 March 1939), emphasis added.

54. Ibid., pp. 3106 and 3132 (22 March 1939), 3276 (24 March 1939), 3319 (25 March 1939), and 4235 (13 April 1939); Public Law 18, *Statutes at Large*, 53: 555–60 (1939).

55. Ibid., pp. 2370–71 (7 March 1939), 2349 and 2363 (8 March 1939), 2824–25 (16 March 1939), 3131–32 (22 March 1939), emphasis added.

56. Routing and referral sheet, Chief of Plans Section to Chief of the Air Corps, 8 April 1939, quoted in Lee, *Employment of Negro Troops*, p. 57.

57. Ibid.

58. Ibid., emphasis added.

59. Lee, *Employment of Negro Troops*, p. 58, quoting the judge advocate general's opinion.

60. Lee, *Employment of Negro Troops*, p. 58; U.S. Congress, House of Representatives, *Supplemental Appropriation Bill for 1940: Hearings on [H.R. 6791]*, pp. 37, 344.

61. U.S. Congress, Senate, *Military Establishment Appropriation Bill for 1940: Hearings Before the Subcommittee of the Committee on Appropriations on H.R. 4630*, 76th Cong., 1st sess., 13–17 March 1939, p. 153.

62. For Brown's statement, together with the statements of J. Finley Wilson and Sen. H. H. Schwartz, regarding a special appropriation for the training of black pilots for the Air Corps, see U.S. Congress, House of Representatives, *Supplemental Appropriation Bill for 1940: Hearings on [H.R. 6791]*, pp. 339–44.

63. Ibid., p. 339.

64. Edgar G. Brown to Rep. Albert J. Engel, 1 June 1939, included in the *Congressional Record*, 76th, Cong., 1st sess., appendix, p. 2421.

65. U.S. Congress, House of Representatives, *Supplemental Appropriation Bill for 1940: Hearings on [H.R. 6791]*, p. 340. Brown's reference to blacks having been trained previously at the North Suburban Flying Corporation is unclear. The most likely conclusion is that the North Suburban Flying Corporation was the old Curtiss Flying School where John C. Robinson and other blacks learned to fly in the early 1930s. This conclusion is supported by the fact that the school was located at the Curtiss Airport (see ibid., p. 37). The name of the school at Glenview apparently changed with some regularity, as it is listed as the Chicago School of Aeronautics in Craven and Cate, *Army Air Forces in World War II*, 6: 456. See also Willard Wiener, *Two Hundred Thousand Flyers: The Story of the Civilian-AAF Pilot Training Program* (Washington, D.C.: The Infantry Journal, 1945), pp. 162–63.

66. U.S. Congress, House of Representatives, *Supplemental Appropriation Bill for 1940: Hearings on [H.R. 6791]*, pp. 339–40.

67. Ibid., p. 342–44.

68. *Congressional Record*, 76th Cong., 1st sess., p. 7649 (21 June 1939).

69. Preliminary Report [regarding black voting patterns and success of black Democratic candidates in 1938 elections], Edgar G. Brown to Irvin H. McDuffie, valet to President Roosevelt, n.d. (ca. November 1938), OF 93, 1936–1937 Folder, FDRL.

70. *Congressional Record*, 76th Cong., 1st sess., pp. 7666–67 (21 June 1939).

71. Ibid., p. 7667.

72. Ibid., p. 7668.

73. Ibid., pp. 7667–68.

74. Ibid., p. 7668. Dirksen's assertion that there were 350 black pilots in

the country does not correspond to Department of Commerce figures on licensed black pilots; according to the January 1939 issue of *Negro Aviators*, only 125 blacks held pilots licenses or were student pilots.

75. *Congressional Record*, 76th Cong., 1st sess., p. 7668 (21 June 1939).

76. Ibid., pp. 7668–69.

77. Ibid., p. 7669.

78. Ibid.

79. Ibid., pp. 7669, 7672–73, 7724.

80. Pisano, "A Brief History of the Civilian Pilot Training Program," pp. 25, 30.

81. U.S. Congress, House of Representatives, *Training of Civil Aircraft Pilots: Hearings on H.R. 50[9]3*; U.S. Congress, Senate, *Training of Civil Aircraft Pilots: Hearing Before a Subcommittee of the Committee on Commerce, United States Senate, on S. 2119*, 76th Cong., 1st sess., 20 April 1939.

82. *Congressional Record*, 76th Cong., 1st sess., p. 4490 (19 April 1939).

83. Ibid.

84. Ibid.

85. Waters, *American Diary*, pp. 205–7; Spencer, *Who Is Chauncey Spencer?* pp. 31–35, 140–44.

86. Spencer, *Who Is Chauncey Spencer?* p. 34.

87. *Congressional Record*, 76th Cong., 1st sess. pp. 7210 (15 June 1939), 7504 (19 June 1939).

88. Civilian Pilot Training Act, *Statutes at Large*, 53: 856 (1939).

6. The Civilian Pilot Training Program at Tuskegee

1. Cornelius R. Coffey to Patterson, 27 January 1939, Co–Cy Folder, GC 1939, Patterson Papers, TUA.

2. Patterson to Coffey, 30 January 1939, Co–Cy Folder, GC 1939, Patterson Papers, TUA.

3. Patterson to Robert H. Hinckley and Patterson to Harry H. Woodring, 22 March 1939, Civil Aeronautics Authority Folder, GC 1939, Patterson Papers, TUA. The reference to landing areas was an exaggeration; one of the greatest problems Tuskegee encountered in establishing its aviation program was the lack of an adequate airfield in the vicinity. Perhaps Patterson was alluding to the availability of land around Tuskegee, but the school possessed no improved landing areas at the time nor did the town of Tuskegee own a municipal airport. The only facility in the area was a private field south of the town that was used by three local white fliers. The institute leased this field the following year.

4. The Adjutant General's Office to Patterson, 29 March 1939, and Hinckley to Patterson, 29 March 1939, Civil Aeronautics Authority Folder, GC 1939, Patterson Papers, TUA.

5. Robinson to Patterson, 1 April 1939, Civil Aeronautics Authority Folder, GC 1939, Patterson Papers, TUA.

6. Patterson to Robinson, 8 April 1939, Civil Aeronautics Authority Folder, GC 1939, Patterson Papers, TUA.

7. *Campus Digest: The Voice of the Tuskegee Student,* 20 May 1939.

8. Ibid.

9. Ibid.; Norfolk *Journal and Guide,* 15 April 1939; Anderson interview, pp. 53–58; George L. Washington, *The History of Military and Civilian Pilot Training of Negroes at Tuskegee, Alabama, 1939–1945* (Washington, D.C.: George L. Washington, 1972), p. 7. The elder Wright first sought to capitalize on the commercial possibilities of aviation when he arranged for an exhibition flight at the 1911 Georgia State Fair (see chapter 3 above). By 1942 another son of Wright's, Bishop R. R. Wright, Jr., had assumed the presidency of Wilberforce University and continued that institution's campaign to secure an army flight training facility on campus.

10. Washington, *History of Military and Civilian Pilot Training at Tuskegee,* p. 7.

11. Robinson to G. L. Washington, 23 May 1939, and Robinson to F. D. Patterson, 29 May 1939, Ro–Ry Folder, GC 1939, Patterson Papers, TUA.

12. Robinson to Patterson, 29 May 1939, Ro–Ry Folder, GC 1939, Patterson Papers, TUA.

13. Ibid.

14. Ibid.

15. Washington to Robinson, 7 June 1939, G. L. Washington Folder, LC 1939, Patterson Papers, TUA.

16. T. M. Campbell to Washington, 4 January 1939, George L. Washington—Civilian Pilot Training Folder, Box 17, Thomas Monroe Campbell Papers, TUA; Washington to Patterson, 6 January 1939, G. L. Washington Folder, LC 1939, Patterson Papers, TUA.

17. Mary McLeod Bethune to Patterson, 2 June 1939, National Youth Administration, Mary McLeod Bethune Folder, GC 1939, Patterson Papers, TUA.

18. Patterson to J. E. Bryan, State Director, NYA for Alabama, 8 June 1939, National Youth Administration, Alabama, Folder, GC 1939, Patterson Papers, TUA.

19. B. L. Balch, Deputy State Director, NYA for Alabama, to Patterson, 10 June and 29 June 1939, National Youth Administration, Alabama, Folder, GC 1939, Patterson Papers, TUA. The NYA officials in West Virginia were

more knowledgeable, or perhaps more cooperative. In November Patterson again wrote to the Alabama NYA director and enclosed a four-page press release that described an NYA-supported aviation mechanics training program at another black school, West Virginia State College. The response he received was not encouraging (see Patterson to Bryan, 4 November 1939, Bryan to Patterson, 14 November 1939, and Patterson to Bryan, 29 November 1939, all in the National Youth Administration, Alabama, Folder, GC 1939, Patterson Papers, TUA).

20. Anderson to Burke, 31 August 1939, Civil Aeronautics Authority Folder, GC 1939, Patterson Papers, TUA.

21. Telegram, James C. Evans to Washington, 13 September 1939, G. L. Washington Folder, LC 1939, Patterson Papers, TUA.

22. Washington to Patterson, 13 September 1939, G. L. Washington Folder, LC 1939, Patterson Papers, TUA.

23. Telegrams, Patterson to Civil Aeronautics Authority, 12 September 1939, and Patterson to Grove Webster, 15 September 1939, Civil Aeronautics Authority Folder, GC 1939, Patterson Papers, TUA.

24. Telegram, Webster to Patterson, 14 September 1939, Civil Aeronautics Authority Folder, GC 1939, Patterson Papers, TUA.

25. Webster to Patterson, 19 and 25 September 1939, Patterson to Webster, 21 September 1939, Civil Aeronautics Authority Folder, GC 1939, Patterson Papers, TUA.

26. Interview with Joseph Wren Allen, Montgomery, Alabama, 24 October 1985.

27. Ibid.

28. *Air Commerce Bulletin* 10 (15 April 1939): 265.

29. Washington, *History of Military and Civilian Pilot Training at Tuskegee*, p. 6; Allen interview, 24 October 1985. Washington's account makes no mention of Patterson's contact with Allen, implying that he made the arrangement with Allen himself. Allen, however, recalled clearly that the initial contact was with Patterson, not with Washington.

30. Barnett to Patterson, 20 September 1939, Barnett Correspondence 1936–1942, LC 1937, Patterson Papers, TUA.

31. Patterson to Barnett, 25 September 1939, Barnett Correspondence 1936–1942, LC 1937, Patterson Papers, TUA.

32. Patterson to Woodring, 4 October 1939, and Woodring to Patterson, 13 October 1939, ROTC Folder, GC 1939, Patterson Papers, TUA.

33. Patterson to Hinckley, 16 October 1939, Civil Aeronautics Authority Folder, GC 1939, Patterson Papers, TUA.

34. Washington, *History of Military and Civilian Pilot Training at Tuskegee*, pp. 15–16, quoting portions of Brown's 6 October 1939 letter to Patterson.

35. Ibid., pp. 16–18.

36. Ibid.; unsigned memorandum, 9 October 1939, and letters, Patterson to Gayle, 16 October 1939, and Walker to Gayle, 21 October 1939, Civil Aeronautics Authority Folder, GC 1939, and telegram, Washington to Patterson, 10 October 1939, G. L. Washington Folder, LC 1939, Patterson Papers, TUA.

37. Hinckley to Patterson, 13 October 1939, Civil Aeronautics Authority Folder, GC 1939, Patterson Papers, TUA.

38. "Tuskegee Institute Civilian Pilot Training Program: Preliminary Announcement," 17 October 1939, G. L. Washington Folder, LC 1940, Patterson Papers, TUA; "Civilian Pilot Training Program: Release No. 2," 19 October [1939], Civil Aeronautics Authority Folder, GC 1939, Patterson Papers, TUA; "Civilian Pilot Training Program, Tuskegee Institute: Release Number 3," 1 November 1939, Civil Aeronautics Authority Folder, GC 1939, Patterson Papers, TUA.

39. See note 38 above.

40. See note 38 above.

41. *The Campus Digest: The Voice of the Tuskegee Student*, 28 October 1939.

42. Memorandum, Washington to Patterson, 28 October 1939, Civil Aeronautics Authority Folder, GC 1939, Patterson Papers, TUA.

43. Ibid.

44. Ibid. G. L. Washington, in his *History of Military and Civilian Pilot Training at Tuskegee*, p. 34, maintains that the scholarship campaign was instituted later, probably for the second CPT class; correspondence and memoranda in the Patterson Papers (Civil Aeronautics Authority Folder, GC 1939) clearly show, however, that the campaign was established in late October 1939 for the first class. The average weekly earnings of American employees in 1939 amounted to $24.31, according to U.S. Bureau of the Census, *Statistical History of the United States From Colonial Times to the Present* (New York: Basic Books, 1976), "Average Annual Earnings of Employees: 1900–1970," Series D, 722–727, p. 164.

45. For a sample of the form letter addressed to Patterson and dated 25 October 1939, see Civil Aeronautics Authority Folder, GC 1939, Patterson Papers, TUA.

46. "Civilian Pilot Training Program Scholarship Drive Report, As of November 28, 1939. Reported by: G. L. Washington," Civil Aeronautics Authority Folder, GC 1939, Patterson Papers, TUA.

47. Harlan, *The Making of a Black Leader*, p. 140.

48. The decision to reject Robinson's offer to serve as Tuskegee's aviation extension school in Chicago may have been a product of the same sentiments; even though Robinson was black, the off-campus location of his

school would have made it difficult for Tuskegee administrators in Alabama to exercise control over his activities.

49. Eldridge Adams, M. D., Chief, CAA Medical Section to Patterson, 19 October 1939, with handwritten note from Patterson: "Mr. Washington[:] Dr. Cary or Dr. Branche"; Washington to Adams, 24 October 1939; and Adams to Washington, 31 October 1939; all in Civil Aeronautics Authority Folder, GC 1939, Patterson Papers, TUA.

50. "Tentative Faculty and Class Assignments, October 17, 1939," prepared by G. L. Washington in his capacity as director, Civilian Pilot Training Program, Civil Aeronautics Authority Folder, GC 1939, Patterson Papers, TUA; Washington, *History of Military and Civilian Pilot Training at Tuskegee*, pp. 22–23.

51. Washington, *History of Military and Civilian Pilot Training at Tuskegee*, p. 22.

52. Ibid., pp. 22–23.

53. Interview with Prof. Robert G. Pitts, Auburn, Alabama, 17 February 1985.

54. Ibid.; Allen interview, 24 October 1985.

55. Washington to Bloomfield M. Cornell, 28 November 1939, and Washington to Patterson, E. H. Burke, Comptroller, and Lloyd Isaacs, Treasurer (with Civilian Pilot Training Program Budget attached), 18 October 1939, Civil Aeronautics Authority Folder, GC 1939, Patterson Papers, TUA.

56. Washington, *History of Military and Civilian Pilot Training at Tuskegee*, p. 18.

57. Washington to Cornell, 28 November 1939, and Washington to Patterson, E. H. Burke, Comptroller, and Lloyd Isaacs, Treasurer (with Civilian Pilot Training Program Budget attached), 18 October 1939, Civil Aeronautics Authority Folder, GC 1939, Patterson Papers, TUA.

58. Allen interview, 24 October 1985.

59. Washington, *History of Military and Civilian Pilot Training at Tuskegee*, p. 27.

60. Grove Webster to Alabama Air Service, 22 December 1939, 836.211, Alabama Air Service Folder, Box 384, General Files, Records of the CAA, RG 237, NA (hereafter cited as CAA General Files, RG 237, NA). On 29 November Allen had received contract approval for twenty CPT students at API; see Webster to Alabama Air Service, 29 November 1939, 836.211, Alabama Air Service Folder, Box 384, CAA General Files, RG 237, NA.

61. Allen interview, 24 October 1985; Washington, *History of Military and Civilian Pilot Training at Tuskegee*, p. 32.

62. Allen interview, 24 October 1985; Washington, *History of Military and Civilian Pilot Training at Tuskegee*, p. 32.

63. Washington to Walker, 2 February 1940, Civil Aeronautics Authority Folder, LC 1940, Patterson Papers, TUA.

64. Ibid.

65. Interview with Forrest Shelton, Atlanta, Georgia, 29 October 1985. According to Shelton, Kennedy Field, built around 1938–39, was originally known as SK&W Field after the initials of the three young fliers who constructed it. It came to be known as Kennedy Field once Tuskegee began operations there, apparently because it was under the control of Stanley Kennedy's father, who leased the property from the owner, John Connor of Tallassee, Alabama.

66. Ibid.

67. "Report of Flying Field Investigation, Tuskegee, Institute," 31 January–1 February 1939, G. L. Washington Folder, LC 1940, Patterson Papers, TUA; Washington, *History of Military and Civilian Pilot Training at Tuskegee*, pp. 37–40.

68. Washington to Walker, 2 February 1939, Civil Aeronautics Authority Folder, LC 1940, Patterson Papers, TUA.

69. Edward C. Nilson to Webster, 3 February 1940, copy in G. L. Washington Folder, LC 1940, Patterson Papers, TUA.

70. Washington, *History of Military and Civilian Pilot Training at Tuskegee*, p. 41.

71. Ibid., p. 43.

72. Ibid., pp. 43–44.

73. W. M. Robertson, Senior Aeronautical Inspector, Atlanta Municipal Airport, to Washington, 27 March 1940, Civil Aeronautics Authority Folder, LC 1940, Patterson Papers, TUA.

74. Washington, *History of Military and Civilian Pilot Training at Tuskegee*, p. 45.

75. *Chattanooga News-Free Press*, 21 April 1940, quoted in Washington, *History of Military and Civilian Pilot Training at Tuskegee*, pp. 44–45.

76. Davis to Patterson, 18 April 1940, D Folder, GC 1940, Patterson Papers, TUA.

77. Telegram, Patterson to Davis, 24 April 1940, D Folder, GC 1940, Patterson Papers, TUA.

78. Allen interview, 24 October 1985.

79. Washington, *History of Military and Civilian Pilot Training at Tuskegee*, p. 45.

80. Allen interview, 24 October 1985.

81. Washington, *History of Military and Civilian Pilot Training at Tuskegee*, pp. 2, 51.

82. G. L. Washington to the Institute of Aeronautical Sciences, New York, New York, 31 July 1940, G. L. Washington Folder, LC 1940, Patterson Papers, TUA; Washington, *History of Military and Civilian Pilot Training at Tuskegee*, pp. 54–55.

83. Allen telephone interview, 26 February 1987; Shelton interview, 29 October 1985.

84. Anderson interview, pp. 72–73.

85. G. L. Washington to the Institute of Aeronautical Sciences, New York, New York, 31 July 1940, G. L. Washington Folder, LC 1940, Patterson Papers, TUA; Washington, *History of Military and Civilian Pilot Training at Tuskegee*, pp. 54–55.

7. Tuskegee Emerges as the Center of Black Aviation

1. Washington to Asa Roundtree [*sic*], Jr., n.d. [ca. 22 January 1940], Civil Aeronautics Authority Folder, LC 1940, Patterson Papers, TUA.

2. Ibid.

3. Washington to Patterson, 22 January 1940, Civil Aeronautics Authority Folder, LC 1940, Patterson Papers, TUA.

4. Ibid.

5. Rountree to Washington, 24 January 1940, Civil Aeronautics Authority Folder, LC 1940, Patterson Papers, TUA.

6. "Report of Flying Field Investigation, Tuskegee Institute," 31 January–1 February 1940, G. L. Washington Folder, LC 1940, Patterson Papers, TUA; Washington, *History of Military and Civilian Pilot Training at Tuskegee*, pp. 37–40. The meeting regarding the selection of a site for Tuskegee's airfield was one of a series of meetings held on 31 January and 1 February 1940. Locating a site for a permanent airfield for Tuskegee Institute was one of two major items under discussion. The other was finding a local flying field that the institute could use in the interim, and it was at this time that the feasibility of conducting CPT operations at Kennedy Field was considered; these developments are discussed in detail in the preceding chapter.

7. "Report of Flying Field Investigation, Tuskegee Institute," 31 January–1 February 1940, G. L. Washington Folder, LC 1940, Patterson Papers, TUA; Washington, *History of Military and Civilian Pilot Training at Tuskegee*, pp. 37–40.

8. Owen Draper to Washington, 3 April 1940, copy in G. L. Washington Folder, LC 1940, Patterson Papers, TUA. Examination of the itemized cost

analysis figures provided by Draper shows that the estimated costs actually amounted to $33,900; due to an arithmetical error, Draper reported the total estimated costs at $22,900. Washington did not detect the error and based his fund-raising campaign on the $22,900 figure. See also Washington, *History of Military and Civilian Pilot Training at Tuskegee*, pp. 46–47.

9. Washington to Alumni Association, Board of Trustees, and Patterson, 5 April 1940, with attached statement on status of aviation training at Tuskegee Institute, G. L. Washington Folder, LC 1940, Patterson Papers, TUA.

10. Ibid. The passage quoted is one of the few instances in which Washington alluded to the threat of white interference; in public statements he frequently cited the support of local whites as a key factor in the success of the aviation program. No evidence has been found of any racial incidents in connection with the flight training at Montgomery. Washington later commented that he did not recall any reports of unpleasant incidents at the municipal airport in Montgomery, and the only problem J. W. Allen recollected was a single derogatory remark by a transient Eastern Airlines pilot (Washington, *History of Military and Civilian Pilot Training at Tuskegee*, p. 34; telephone interview with J. W. Allen, 26 February 1987). The concern Washington expressed to the Board of Trustees in April 1940 may have nevertheless been sincere, as there is evidence which suggests that some prominent local whites believed that federal officials were predisposed to give Tuskegee's flight training program priority. Robert G. Pitts, one of the Auburn professors who taught the ground school for the early CPT classes, recalled that API president L. N. Duncan, concerned that Tuskegee might be given control of the airport at Auburn, sent Pitts and Cornell to Washington, D.C., with instructions to lobby against such action (Pitts interview, 17 February 1985).

11. Apparently, Washington and Neely's proposal was never formally presented to the Board of Trustees, based on the minutes of the 6 April meeting, which contain no mention of the airfield project (*Minutes of the Annual Meeting of the Board of Trustees of the Tuskegee Institute, April 6, 1940*).

12. Washington to Alumni Association, Board of Trustees, and Patterson, 5 April 1940, with attached statement on status of aviation training at Tuskegee Institute, G. L. Washington Folder, LC 1940, Patterson Papers, TUA.

13. Patterson to Harry H. Woodring, Secretary of War, 4 October 1939, R.O.T.C. Folder, GC 1939, Patterson Papers, TUA.

14. Patterson to J. E. Bryan, State Administrator, NYA for Alabama, 4 November 1939, and Bryan to Patterson, 14 November 1939, National Youth Administration, Alabama, Folder, GC 1939, Patterson Papers, TUA.

15. Nilson to Washington, 3 February 1940, and Washington to Patterson, 7 February 1940, Civil Aeronautics Authority Folder, LC 1940, Patterson Papers, TUA.

16. Nilson to Grove Webster, Chief, Private Flying Development Division, CAA, 3 February 1940, copy in G. L. Washington Folder, LC 1940, Patterson Papers, TUA.

17. Allen to Washington, 20 April 1940, G. L. Washington Folder, LC 1940, Patterson Papers, TUA; Washington to Patterson, 15 April 1940, Civil Aeronautics Authority Folder, LC 1940, Patterson Papers, TUA; Allen interview, 24 October 1985.

18. Washington to Patterson, 15 April 1940, Civil Aeronautics Authority Folder, LC 1940, Patterson Papers, TUA.

19. Ibid.

20. Patterson to Jackson Davis, Associate Director, General Education Board, 25 May 1940, General Education Board: Jackson Davis Folder, GC 1940, Patterson Papers, TUA.

21. Davis to Patterson, 29 May 1940, General Education Board: Jackson Davis Folder, GC 1940, Patterson Papers, TUA.

22. Washington to Patterson, 25 October 1939, Civil Aeronautics Authority Folder, GC 1939, Patterson Papers, TUA.

23. Washington, *History of Military and Civilian Pilot Training at Tuskegee,* p. 21.

24. "Report on Civilian Pilot Training Program, Summer 1940," prepared by G. L. Washington, 1 October 1940, Air Corps—General Folder, Subject File, 1940–1947, Civilian Aide Records, RG 107, NA. See also, Washington, *History of Military and Civilian Pilot Training at Tuskegee,* p. 62.

25. Washington to Patterson, 31 April 1940, G. L. Washington Folder, LC 1940, Patterson Papers, TUA.

26. Washington to E. C. Nilson, Regional Manager, Civil Aeronautics Authority, Atlanta, Georgia, 17 May 1940, and Washington to Patterson, 12 September 1940, G. L. Washington Folder, LC 1940, Patterson Papers, TUA.

27. Washington to Patterson, 15 April 1940, Civil Aeronautics Authority Folder, LC 1940, Patterson Papers, TUA.

28. Strickland, *Putt-Putt Air Force,* pp. 12, 18–19; Civil Aeronautics Administration, "Wartime History of the Civil Aeronautics Administration," unpublished report, Federal Aeronautics Administration Library, Washington, D.C., n.d., pp. CPT–18 and CPT–19.

29. U.S. Congress, Senate, *Training of Civil Aircraft Pilots: Hearing on S. 2119,* 20 April 1939, p. 76.

30. Pisano, "Brief History of the Civilian Pilot Training Program," p. 27.

31. Washington to Patterson, 31 April 1940, G. L. Washington Folder, LC 1940, Patterson Papers, TUA.

32. Memorandum for Pres. Franklin D. Roosevelt from Harold Smith, Director of the Budget, 2 December 1940, on the question "Should the

Civilian Pilot Training program be continued during the coming year on the present level?" July–December 1940 Folder, OF 2955, FDRL.

33. Washington to Patterson, 15 April 1940, Civil Aeronautics Authority Folder; Washington to Patterson, 20 April 1940, G. L. Washington Folder; in LC 1940, Patterson Papers, TUA.

34. Washington to Patterson, 31 [*sic*] April 1940, G. L. Washington Folder, LC 1940, Patterson Papers, TUA.

35. Patterson to Civil Aeronautics Authority, attention Grove Webster, 11 May 1940, with unsigned draft attached, n.d., Civil Aeronautics Authority Folder, LC 1940, Patterson Papers, TUA. Patterson's proposal of 11 May 1940 was taken verbatim from the attached draft. The initials on the draft indicate that it was typed by G. L. Washington's secretary and was no doubt prepared by Washington.

36. Grove Webster to Patterson, 15 May 1940, Civil Aeronautics Authority Folder, LC 1940, Patterson Papers, TUA.

37. Washington, *History of Military and Civilian Pilot Training at Tuskegee*, p. 58.

38. Washington, *History of Military and Civilian Pilot Training at Tuskegee*, pp. 58–61. The activities of the Coffey School of Aeronautics during 1939–40 are described in detail in the following chapters. Although Coffey and Brown had both contacted Tuskegee to inquire about job opportunities, they ultimately became vocal critics of Tuskegee's aviation program; see Coffey to Patterson, 27 January 1939, Co–Cy Folder, and Brown to Patterson, 5 October 1939, Civil Aeronautics Authority Folder, both in GC 1939, Patterson Papers, TUA.

39. Washington, *History of Military and Civilian Pilot Training at Tuskegee*, pp. 63–65.

40. Ibid., p. 66, emphasis added.

41. "G. L." to "Pat" [i.e., Washington to Patterson], 11 July 1940, G. L. Washington Folder, LC 1940, Patterson Papers, TUA.

42. Washington, *History of Military and Civilian Pilot Training at Tuskegee*, pp. 68–70; Anderson interview, p. 69.

43. "G. L." to "Pat" [i.e., Washington to Patterson], 11 July 1940, G. L. Washington Folder, LC 1940, Patterson Papers, TUA.

44. "G. L." to "Pat" [i.e., Washington to Patterson], 11 July 1940, and Washington to Lewis A. Jackson, 20 July 1940, G. L. Washington Folder, LC 1940, Patterson Papers, TUA; Washington, *History of Military and Civilian Pilot Training at Tuskegee*, p. 93.

45. "G. L." to "Pat" [i.e., Washington to Patterson], 11 July 1940, G. L. Washington Folder, LC 1940, Patterson Papers, TUA.

46. Washington, *History of Military and Civilian Pilot Training at Tuskegee*, pp. 69–70. Washington used the military designation YPT–14 rather than the

company nomenclature of UPF–7. Waco UPF–7s were used extensively in the CPT Program; over six hundred were purchased by CPT operators, at a cost of approximately $7,500 each (see Herm Schreiner, "The Waco Story: Part II—Expansion with the 'F' Series," *American Aviation Historical Society Journal* 29 [Fall 1984]: 214–27).

47. "G. L." to "Pat" [i.e., Washington to Patterson], 11 July 1940, G. L. Washington Folder, LC 1940, Patterson Papers; TUA.

48. Washington, *History of Military and Civilian Pilot Training at Tuskegee*, p. 70.

49. "G. L." to "Pat" [i.e., Washington to Patterson], 11 July 1940, G. L. Washington Folder, LC 1940, Patterson Papers, TUA.

50. Washington, *History of Military and Civilian Pilot Training at Tuskegee*, unpaginated leaf inserted between pp. 69 and 70.

51. Robert T. Warner and Harold R. Decker, "Auburn–Opelika Airport History," 24 May 1974, copy in Historical Collection, Auburn University Archives, Auburn, Alabama.

52. Washington, *History of Military and Civilian Pilot Training at Tuskegee*, pp. 70–71; "G. L." to "Pat" [i.e., Washington to Patterson], 11 July 1940, G. L. Washington Folder, LC 1940, Patterson Papers, TUA. The API students may have been pressured by Cornell into voting in favor of the proposal; Washington recounted that "Cornell told me in reporting that woe be unto the student that disagreed, but none did" (Washington, *History of Military and Civilian Pilot Training at Tuskegee*, p. 71).

53. Patterson to Pres. L. N. Duncan, API, D Folder, GC 1940, Patterson Papers, TUA.

54. "G. L." to "Pat" [i.e., Washington to Patterson], 11 July 1940, G. L. Washington Folder, LC 1940, Patterson Papers, TUA.

55. Washington, *History of Military and Civilian Pilot Training at Tuskegee*, p. 71.

56. Ibid., pp. 77–78.

57. Ibid., pp. 72–73.

58. J. L. Patzolcl to Washington, 1 August 1940, G. L. Washington Folder, LC 1940, Patterson Papers, TUA.

59. Ibid. Washington's recollection was that Anderson arrived on 29 July but Patzolcl's letter clearly states that Anderson "left today" (Washington, *History of Military and Civilian Pilot Training at Tuskegee*, p. 73). See also Washington to Lloyd Isaacs, Tuskegee Institute Treasurer, G. L. Washington Folder, LC 1940, Patterson Papers, TUA, which indicates that Anderson entered the employ of the Institute "on August 1, 1940,—the day he left Chicago with the aeroplane."

60. Anderson interview, pp. 69–70.

61. Ibid., p. 70.

62. Anderson interview, p. 70; Washington, *History of Military and Civilian Pilot Training at Tuskegee*, pp. 73, 81.

63. Washington, *History of Military and Civilian Pilot Training at Tuskegee*, p. 81.

64. Ibid., p. 84.

65. Telegram, Lewis Jackson to Washington, [4 August 1940], Jackson to Patterson, 20 August 1940, G. L. Washington Folder; Patterson to Grove Webster, 30 September 1940, Civil Aeronautics Authority Folder; all in LC 1940, Patterson Papers, TUA.

66. Washington, *History of Military and Civilian Pilot Training at Tuskegee*, pp. 84–85. Washington was generous in his assessment of the attitudes of the white spectators. He took no offense at the "expressions of the whites, when a perfect landing was made,—'Did you see that nigger land that plane!'" Such a comment, he asserted, was "the same enthusiastic exclamation which would have been made [about a white trainee], with boy substituted for nigger."

67. Washington, *History of Military and Civilian Pilot Training at Tuskegee*, p. 89. As late as 5 October 1940, seven of the ten secondary students still had not passed all five sections of the written examination, most having failed only one section (Washington to A. and T. College et al., 5 October 1940, G. L. Washington Folder, LC 1940, Patterson Papers, TUA).

68. Washington to the Budget Committee, Tuskegee Institute, 29 July 1940, G. L. Washington Folder, LC 1940, Patterson Papers, TUA.

69. Washington to Patterson, 1 August 1940; "G. L." to "Pat" [i.e., Washington to Patterson], [8 August 1940]; and Patterson to Washington, 12 August 1940; all in G. L. Washington Folder, LC 1940, Patterson Papers, TUA.

70. Patterson to CAA, attention Grove Webster, 2 September 1940, Civil Aeronautics Authority Folder, LC 1940, Patterson Papers, TUA.

71. "Division of Aeronautics," [presented 9 September 1940], G. L. Washington Folder, LC 1940, Patterson Papers, TUA. Although undated and unsigned, the analysis was definitely prepared by Washington and presented to Patterson on Monday, 9 September 1940, as shown by Washington's letter to Patterson of 12 September 1940, in the same folder, in which Washington states "*We must close out items . . .* on 'Immediate Considerations' on yellow sheet exhibited in our conference of Monday." Portions of the analysis were also presented at a 14 September meeting of the Executive Committee (Southern Members) of the Board of Trustees and described as "Mr. Washington's report" (see *Minutes of the Annual Meeting of the Board of Trustees of the Tuskegee Institute*, 24 October 1940, pp. 28–31). The figures cited above have been rounded to the nearest thousand.

72. "Division of Aeronautics," [presented 9 September 1940], G. L. Washington Folder, LC 1940, Patterson Papers, TUA.

73. Washington to Patterson, 12 September 1940, G. L. Washington Folder, LC 1940, Patterson Papers, TUA.

74. Benjamin Russell, Executive Committee member, to Patterson, 12 September 1940, G. L. Washington Folder, LC 1940, Patterson Papers, TUA.

75. "Minutes of the Meeting of the Executive Committee (Southern Members) of the Board of Trustees of Tuskegee Institute, September 14, 1940," in *Minutes of the Annual Meeting of the Board of Trustees of the Tuskegee Institute*, 24 October 1940, pp. 28–31.

76. "Milestones," n.d., 1946 Moton Field Folder, Series 2, Box 2, Patterson Papers, TUA; Washington to Patterson, 31 October 1940, G. L. Washington Folder, LC 1940, Patterson Papers, TUA; Washington, *History of Military and Civilian Pilot Training at Tuskegee*, pp. 94–95; Shelton interview, 29 October 1985. The three black instructors were Anderson, Lewis A. Jackson, and George Allen; the whites, Forrest Shelton, Joseph T. Camilleri, and Frank Rosenberg.

77. Washington to Patterson, 6 November 1940, G. L. Washington Folder, LC 1940, Patterson Papers, TUA.

78. "Statements and Statistics on 1940–1941 Aviation Program as of October 18, 1940," by G. L. Washington, Director, Division of Aeronautics, Civil Aeronautics Authority Folder, LC 1940, Patterson Papers, TUA.

79. "G. L" to "Pat" [i.e., Washington to Patterson], 11 July 1940, G. L. Washington Folder, LC 1940, Patterson Papers, TUA.

80. Patterson to Washington, 13 June 1940, G. L. Washington Folder, LC 1940, Patterson Papers, TUA.

81. Washington to Patterson, 15 June 1940, G. L. Washington Folder, LC 1940, Patterson Papers, TUA.

82. Patterson to Percy R. Hines, President, Chicago Tuskegee Club, 13 July 1940, He–Hi Folder, GC 1940, Patterson Papers, TUA.

83. Patterson to R. Hayne King, Pierce and Headrick, Inc., 15 August 1940, and King to Patterson, 26 August 1940, K Folder, GC 1940, Patterson Papers, TUA.

84. During the first week of November, Patterson sent personal letters to each alumni who agreed to serve on the Citizens Committee; twenty-eight letters for Chicago and thirty-four letters for Cleveland were found scattered throughout GC 1940 of the Patterson Papers, TUA, but none were found for Detroit. See, for example, Patterson to John C. Robinson, 7 November 1940, Ro–Ry Folder, and Patterson to W. H. McKinney, Cleveland, 2 November 1940, Ma–Mc Folder.

85. Washington to Robert P. Morgan, 28 October 1940, with attached

statement by Washington, G. L. Washington Folder, LC 1940, Patterson Papers, TUA. Morgan subsequently printed Washington's statement verbatim on publicity brochures in support of Cleveland's campaign; copy in the same folder.

86. Patterson to Brown, 30 September 1940, Civilian Conservation Corps —Edgar G. Brown Folder; telegrams, Patterson to Hill, 26 September 1940, He–Hi Folder, and Patterson to Bankhead, 26 September 1940, Ba–Be Folder; all in GC 1940, Patterson Papers, TUA.

87. Sen. Lister Hill to Patterson, 26 September 1940, He–Hi Folder, GC 1940; and Patterson to Webster, 30 September 1940, Civil Aeronautics Authority Folder, LC 1940; both in Patterson Papers, TUA.

88. Patterson to Webster, 30 September 1940, Civil Aeronautics Authority Folder, LC 1940, Patterson Papers, TUA.

89. *Minutes of the Board of Trustees of the Tuskegee Institute*, 24 October 1940, pp. 21–22.

90. "Statements and Statistics on 1940–1941 Aviation Program as of October 18, 1940," by G. L. Washington, Director, Division of Aeronautics, Civil Aeronautics Authority Folder, LC 1940, Patterson Papers, TUA.

8. The Campaign for Air Corps Participation Broadens

1. The microfilm edition of the *Tuskegee Clippings File* is a convenient source for documenting the interest of the black press in the CPT Program during the prewar years; see the entries under "Aviation" for 1939 (60: 598–644) and 1940 (65: 385–430).

2. Memorandum, Wright to Howard Rough, Director of Regional Offices, CAA, 28 October 1940, with attached letter, Wright to Hennings, 21 October 1940, 836.17, Negro Pilot Training, CAA General Files, RG 237, NA.

3. *Pittsburgh Courier*, 21 October 1939, quoted in Osur, *Blacks in the Army Air Forces*, p. 13.

4. Public Law 18, *Statutes at Large*, 53: 556 (1939). The origins and development of the provision of P.L. 18 relating to the training of black pilots are discussed fully in chapter 5 above.

5. Another Illinois school was Parks Air College in East St. Louis; five were in areas outside the South, three in southern California, and one each in Oklahoma and Nebraska. The two southern schools were in Texas and Alabama (Craven and Cate, *Army Air Forces in World War II*, 6: 456). Harold S. Darr was awarded a contract for primary training on 28 June 1939, to be conducted by the Chicago School of Aeronautics, at the Curtiss Airport in Glenview (see "Historical Record, Air Corps Training Detachment, Chicago

School of Aeronautics, Glenview, Illinois," 234.142-1 [28 June 1939–30 March 1940], USAF Collection, USAF-HRC).

6. John C. Robinson had learned to fly at the Curtiss Airport, which was established by Darr and Maj. R. W. Schroeder in 1928 (see Scott, "Colonel John C. Robinson," pp. 60–61; and Wiener, *Two Hundred Thousand Flyers*, p. 163).

7. Memorandum summarizing statement of John C. Robinson, 30 November [1940], Folder 1, Box 171, Barnett Papers, CHS.

8. Four reports of the meeting have been found and the following description is based on those accounts, which vary in length but generally agree as to the tone and conclusions of the proceedings. The longest and most candid report was that submitted by the Air Corps representative to General Arnold: memorandum, Maj. R. M. Webster, Chief, Training Section, to Chief of the Air Corps, 18 January 1940, "Reports of Conferences Relative to the Training of Negro Pilots, January 1940," 220.765-5 (1940), USAF Collection, USAF-HRC. Other reports of the conference include memorandum, Grove Webster, Chief, Private Flying Development Division, to Chief, Civil Aeronautics Authority, 23 January 1940, "Reports of Conferences Relative to the Training of Negro Pilots, January 1940," 220.765-5 (1940), USAF Collection, USAF-HRC; memorandum, Willa B. Brown to M. O. Bousefield, 19 February 1940, Brown, Willa B.—Coffey School of Aeronautics Folder, Civilian Aide Records, RG 107, NA; memorandum summarizing statement of John C. Robinson, 30 November [1940], Folder 1, Box 171, Barnett Papers, CHS.

9. Memorandum summarizing statement of John C. Robinson, 30 November [1940], Folder 1, Box 171, Barnett Papers, CHS.

10. Ibid.

11. Ibid.

12. Memorandum, Maj. R. M. Webster, Chief, Training Section, to Chief of the Air Corps, 18 January 1940, "Reports of Conferences Relative to the Training of Negro Pilots, January 1940," 220.765-5 (1940), USAF Collection, USAF-HRC.

13. Memorandum, Grove Webster, Chief, Private Flying Development Division, to Chief, Civil Aeronautics Authority, 23 January 1940, "Reports of Conferences Relative to the Training of Negro Pilots, January 1940," 220.765-5 (1940), USAF Collection, USAF-HRC.

14. Memorandum, Maj. R. M. Webster, Chief, Training Section, to Chief of the Air Corps, 18 January 1940, "Reports of Conferences Relative to the Training of Negro Pilots, January 1940," 220.765-5 (1940), USAF Collection, USAF-HRC.

15. Ibid.

16. Memorandum, Willa B. Brown to M. O. Bousefield, 19 February 1940,

Brown, Willa B.—Coffey School of Aeronautics Folder, Civilian Aide Records, RG 107, NA.

17. *Congressional Record*, 76th Cong., 3rd sess., pp. 671–72 (25 January 1940).

18. "Annual Inspection of Chicago School of Aeronautics," 14 March 1940, 168.7032–53 (FY 1940), Harold A. McGinnis Papers, USAF-HRC.

19. Memorandum, Willa B. Brown to M. O. Bousefield, 19 February 1940, Brown, Willa B.—Coffey School of Aeronautics Folder, Civilian Aide Records, RG 107, NA.

20. Ibid.; for an example of the newspaper announcement, see *Pittsburgh Courier*, 17 February 1940, p. 4.

21. A copy of the flyer is attached to memorandum, Willa B. Brown to M. O. Bousefield, 19 February 1940, Brown, Willa B.—Coffey School of Aeronautics Folder, Civilian Aide Records, RG 107, NA.

22. Ibid.; *Pittsburgh Courier*, 17 February 1940, p. 4.

23. Memorandum, Yount to Marshall, 4 April 1940, xerox 2457, George C. Marshall Foundation, Lexington, Virginia (hereafter cited as Marshall Foundation).

24. *Pittsburgh Courier* (nat. ed.), 16 March 1940, p. 4.

25. Memorandum for record, Maj. R. M. Webster, Chief, Training Section, 21 March 1940, 220.765–6 (1940), USAF Collection, USAF-HRC.

26. Atlanta *Daily World*, 21 July 1940, in *Tuskegee Clippings File* 65: 388.

27. Oral History Interview of Brig. Gen. Noel F. Parrish by Dr. James C. Hasdorff, 10–14 June 1974, pp. 89–90, K239.0512–744, USAF Collection, USAF-HRC.

28. U.S. Congress, House of Representatives, *Military Establishment Appropriation Bill for 1941: Hearings before the Subcommittee of the Committee on Appropriations*, 76th Cong., 3rd sess., 1940, pp. 890–95.

29. *Congressional Record*, 76th Cong., 3rd sess., pp. 4017–18 (4 April 1940).

30. *Pittsburgh Courier*, 20 April 1940, in *Tuskegee Clippings File* 65: 391.

31. *Pittsburgh Courier* (nat. ed.), 4 May 1940, pp. 1, 4; Osur, *Blacks in the Army Air Forces*, p. 13.

32. U.S. Congress, Senate, *Military Establishment Appropriation Bill for 1941: Hearings on H.R. 9209*, p. 367.

33. Osur, *Blacks in the Army Air Forces*, p. 13; *Papers of the NAACP, Part 1*, 2: 1078.

34. U.S. Congress, Senate, *Military Establishment Appropriation Bill for 1941: Hearings on H.R. 9209*, p. 368.

35. Ibid.

36. Ibid., p. 369.

37. Lee, *Employment of Negro Troops*, pp. 68–71.

38. Arnold to Assistant Chief of Staff, G–3, 31 May 1940, in MacGregor and Nalty, *Basic Documents*, 5: 17.

39. For documentation of objections within the General Staff to the Air Corps' refusal to admit blacks, see ibid., 5: 9, 19–20.

40. *Papers of the NAACP, Part 1*, 10: 912–16.

41. Ibid., 10: 918.

42. Ibid., 10: 991, 993.

43. Ibid., 10: 1044.

44. Ibid., 10: 741.

45. Forrester B. Washington to Will W. Alexander, 21 June 1940, Commission on Interracial Cooperation Folder, GC 1940, Patterson Papers, TUA.

46. James Middleton, Port Morris Community Council, Bronx, New York, to President Roosevelt, 13 June 1940, 1940 Folder, OF 93, FDRL.

47. Collier Anderson to Roosevelt, 16 August 1940, 1940 Folder, OF 93, FDRL.

48. White House memorandum, 12 July 1940, with "Memorandum on Negro Participation in the Armed Forces," n.d., attached, 1940 Folder, OF 93, FDRL.

49. U.S. Congress, House of Representatives, *Second Supplemental National Defense Appropriation Bill for 1941: Hearings before the Subcommittee of the Committee on Appropriations . . .* , 76th Cong., 3rd sess., 1940, pp. 132–33.

50. *Pittsburgh Courier* (nat. ed.), 17 August 1940, pp. 1, 4.

51. Julian S. Rammelkamp, "McCormick, Robert Rutherford," *Dictionary of American Biography: Supplement Five, 1951–1955* (N.Y.: Charles Scribner's Sons, 1977), pp. 447–51.

52. Editorial, *Chicago Tribune*, 8 August 1940, in *Tuskegee Clippings File* 65: 400.

53. *Pittsburgh Courier* (nat. ed.), 17 August 1940, pp. 1, 4.

54. *Chicago Bee*, 11 August 1940, in *Tuskegee Clippings File* 65: 401.

55. *Washington Tribune*, 17 August 1940, in *Tuskegee Clippings File* 65: 401; see also *Boston Guardian*, 10 August 1940, in *Tuskegee Clippings File* 65: 401.

56. Quoted in Sitkoff, *New Deal for Blacks*, p. 304.

57. Ibid., p. 303.

58. Otis L. Graham, Jr., and Meghan Robinson Wander, eds., *Franklin D. Roosevelt: His Life and Times, An Encyclopedic View* (Boston: G. K. Hall, 1985), pp. 384–85.

59. U.S. Congress, House of Representatives, *Selective Compulsory Military*

Training and Service: Hearings before the Committee on Military Affairs . . . on H.R. 10132, 76th Cong., 3rd sess., 1940, pp. 585–90.

60. Walter White to Sen. Robert F. Wagner, 8 August 1940, quoted in *Congressional Record*, 76th Cong., 3rd sess., pp. 10889–90 (26 August 1940).

61. *Congressional Record*, 76th Cong., 3rd sess., p. 10888 (26 August 1940).

62. Ibid.

63. Ibid., p. 10890.

64. Ibid.

65. Ibid., p. 10891.

66. Ibid., p. 10892.

67. Ibid., pp. 10892–93.

68. Ibid., p. 10893.

69. Ibid.

70. Ibid., pp. 10893–94.

71. Ibid., p. 10894.

72. Ibid.

73. Ibid., p. 10895.

74. Selective Training and Service Act of 1940, *Statutes at Large*, 54: 885–86, 896.

9. Fruits of the Campaign: The Ninety-ninth Pursuit Squadron

1.   Captain J. T. Cumberpatch, Assistant Chief, Personnel Division, to Howard Williams, 6 March 1939, and Williams to Henry Stimson, Franklin D. Roosevelt, and Eleanor Roosevelt, 16 September 1940, in U.S. Army Air Corps Cadet Program Information Folder, Box IIA–647, NAACP Papers, Lib. Cong.

2.   Adjutant General Adams to Williams, 1 October 1940, U.S. Army Air Corps Cadet Program Information Folder, Box IIA–647, NAACP Papers, Lib. Cong.

3.   Memorandum, Lt. Col. Orlando Ward, Secretary, General Staff, to Assistant Chief of Staff, G–1, 5 September 1940, in MacGregor and Nalty, *Basic Documents*, 5: 24.

4.   Memorandum, George C. Marshall to General Shedd, 14 September 1940, in MacGregor and Nalty, *Basic Documents*, 5: 25.

5.   White House press release, 16 September 1940, 1940 Folder, OF 93, FDRL.

6.   Lee, *Employment of Negro Troops*, p. 116.

7. Henry Lewis Stimson Diaries, 30: 200 (microfilm 6: 328), YUL.

8. The 27 September 1940 meeting and its aftermath are recounted in a number of sources. See, for example, Dalfiume, *Desegregation of the U.S. Armed Forces*, pp. 37–40; Lee, *Employment of Negro Troops*, pp. 74–82; MacGregor, *Integration of the Armed Forces*, p. 15; Nalty, *Strength for the Fight*, pp. 138–40; and Sitkoff, *New Deal for Blacks*, pp. 306–9.

9. Stimson Diaries, 30: 200 (microfilm 6: 328), YUL.

10. Quoted in Sitkoff, *New Deal for Blacks*, p. 306.

11. NAACP memorandum, "CONFERENCE AT THE WHITE HOUSE, Friday–September 27, 1940—11:35 A.M.–12:10 P.M. Subject: *Discrimination Against Negroes in the Armed Forces of the United States*," in MacGregor and Nalty, *Basic Documents*, 5: 26–27; the memorandum presented to the administration official at the conclusion of the meeting is printed in *Crisis* 47 (November 1940): 351.

12. *Crisis* 47 (November 1940): 351.

13. Memorandum, Robert P. Patterson, Assistant Secretary of War, to Roosevelt, 8 October 1940, and Stephen Early, Secretary to the President, to Patterson, 9 October 1940, in MacGregor and Nalty, *Basic Documents*, 5: 29–31.

14. Ibid.

15. Ibid.

16. NAACP press release, 11 October 1940, quoting from 10 October 1940 telegram, NAACP to President Roosevelt, 1940 Folder, OF 93, FDRL.

17. War Department, Adjutant General's Office, to Commanding Generals of all Armies, Corps Areas and Departments, and Chiefs of the Arms and Services, 16 October 1940, in MacGregor and Nalty, *Basic Documents*, 5: 32–33.

18. *New York Times*, 16 October 1940, p. 9.

19. Lee, *Employment of Negro Troops*, p. 117, quoting marginal note, signed HHA, on memorandum, Training and Operations Division for Executive Officer, 15 October 1940.

20. *Pittsburgh Courier*, 19 October 1940, in *Tuskegee Clippings File* 65: 399.

21. Ibid.

22. Memorandum summarizing statement of John C. Robinson, 30 November [1940], Folder 1, Box 171, Barnett Papers, CHS.

23. "Report of the Findings Committee of 'Military and Naval Defense': Conference on the participation of the Negro in the national defense— Hampton Institute, November 25–26, 1940," Folder 102–13, William H. Hastie Papers, Harvard Law School Library, Cambridge, Massachusetts.

24. Edgar G. Brown to Pres. Franklin D. Roosevelt, 19 November 1940, 1940 Folder, OF 93, FDRL.

25. *Crisis* 47 (December 1940).

26. Ibid., pp. 376–78, 388.

27. Peck was reputedly one of a handful of American pilots who flew with government forces during the Spanish Civil War; see *Current Biography* 3 (August 1942): 40–42. This claim is questioned by at least one researcher; see Allen Herr's articles, "American Pilots in the Spanish Civil War," *American Aviation Historical Society Journal* 22 (Fall 1977): 171–72, and "American Pilots in The Spanish Civil War: Addenda," *American Aviation Historical Society Journal* 23 (Fall 1978): 235.

28. James L. H. Peck, "When Do *We* Fly?" *Crisis* 47 (December 1940): 375–78, 388.

29. Ibid.

30. Ibid.

31. Ibid.

32. Press release, 29 November 1940, U.S. Army Air Corps General Folder, IIA–647, NAACP Papers, Lib. Cong.

33. Editorial, *Pittsburgh Courier* (nat. ed.), 7 December 1940, p. 6.

34. Robert P. Patterson to Arnold, 26 December 1940, 291.2, Classified Decimal File, 1940–1945, Office of Assistant Secretary of War for Air, RG 107, NA.

35. Lee, *Employment of Negro Troops*, p. 116; quotation from memorandum, Training and Operations Division, Office of the Chief of the Air Corps to General Staff, G–3, 3 October 1940.

36. Lee, *Employment of Negro Troops*, p. 116.

37. Ibid., pp. 116–17; quotations from memorandum, Chief of Air Plans Division to Chief of Training and Operations Division, 5 October 1940.

38. Lee, *Employment of Negro Troops*, p. 117, quoting marginal note, signed HHA, on memorandum, Training and Operations Division for Executive Officer, 15 October 1940, emphasis in original.

39. Lee, *Employment of Negro Troops*, p. 117, quotations from memorandum, Office of the Chief of the Air Corps for G–3, 22 October 1940.

40. Stimson Diaries, 31: 71–72 (microfilm 6: 408–9), YUL.

41. Lee, *Employment of Negro Troops*, p. 117.

42. Ibid., p. 121.

43. Brig. Gen. Davenport Johnson, Office of the Chief of the Air Corps, Operations and Training Division, to Commanding General, SEACTC, 8 November 1940, copy in "History of Tuskegee Army Air Field," 289.28–1, vol. 1 (23 July–6 December 1941), USAF Collection, USAF-HRC.

44. Ibid.

45. Ibid.

46. Ibid.

47. Ibid.

48. Ibid.

49. Memorandum, Col. C. W. Howard, G–3, Miscellaneous Branch, to Chief, G–3, Mobilization Branch, 12 November 1940, 220.765–2, vol. 1 (1940), USAF Collection, USAF-HRC.

50. Memorandum, Col. C. W. Howard, G–3, Miscellaneous Branch, to General [Frank] Andrews, 14 November 1940, 220.765–2, vol. 1 (1940), USAF Collection, USAF-HRC.

51. Memorandum, Gen. Davenport Johnson, Assistant Chief of the Air Corps, to Manning, n.d., 220.765–2, vol. 1 (1940), USAF Collection, USAF-HRC.

52. Telegram, Enoch Waters to Walter White, NAACP, 15 November [1940], United States Army Air Corps Folder, Box IIA–647, NAACP Papers, Lib. Cong.

53. Howard Gould to White, 15 November 1940, United States Army Air Corps Folder, Box IIA–647, NAACP Papers, Lib. Cong.; Gould to William H. Hastie, 15 November [1940], Air Corps—General Folder, Civilian Aide Records, RG 107, NA; this quotation and the two following are taken from Gould to White, 15 November 1940.

54. Ibid.

55. Ibid.

56. Gould was referring to C. Alfred Anderson and Lewis Jackson. Anderson was teaching CPT courses at Howard University when President Patterson recruited him for Tuskegee; thus Gould's charge appears to be groundless in his case. But in the case of Jackson, the assertion seems more plausible, because he was teaching at the Coffey School of Aeronautics before he accepted employment at Tuskegee Institute.

57. Gould to White, 15 November 1940, United States Army Air Corps Folder, Box IIA–647, NAACP Papers, Lib. Cong.; Gould to Hastie, 15 November [1940], Air Corps—General Folder, Civilian Aide Records, RG 107, NA; quotation taken from Gould to Hastie, 15 November [1940].

58. Gould to White, 15 November 1940, United States Army Air Corps Folder, Box IIA–647, NAACP Papers, Lib. Cong.

59. Ibid.

60. Roy Wilkins to Hastie, 15 November 1940, United States Army Air Corps Folder, Box IIA–647, NAACP Papers, Lib. Cong.

61. Memorandum, "IN RE: *ARMY DISCRIMINATION*," 10 October 1940, U.S. Army Air Corps, Yancey Williams Case, 1940–1941 Folder, Series IIB, NAACP Papers, Lib. Cong.

62. *Papers of the NAACP, Part 1*, 2: 1076.

63. Alfred Baker Lewis to White, 16 and 28 August, U.S. Army Air Corps 1940, Box IIA–647, NAACP Papers, Lib. Cong..

64. Lewis to White, 5 September 1940, U.S. Army Air Corps 1940, Box IIA–647, NAACP Papers, Lib. Cong.

65. W. Robert Ming to Thurgood Marshall, 27 November 1940, U.S. Army Air Corps, Yancey Williams Case, 1940–1941 Folder, Series IIB, NAACP Papers, Lib. Cong.

66. Major Floyd W. Ferree to Yancey Williams, 23 November 1940, U.S. Army Air Corps, Yancey Williams Case, 1940–1941 Folder, Series IIB, NAACP Papers, Lib. Cong.

67. Ming to Marshall, 27 November 1940, U.S. Army Air Corps, Yancey Williams Case, 1940–1941 Folder, Series IIB, NAACP Papers, Lib. Cong.

68. Marshall to Ming, 10 December 1940, and Ming to Marshall, 12 December 1940, U.S. Army Air Corps, Yancey Williams Case, 1940–1941 Folder, Series IIB, NAACP Papers, Lib. Cong.

69. Telegram, [George H.] Brett to Commanding General, SEACTC, 30 November 1940, 220.765–2, vol. 1 (1940), USAF Collection, USAF-HRC.

70. Telegram, Weaver to Chief of the Air Corps, 3 December 1940, 220.765–2, vol. 1 (1940), USAF Collection, USAF-HRC.

71. Telegram, Brett to Commanding General, SEACTC, 5 December 1940, 220.765–2, vol. 1 (1940), USAF Collection, USAF-HRC.

72. 1st Indorsement, HQ, SEACTC, to Chief of the Air Corps, 6 December 1940, 220.765–2, vol. 1 (1940), USAF Collection, USAF-HRC.

73. Ibid.

74. Ibid.

75. Ibid.

76. Ibid.

77. See Airfield Options Folder, GC 1941, Patterson Papers, TUA.

78. Telegram, Patterson to John H. Bankhead, Lister Hill, and Henry B. Steagall, 11 December 1940, Ba–Be Folder, GC 1940, Patterson Papers, TUA. An identical telegram was also sent to Congresswoman Frances P. Bolton, 11 December 1940, Ba–Be Folder, GC 1940, Patterson Papers, TUA.

79. Bankhead to Patterson, 14 December 1940, Ba–Be Folder, GC 1940, Patterson Papers, TUA.

80. Patterson to Bankhead, 17 December 1940, Ba–Be Folder, GC 1940, Patterson Papers, TUA.

81. Ibid.

82. G. B. Edwards, Mayor, City of Tuskegee to Patterson, 3 January 1941, quoting telegram sent to Bankhead, Hill, and Steagall, E Folder, GC 1941, Patterson Papers, TUA.

83. Telegram, Patterson to White, 11 December 1940, NAACP Folder, GC 1940, Patterson Papers, TUA.

84. Telegram, Patterson to White, 3 January 1941, Tuskegee Air Base 1940–1945 Folder, Series IIB, NAACP Papers, Lib. Cong.

85. Telegrams, White to Hastie and White to Patterson, 3 January 1941, Tuskegee Air Base 1940–1945 Folder, Series IIB, NAACP Papers, Lib. Cong.

86. Telegram, White to Patterson, 4 January 1941, Tuskegee Air Base 1940–1945 Folder, Series IIB, NAACP Papers, Lib. Cong.

87. Telegram, Patterson to White, 4 January 1941, Tuskegee Air Base 1940–1945 Folder, Series IIB, NAACP Papers, Lib. Cong.

88. White to Hastie, 4 January 1941, Tuskegee Air Base 1940–1945 Folder, Series IIB, NAACP Papers, Lib. Cong.

89. Telegram, White to Patterson, 6 January 1941, Tuskegee Air Base 1940–1945 Folder, Series IIB, NAACP Papers, Lib. Cong.

90. "Plan for Flying Training and Establishment of Pursuit Squadron (C) [Colored] SE [Single Engine]," 10 December 1940, "Plan for Technical Training, Air Corps Squadron (Colored), Pursuit, Single-Engine, Single-Seat, and Supporting Air Base Enlisted Personnel, Air Corps (Colored)," n.d., in 220.765–2, vol. 1 (1940), USAF Collection, USAF-HRC.

91. Memorandum, H. H. Arnold to George H. Brett, 16 December 1940, 220.765–2, vol. 1 (1940), USAF Collection, USAF-HRC.

92. Office of the Chief of the Air Corps to the Adjutant General, 18 December 1940, with inclosures, 220.765–2, vol. 1 (1940), USAF Collection, USAF-HRC. (For a copy of the letter, without enclosures, see MacGregor and Nalty, *Basic Documents*, 5: 43–45.)

93. Ibid.

94. Ibid.

95. Ibid.

96. Memorandum, Hastie to the Under Secretary of War, 31 December 1940, item 6063, reel 656, Marshall Foundation. (Also in MacGregor and Nalty, *Basic Documents*, 5: 46–47.)

97. Ibid.

98. Ibid.

99. Ibid.

100. Ibid.

101. MacGregor and Nalty, *Basic Documents*, 5:32–33.

102. Memorandum, Arnold to Robert A. Lovett, 6 January 1941, Negro Pilot Training Folder, Box 44, Official File, Henry H. Arnold Papers, Lib. Cong.

103. Ibid.

104. Memorandum, Arnold to Colonel Ward, 8 January 1941, Negro Pilot Training Folder, Box 44, Official File, Arnold Papers, Lib. Cong.

105. 1st Indorsement, Adjutant General to Chief of the Air Corps, 9 January 1941, in MacGregor and Nalty, *Basic Documents*, 5: 45.

10. The Reaction to the Ninety-ninth

1. Telegram, Malcolm S. MacLean to Patterson, 30 December [1940], Ma–Mc Folder, GC 1940, Patterson Papers, TUA.

2. Telegrams, Patterson to MacLean, 31 December 1940, and MacLean to Patterson, n.d., Ma–Mc Folder, GC 1940, Patterson Papers, TUA.

3. Washington, *History of Military and Civilian Pilot Training at Tuskegee*, p. 135.

4. Some thirty-two letters of invitation, all dated 4 January 1941, to a wide variety of black schools and colleges are scattered throughout the 1940 General Correspondence series of the Patterson Papers, generally filed alphabetically under the name of the institution's president. Since only some half-dozen black schools had CPT programs, most of the institutions did not have active aviation programs but many expressed an interest in entering the field of aviation education. Twenty-eight of the invitations went to institutions in southern or border states, and one each to schools in Pennsylvania, Ohio, Oklahoma, and the District of Columbia. No evidence was found to indicate that the Coffey School of Aeronautics was invited to participate in the conference. For samples of the letter of invitation see Patterson to D. Ormande Walker, Wilberforce University, Ohio, Wa–We Folder; T. E. Jones, Fisk University, Tennessee, Fisk University Folder; R. S. Grossley, State College for Colored Students, Dover, Delaware, Go–Gy Folder; and Mordecai W. Johnson, Howard University, Washington, D.C., J Folder; all in GC 1940, Patterson Papers, TUA.

5. If Marshall knew the status of the Air Corps' plans for establishing a black air unit, he may well have obtained his information from Hastie, though no evidence has been found to support this theory. Such a conclusion is plausible, however, because Hastie had been active in the NAACP since the early 1930s and had worked with Marshall on a number of NAACP-sponsored civil rights suits (see Gilbert Ware, *William Hastie: Grace Under Pressure* [New York: Oxford Univ. Press, 1984], pp. 35–94).

6. Telegram, Marshall to Ming, 8 January 1941, U.S. Army Air Corps, Yancey Williams Case, 1940–1941 Folder, Series IIB, NAACP Papers, Lib. Cong.

7. Telegram, Ming to Marshall, 8 January 1941, U.S. Army Air Corps, Yancey Williams Case, 1940–1941 Folder, Series IIB, NAACP Papers, Lib. Cong.

8. Ming to Marshall, 9 January 1941, U.S. Army Air Corps, Yancey Williams Case, 1940–1941 Folder, Series IIB, NAACP Papers, Lib. Cong.

9. Telegram, Marshall to Ming, 9 January 1941, U.S. Army Air Corps, Yancey WIlliams Case, 1940–1941 Folder, Series IIB, NAACP Papers, Lib. Cong.

10. Truman Gibson to Claude A. Barnett, 11 January 1941, *The Claude A. Barnett Papers*, microfilm ed. (Frederick, Md.: University Publications of America, 1986), part 3, series F, 2: 1325–26.

11. Ibid.

12. Barnett to Patterson, 13 January 1941, Barnett Correspondence, see earlier examples 1936–1942, LC 1937, Patterson Papers, TUA.

13. Memorandum, Washington to Patterson, 16 January 1940 [*sic*, 1941], and agenda, "Conference on Aviation," 16 January 1941, G. L. Washington Folder, Box 1, Series 2, Patterson Papers, TUA.

14. Telegram, Patterson to Colonel McDaniel, Aviation Section, War Department, 16 January 1941, G. L. Washington Folder, Box 1, Series 2, Patterson Papers, TUA.

15. Telegram, Arnold to Patterson, 16 January 1941, G. L. Washington Folder, Box 1, Series 2, Patterson Papers, TUA.

16. Minutes, Conference on Aviation, Tuskegee Institute, 16 January 1941, G. L. Washington Folder, Box 1, Series 2, Patterson Papers, TUA.

17. Ibid.

18. Ibid.

19. Memorandum, Washington to Patterson, 16 January 1940 [*sic*, 1941], G. L. Washington Folder, Box 1, Series 2, Patterson Papers, TUA.

20. Ibid. Washington's comment on the Veterans Hospital refers to the establishment of a segregated facility adjacent to the Tuskegee campus during the 1920s (see Daniel, "Black Power in the 1920s," pp. 368–88).

21. Minutes, Conference on Aviation, Tuskegee Institute, 16 January 1941, G. L. Washington Folder, Box 1, Series 2, Patterson Papers, TUA.

22. Neither Ming nor Marshall could represent Williams directly as they were not members of the District of Columbia Bar; consequently, they obtained the cooperation of Wendell L. McConnell, who served as Williams's counsel and requested the assistance of the NAACP National Legal Committee in prosecuting the case. See Ming to Marshall, 9 January 1941; telegram, Marshall to Ming, 10 January 1941; McConnell to National Legal Committee, 11 January 1941; Marshall to McConnell, 13 January 1941; all in U.S. Army Air Corps, Yancey Williams Case, 1940–1941 Folder, Series IIB, NAACP Papers, Lib. Cong.

23. Civil Action Suit No. 9763, 16 January 1941, Civil Docket book, p. 382, District Court of the United States for the District of Columbia, Washington,

D.C. Records for the case are in the custody of the Washington National Records Center, Suitland, Maryland, File No. CA 9763, Accession No. 21–66–A–1059, Box No. 152, Location No. 16–78–4–5. The date of filing has been incorrectly reported in various sources. An NAACP press release on the case, dated 17 January 1941, announced that the suit was filed "today" and was apparently the source for articles in the *NAACP Annual Reports* for 1940 and 1941 that cited the date of filing as 17 January (Press release, "Barred from Air Corps: Howard Student Sues U.S. War Department," 17 January 1941, U.S. Army Air Corps, Yancey Williams Case, 1940–1941 Folder, Series IIB, NAACP Papers, Lib. Cong.; *NAACP Annual Report for 1940*, p. 5; *NAACP Annual Report for 1941*, p. 27). Osur cites the date of filing as 15 January 1941 (*Blacks in the Army Air Forces*, p. 24).

24. Civil Action Suit No. 9763, 16 January 1941, Civil Docket book, p. 382, District Court of the United States for the District of Columbia, Washington, D.C. Applicants for aviation cadet appointment were required to submit their application through the corps area commander, in this case the Third Corps Area, which included the District of Columbia; thus General Grant was included as a codefendant.

25. *NAACP Annual Report for 1941*, pp. 27–28.

26. Roy Wilkins and Tom Matthews, *Standing Fast: The Autobiography of Roy Wilkins* (New York: Viking Press, 1982), p. 181.

27. Barnett to Patterson, 18 January 1941, Barnett Correspondence 1936–1942, LC 1937, Patterson Papers, TUA.

28. Ibid.

29. Ibid. For a discussion of the contacts between Waters, Gould, and the NAACP, see chapter 9 above.

30. Telegram, White to Hastie, 17 January 1941, Tuskegee Air Base 1940–1945 Folder, Series IIB, NAACP Papers, Lib. Cong.

31. Hastie to White, 24 January 1941, Tuskegee Air Base 1940–1945 Folder, Series IIB, NAACP Papers, Lib. Cong.

32. White to Hastie, 28 January 1941, and press release, "N.A.A.C.P. Protests War Department Plan for Segregated Air Squadron at Tuskegee," 14 February 1941, Tuskegee Air Base 1940–1945 Folder, Series IIB, NAACP Papers, Lib. Cong.

33. Marshall to Hastie, 17 January 1941, Tuskegee Air Base 1940–1945 Folder, Series IIB, NAACP Papers, Lib. Cong.

34. Ibid.

35. Telegram, George Murphy, NAACP Director of Publicity and Promotion, to Peck, 11 February 1941, Tuskegee Air Base 1940–1945 Folder, Series IIB, NAACP Papers, Lib. Cong.

36. Memorandum, White to Murphy, 27 January 1941, with attached

draft, Tuskegee Air Base 1940–1945 Folder, Series IIB, NAACP Papers, Lib. Cong.

37. White to Hastie, 28 January 1941, Tuskegee Air Base 1940–1945 Folder, Series IIB, NAACP Papers, Lib. Cong.

38. National Airmen's Association of America press release by Enoch P. Waters, Jr., 18 January 1941, Air Corps—General Folder, Civilian Aide Records, RG 107, NA.

39. Ibid.

40. Waters to Walter White, 1 February 1941, Tuskegee Air Base 1940–1945 Folder, Series IIB, NAACP Papers, Lib. Cong.

41. Barnett to Patterson, 20 January 1941, Barnett Correspondence 1936–1942, LC 1937, Patterson Papers, TUA; Waters to Truman K. Gibson, 20 January 1941, Air Corps—General Folder, Civilian Aide Records, RG 107, NA.

42. Memorandum, "Conversation with Willa Brown, of Harlem Air Port, December 10, 1940," Air Corps—General Folder, Civilian Aide Records, RG 107, NA.

43. Washington, *History of Military and Civilian Pilot Training at Tuskegee*, p. 166.

44. Barnett to Patterson, 3 February 1941, Barnett Correspondence 1936–1942, LC 1937, Patterson Papers, TUA.

45. Ibid.

46. Hastie to Lovett, 14 May 1941, Air Corps—Departmental Folder, Civilian Aide Records, RG 107, NA.

47. Lovett to Hastie, 19 May 1941, Air Corps—Departmental Folder, Civilian Aide Records, RG 107, NA.

48. Washington, *History of Military and Civilian Pilot Training at Tuskegee*, p. 166.

49. Ibid.

50. Report, Washington to Patterson, "Conference at Maxwell Field with Major L. S. Smith, Director of Training," 15 February 1941, appendix 1 to "History of 66th AAF Flying Training Detachment, Moton Field," 234.821 (February–7 December 1941), USAF Collection, USAF-HRC.

51. Telegrams, Patterson to Gen. George C. Marshall, 14 March 1941, and Gen. George H. Brett to Patterson, 15 March 1941, item 6063, microfilm reel 656, Marshall Foundation.

52. For a copy of the report, see *Barnett Papers*, microfilm ed., part 3, series B, 7: 590, 611–12, 715–16.

53. Barnett to Patterson, 2 May 1941, Barnett Correspondence 1936–1942, LC 1937, Patterson Papers, TUA.

54. Ibid.

55. Ibid.

56. Patterson to Webster, 8 May 1941, Civil Aeronautics Authority Folder, Box 1, Series 2, Patterson Papers, TUA.

57. Patterson to Marshall, 8 May 1941, U.S. War Department, January–July 1941 Folder, Box 2, Series 2, Patterson Papers, TUA.

58. Memorandum, Arnold to Marshall, "Subject: Chicago–Tuskegee Difficulty," 20 May 1941, Negro Pilot Training Folder, Box 44, Official File 1932–1946, Arnold Papers, Lib. Cong.

59. Ibid.

60. Marshall to Patterson, 24 May 1941, Negro Pilot Training Folder, Box 44, Official File 1932–1946, Arnold Papers, Lib. Cong. For a copy of Arnold's draft letter, dated 20 May 1941, see the same file.

61. Nalty, *Strength for the Fight*, pp. 109–10.

62. Telegram, White to Lovell, 7 February 1941, and letter White to Lovell, 14 February 1941, U.S. Army Air Corps Cadet Program Information 1940–1941 Folder, IIA–647, NAACP Papers, Lib. Cong.

63. For a copy of *Negro Aviators*, September 1940, with annotation "letter Feb. 21, 1940 [*sic*, 1941]," and letters from Delaware State College for Colored Students, Howard University, Tuskegee Institute, A. and T. College of North Carolina, West Virginia State College, Cornell University, University of Minnesota, Carroll College, and New York City College, see U.S. Army Air Corps Cadet Program Information 1940–1941 Folder, IIA–647, NAACP Papers, Lib. Cong.

64. For a copy of the 21 February 1941 form letter, signed by White, see U.S. Army Air Corps Cadet Program Information 1940–1941 Folder, IIA–647, NAACP Papers, Lib. Cong.

65. Telegrams, Marshall to Hastie and Hastie to Marshall, 25 February 1941, U.S. Army Air Corps Cadet Program Information 1940–1941 Folder, IIA–647, NAACP Papers, Lib. Cong.

66. Press release, "N.A.A.C.P. Asks Letters from Prospective Air Cadets," 7 March 1941; newspaper clippings from various black papers that published the release; and correspondence citing similar articles; all in U.S. Army Air Corps Cadet Program Information 1940–1941 Folder, IIA–647, NAACP Papers, Lib. Cong.

67. See the folder cited in note 66 for four consolidated lists of respondents, arranged alphabetically and including a synopsis of each letter of response.

68. For a sample of the form letter, dated 9 May 1941, see the folder cited in note 66.

69. Press release, "Applications for Air Corps Available at NAACP Office," 9 May 1941, also in the folder cited in note 66.

70. Truman K. Gibson to Claude A. Barnett, 25 March 1941, Air Corps— Enlistments, A–H Folder, Civilian Aide Records, RG 107, NA.

71. Charles W. Dryden to White, 25 February 1941, U.S. Army Air Corps Cadet Program Information 1940–1941 Folder, IIA–647, NAACP Papers, Lib. Cong.

72. Oscar Oliver to the NAACP, 24 May 1941, U.S. Army Air Corps Cadet Program Information 1940–1941 Folder, IIA–647, NAACP Papers, Lib. Cong.

73. Washington, *History of Military and Civilian Pilot Training at Tuskegee*, p. 44.

74. Alexander Anderson to White, 26 March 1941, U.S. Army Air Corps Cadet Program Applications Folder, IIA–646, NAACP Papers, Lib. Cong.

75. Anderson to White, 14 May 1941, U.S. Army Air Corps Cadet Program Applications Folder, IIA–646, NAACP Papers, Lib. Cong.

76. Anderson to White, 8 November 1941, U.S. Army Air Corps Cadet Program Applications Folder, IIA–646, NAACP Papers, Lib. Cong.

77. Ibid.

11. Military Flight Training Begins

1. Washington, *History of Military and Civilian Pilot Training at Tuskegee*, pp. 105–6.

2. Ibid., p. 120A.

3. "History of 66th AAF Flying Training Detachment, Moton Field, Tuskegee Institute, Alabama," p. 14, 234.821 (February–7 December 1941), USAF Collection, USAF-HRC.

4. Washington, *History of Military and Civilian Pilot Training at Tuskegee*, p. 159.

5. For a discussion of the decision to abandon plans to use CPT secondary and use instead a primary flying school, see chapter 10.

6. G. L. Washington to F. D. Patterson, "Report on: Conference at Maxwell Field with Major L. S. Smith, Director of Training," 15 February 1941, appendix 1, "History of 66th AAF Flying Training Detachment, Moton Field, Tuskegee Institute, Alabama," 234.821 (February–7 December 1941), USAF Collection, USAF-HRC.

7. Ibid.

8. Ibid.

9. Ibid.

10. Washington, *History of Military and Civilian Pilot Training at Tuskegee*, p. 182; Wiener, *Two Hundred Thousand Flyers*, pp. 162–63.

11. Washington, *History of Military and Civilian Pilot Training at Tuskegee*, p. 182; Washington to Patterson, 18 February 1941, G. L. Washington Folder, Box 1, Series 2, and Patterson to Claude A. Barnett, 11 March 1941, Barnett Correspondence 1936–1942, LC 1937, Patterson Papers, TUA.

12. An exact accounting of the amounts raised by alumni in Chicago and Cleveland was not found, and judging from the correspondence between Patterson and campaign managers in both cities, it did not amount to much over $1,000; by June 1941 the campaign in both cities had been abandoned. See Patterson–Daniel J. Faulkner correspondence and Patterson–Robert P. Morgan correspondence, F and Mo–My Folders, GC 1941, Patterson Papers, TUA.

13. Washington, *History of Military and Civilian Pilot Training at Tuskegee*, p. 182; unsigned memorandum, n.d. (ca. 1 March 1941), Carnegie Corporation of New York Folder, GC 1941, Patterson Papers, TUA. The RFC could not lend funds directly to Tuskegee Institute because it was a nonprofit organization. To qualify for an RFC loan for a primary school, Tuskegee would have had to set up a special corporation, subject to federal taxes. The quotation is from Patterson to Claude A. Barnett, 11 March 1941, Barnett Correspondence 1936–1942, LC 1937, Patterson Papers, TUA.

14. *Minutes of the Spring Meeting of the Board of Trustees of the Tuskegee Institute, April 5, 1941*, pp. 16–18.

15. Ibid.; "History of 66th AAF Flying Training Detachment, Moton Field, Tuskegee Institute, Alabama," p. 28, 234.821 (February–7 December 1941), USAF Collection, USAF-HRC.

16. "History of 66th AAF Flying Training Detachment, Moton Field, Tuskegee Institute, Alabama," p. 28, 234.821 (February–7 December 1941), USAF Collection, USAF-HRC; Washington, *History of Military and Civilian Pilot Training at Tuskegee*, pp. 183–84; *Minutes of the Spring Meeting of the Board of Trustees of the Tuskegee Institute, April 5, 1941*, p. 18.

17. Unsigned memorandum, n.d. (ca. 1 March 1941), Carnegie Corporation of New York Folder, GC 1941, Patterson Papers, TUA.

18. "History of 66th AAF Flying Training Detachment, Moton Field, Tuskegee Institute, Alabama," p. 28, 234.821 (February–7 December 1941), USAF Collection, USAF-HRC; *Minutes of the Spring Meeting of the Board of Trustees of the Tuskegee Institute, April 5, 1941*, p. 18.

19. "History of 66th AAF Flying Training Detachment, Moton Field, Tuskegee Institute, Alabama," p. 29, 234.821 (February–7 December 1941), USAF Collection, USAF-HRC.

20. Patterson to Carnegie Corporation, 7 March 1941, and Robert M.

Lester, Secretary, Carnegie Corporation, to Patterson, 18 April 1941, Carnegie Corporation of New York Folder, GC 1941, Patterson Papers, TUA.

21. Washington, *History of Military and Civilian Pilot Training at Tuskegee*, p. 186.

22. *Minutes of the Spring Meeting of the Board of Trustees of the Tuskegee Institute, April 5, 1941*, p. 18.

23. Eleanor Roosevelt, "My Day" newspaper column, for release 31 March 1941, My Day Bound Copies, Eleanor Roosevelt Papers, FDRL.

24. Washington to Patterson, 24 March 1941, and Patterson to Washington, 25 March 1941, G. L. Washington Folder, Box 1, Series 2, Patterson Papers, TUA.

25. Eleanor Roosevelt, "My Day" newspaper column, for release 1 April 1941, My Day Bound Copies, Eleanor Roosevelt Papers, FDRL.

26. Anderson interview, p. 68.

27. Ibid.

28. Chief Anderson maintained in his interview (pp. 68–69) and in numerous public statements that after her flight Mrs. Roosevelt became a staunch advocate of black participation in the air force, and used her influence to force the War Department to establish a black unit: "I think she was in our corner then 100 percent, seeing that blacks got into the Army Air Forces. I know that had to be. I know a lot of good came out of it [i.e., the flight]. It wasn't long after that that it was announced that they would be starting this flight program here. I imagine not more than 2 or 3 weeks at the most." However, by the time that Mrs. Roosevelt visited Tuskegee and took her flight with Anderson, the decision to establish and train a black air unit was several months old. Based on the chronology of events as outlined in this book, the announcement Anderson is referring to is almost certainly the establishment of the primary flying school, not the overall decision to establish and train a segregated unit that was released on 16 January 1941.

29. *Minutes of the Spring Meeting of the Board of Trustees of the Tuskegee Institute, April 5, 1941*, p. 20.

30. Ibid. See also Patterson's telegrams of thanks to Lessing Rosenwald, Eleanor Roosevelt, and Will W. Alexander for their support in securing the loan from the Rosenwald Fund; telegrams, Patterson to Rosenwald and Roosevelt, 7 April 1941, Ro–Ry Folder, and telegram, Patterson to Alexander, 7 April 1941, Julius Rosenwald Fund—W. W. Alexander Folder, all in GC 1941, Patterson Papers, TUA.

31. Washington, *History of Military and Civilian Pilot Training at Tuskegee*, p. 184.

32. Des Moines *Iowa Bystander*, 27 February 1941, p. 1; A. A. Alexander to Chief of Engineers, Engineering Corps, Washington, D.C., 27 March 1941, Alexander and Repass Folder, GC 1941, Patterson Papers, TUA. For addi-

tional biographical details on Alexander, see Charles E. Wynes, "'Alexander the Great,' Bridge Builder," *Pamlimpsest* 66 (May/June 1985): 78–86; and Raymond A. Smith, Jr., "'He Opened Holes Like Mountain Tunnels,'" *Palimpsest* 66 (May/June 1985): 87–100.

33. Washington, *History of Military and Civilian Pilot Training at Tuskegee*, p. 184; Wiener, *Two Hundred Thousand Flyers*, pp. 165, 179.

34. Washington, *History of Military and Civilian Pilot Training at Tuskegee*, p. 184.

35. Capt. Leland S. Stranathan, Assistant Director of Training, SEACTC, to Patterson, 20 February 1941, Maxwell Field Folder, GC 1941, Patterson Papers, TUA.

36. "History of 66th AAF Flying Training Detachment, Moton Field, Tuskegee Institute, Alabama," pp. 2–3, 234.821 (February–7 December 1941), USAF Collection, USAF-HRC; Washington, *History of Military and Civilian Pilot Training at Tuskegee*, p. 185; Lloyd Isaacs, Treasurer, Tuskegee Institute, to S. M. Eich, 2 June 1941, Airfield Options Folder, GC 1941 Patterson Papers, TUA.

37. Moton Field, named after Tuskegee Institute's second president, Robert Russa Moton, has remained an active airport since its construction in 1941 and currently serves as the municipal airport for Tuskegee, Alabama.

38. Telegram, Patterson to Gen. George C. Marshall, 14 March 1941, 220.1645–6 (1941), USAF Collection, USAF-HRC.

39. Office of the Chief of the Air Corps to the Adjutant General, 11 March 1941, 220.1645–6 (1941), USAF Collection, USAF-HRC.

40. For a full discussion of the establishment and development of the preflight schools, see "The Preflight Schools in World War II," U.S. Air Force Historical Study no. 90, 101–90 (1941–1953), USAF Collection, USAF-HRC. The following discussion of the preflight schools is taken from pp. 15–24.

41. At first the preflight schools were officially known as "Replacement Training Centers," but in April 1942 the term *Preflight Schools* was adopted as the official designation ("The Preflight Schools in World War II," p. 23).

42. Office of the Chief of the Air Corps to the Adjutant General, 11 March 1941, 220.1645–6 (1941), USAF Collection, USAF-HRC.

43. Telegram, Patterson to Marshall, 14 March 1941, 220.1645–6 (1941), USAF Collection, USAF-HRC.

44. Telegrams, Brett to Weaver and Brett to Patterson, 15 March 1941, 220.1645–6 (1941), USAF Collection, USAF-HRC.

45. 1st Indorsement, War Department, Adjutant General's Office to Chief of the Air Corps, 27 March 1941, 220.1645–6 (1941), USAF Collection, USAF-HRC.

46. "History of 66th AAF Flying Training Detachment, Moton Field,

Tuskegee Institute, Alabama," pp. 3–4, 234.821 (February–7 December 1941), USAF Collection, USAF-HRC.

47. Ibid.

48. Johnson to Commanding General, SEACTC, 23 April 1941, "History of 66th AAF Flying Training Detachment, Moton Field, Tuskegee Institute, Alabama," appendix 2, 234.821 (February–7 December 1941), USAF Collection, USAF-HRC.

49. Washington, *History of Military and Civilian Pilot Training at Tuskegee*, p. 188.

50. Ibid., pp. 188–89.

51. Patterson to Smith, 7 May 1941, Maxwell Field Folder, GC 1941, Patterson Papers, TUA.

52. Washington to Smith, 12 May 1941, G. L. Washington Folder, Box 1, Series 2, Patterson Papers, TUA.

53. Capt. J. P. McConnell to Washington, 14 May 1941, Maxwell Field Folder, GC 1941, Patterson Papers, TUA.

54. Routing and referral sheet, Training and Operations Division to Materiel Division, 15 May 1941, 220.1645–6 (1941), USAF Collection, USAF-HRC.

55. Ibid.

56. Ibid.

57. Routing and referral sheet, Materiel Division to Intelligence Division, 11 June 1941, 220.1645–6 (1941), USAF Collection, USAF-HRC.

58. "History of 66th AAF Flying Training Detachment, Moton Field, Tuskegee Institute, Alabama," p. 22, 234.821 (February–7 December 1941), USAF Collection, USAF-HRC.

59. Washington to Patterson, 6 June 1941, G. L. Washington Folder, Box 1, Series 2, Patterson Papers, TUA.

60. Patterson to Washington, 9 June 1941, G. L. Washington Folder, Box 1, Series 2, Patterson Papers, TUA.

61. Washington, *History of Military and Civilian Pilot Training at Tuskegee*, p. 193.

62. For biographical information on Parrish, see James C. Hasdorff, "Reflections on the Tuskegee Experiment: An Interview with Brig. Gen. Noel F. Parrish, USAF (Ret.)," *Aerospace Historian* 24 (Fall, September 1977): 173–80; and Oral History Interview of Brig. Gen. Noel F. Parrish by Dr. James C. Hasdorff, 10–14 June 1974, K239.0512–744, USAF Collection, USAF-HRC. Parrish was assigned to the Chicago School of Aeronautics on 25 June 1939; see "Historical Record, Air Corps Training Detachment, Chicago School of

Aeronautics," 234.142–1 (28 June 1939–30 March 1940), USAF Collection, USAF-HRC.

63. Parrish interview, pp. 89–90.

64. Ibid., p. 92.

65. Washington, *History of Military and Civilian Pilot Training at Tuskegee*, p. 195.

66. "History of 66th AAF Flying Training Detachment, Moton Field, Tuskegee Institute, Alabama," pp. 8–9, 234.821 (February–7 December 1941), USAF Collection, USAF-HRC.

67. Washington, *History of Military and Civilian Pilot Training at Tuskegee*, p. 194.

68. "No. 1 Graduate of the Nation," *Crisis* 43 (August 1936), cover photograph and pp. 239, 254.

69. Alan L. Gropman, "Benjamin O. Davis, Jr.: History on Two Fronts," in *Makers of the United States Air Force*, ed. John L. Frisbee (Washington, D.C.: Office of Air Force History, 1987), p. 231.

70. Ibid., pp. 231–32.

71. Washington, *History of Military and Civilian Pilot Training at Tuskegee*, pp. 127, 141.

72. "History of 66th AAF Flying Training Detachment, Moton Field, Tuskegee Institute, Alabama," p. 19, 234.821 (February–7 December 1941), USAF Collection, USAF-HRC.

73. The army's flight training program was standardized nationwide so that each school followed the same schedule and graduated students according to a carefully coordinated plan. When a cadet entered training, he was assigned a "class" number based on his projected graduation date. For example, cadets who entered training in early spring 1941 were part of the first increment scheduled to complete advanced training in 1942 and were therefore designated class 42–A. The second increment was designated 42–B, the third 42–C, and so on. Thus as part of class 42–C, Captain Davis and the first group of black aviation cadets were part of the overall group of flying students scheduled to graduate in March 1942, the third increment for the year.

74. No record has been found to indicate the rationale for the class overage; G. L. Washington recalled that when the discrepancy was brought to the attention of the contracting officer, a change order, dated 10 September 1941, was issued, authorizing a class size of thirteen for class 42–C. A similar problem occurred with class 42–D, which consisted of eleven cadets and it was handled likewise, with a change order dated 6 October 1941 (see Washington, *History of Military and Civilian Pilot Training at Tuskegee*, pp. 249–50).

75. Report to Dr. F. D. Patterson, "*FINAL PLANS*: Inaugural Ceremony initiating Training of Negroes as Military Pilots for United States Army Air Corps, Saturday afternoon, 4:00, Booker T. Washington Monument," 17 July 1941, U.S. War Department, January–July 1941 Folder, Box 2, Series 2, Patterson Papers, TUA; Washington, *History of Military and Civilian Pilot Training at Tuskegee*, p. 229.

76. Program, "Inaugural Ceremony Initiating the Training of Negroes as Military Aviators for the United States Army Air Corps," U.S. War Department, January–July 1941 Folder, Box 2, Series 2, Patterson Papers, TUA.

77. Washington, *History of Military and Civilian Pilot Training at Tuskegee*, pp. 229–31.

78. Telegram, Patterson to Col. James M. Lockett, 15 July 1941, Li–Ly Folder, GC 1941, Patterson Papers, TUA.

79. Telegrams, Patterson to Marshall and Arnold, 15 July 1941, U.S. War Department, January–July 1941 Folder, Box 2, Series 2, Patterson Papers, TUA.

80. Telegram, Patterson to Webster, 15 July 1941, Civil Aeronautics Authority Folder, Patterson Papers, TUA.

81. "History of 66th AAF Flying Training Detachment, Moton Field, Tuskegee Institute, Alabama," p. 18, 234.821 (February–7 December 1941), USAF Collection, USAF-HRC.

82. Telegram, Marshall to Patterson, 18 July 1941, U.S. War Department Folder, Box 1, Series 2, Patterson Papers, TUA.

83. Telegram, Webster to Patterson, n.d., Civil Aeronautics Authority Folder, Box 1, Series 2, Patterson Papers, TUA.

84. Telegram, Arnold to Patterson, 17 July 1941, U.S. War Department—99th Pursuit Squadron Folder 1, Box 1, Series 2, Patterson Papers, TUA.

85. "Remarks by Doctor F. D. Patterson . . . ," 19 July 1941, U.S. War Department, January–July 1941 Folder, Box 2, Series 2, Patterson Papers, TUA.

86. "Address by Major General Walter R. Weaver . . . ," 19 July 1941, U.S. War Department, January–July 1941 Folder, Box 2, Series 2, Patterson Papers, TUA.

87. "History of 66th AAF Flying Training Detachment, Moton Field, Tuskegee Institute, Alabama," p. 18, 234.821 (February–7 December 1941), USAF Collection, USAF-HRC.

88. Washington to Patterson, "Report on: Conference at Maxwell Field with Major L. S. Smith, Director of Training," 15 February 1941, appendix 1, "History of 66th AAF Flying Training Detachment, Moton Field, Tuskegee Institute, Alabama," 234.821 (February–7 December 1941), USAF Collection, USAF-HRC.

89. Robinson to Washington, 20 February 1941, G. L. Washington Folder, Box 1, Series 2, Patterson Papers, TUA.

90. Ibid.

91. Robinson to Bethune, 19 March 1941; Patterson to Robinson, 3 April 1941; Robinson to Patterson, 8 April 1941; all in Ro–Ry Folder, GC 1941, Patterson Papers, TUA. Patterson to J. E. Bryan, Alabama NYA, 10 April and 13 May 1941, National Youth Administration, Alabama Folder, GC 1941, Patterson Papers, TUA. See also Patterson to Washington, 7 May 1941, G. L. Washington Folder, Box 1, Series 2, Patterson Papers, TUA.

92. Washington to Patterson, 5 May 1941 (with attached application forms) and 20 May 1941, G. L. Washington Folder, Box 1, Series 2, Patterson Papers, TUA.

93. Robinson to Washington, 20 May 1941, Ro–Ry Folder, GC 1941, Patterson Papers, TUA.

94. Washington to Robinson, 22 May 1941, G. L. Washington Folder, Box 1, Series 2, Patterson Papers, TUA.

95. Barnett to Washington, 27 May 1941, G. L. Washington Folder, Box 1, Series 2, Patterson Papers, TUA.

96. Barnett to Patterson, 27 February 1941, Barnett Correspondence 1936–1942, LC 1937, Patterson Papers, TUA.

97. Washington to Patterson, 25 June 1941, U.S. War Department, January–July 1941 Folder, Box 2, Series 2, Patterson Papers, TUA.

98. Ibid.; Washington, *History of Military and Civilian Pilot Training at Tuskegee*, pp. 94, 150–52.

99. Washington to Patterson, 25 June 1941, U.S. War Department, January–July 1941 Folder, Box 2, Series 2, Patterson Papers, TUA.

100. Ibid.

101. Washington to Patterson, 5 May 1941 (with attached application forms) and 20 May 1941, G. L. Washington Folder, Box 1, Series 2, Patterson Papers, TUA; "History of 66th AAF Flying Training Detachment, Moton Field, Tuskegee Institute, Alabama," p. 13, 234.821 (February–7 December 1941), USAF Collection, USAF-HRC.

102. "History of 66th AAF Flying Training Detachment, Moton Field, Tuskegee Institute, Alabama," p. 13, 234.821 (February–7 December 1941), USAF Collection, USAF-HRC.

103. Ibid., pp. 8–9.

104. Washington, *History of Military and Civilian Pilot Training at Tuskegee*, p. 194.

105. Ibid., pp. 194–95.

106. Ibid., p. 195.

107. Ibid., p. 246.

108. Ibid.; "History of 66th AAF Flying Training Detachment, Moton Field, Tuskegee Institute, Alabama," p. 19, 234.821 (February–7 December 1941), USAF Collection, USAF-HRC.

109. Washington, *History of Military and Civilian Pilot Training at Tuskegee*, pp. 246–47.

110. "History of 66th AAF Flying Training Detachment, Moton Field, Tuskegee Institute, Alabama," p. 19, 234.821 (February–7 December 1941), USAF Collection, USAF-HRC.

111. Washington to Patterson, 14 November 1941, U.S. War Department, August-December 1941 Folder, Box 2, Series 2, Patterson Papers, TUA.

112. Craven and Cate, *Army Air Forces in World War II*, 6: 577–78.

113. "History of 66th AAF Flying Training Detachment, Moton Field, Tuskegee Institute, Alabama," p. 17, 234.821 (February–7 December 1941), USAF Collection, USAF-HRC.

114. Craven and Cate, *Army Air Forces in World War II*, 6: 569; William Wolf, "U.S.A.A.F. Pilot Training in World War II," pt. 1, *The Historical Aviation Album* 16 (1980): 60.

115. Anderson interview, pp. 77–78. G. L. Washington, in his *History of Military and Civilian Pilot Training at Tuskegee*, pp. 372–73, contends that Davis had difficulty in primary and that Anderson provided additional—and unauthorized—flying instruction to the young officer after hours at Kennedy Field. When asked about Washington's account (in the interview cited above), Anderson discounted it, declaring that "Davis did not have a problem." Since Anderson was Davis's primary instructor, his account seems more plausible.

116. The seven eliminees in the first class were John C. Anderson, Jr., Charles D. Brown, Theodore E. Brown, Marion A. B. Carter, Ulysses S. Pannell, William H. Slade, and Roderick C. Williams ("History of 66th AAF Flying Training Detachment, Moton Field, Tuskegee Institute, Alabama," pp. 11–12 and appendix 13, 234.821 [February–7 December 1941], USAF Collection, USAF-HRC).

12. Making the Dream a Reality: The Air Corps' First Black Pilots

1. Memorandum, Judge Hastie to Robert A. Lovett, Assistant Secretary of War (Air), 23 April 1941, Air Corps Departmental Folder, Civilian Aide Records, RG 107, NA.

2. Washington, *History of Military and Civilian Pilot Training at Tuskegee*, p. 134.

3. Ibid., pp. 102–3.

4. G. B. Edwards, Mayor, City of Tuskegee to Patterson, 3 January 1941, quoting telegram sent to Sen. John Bankhead, Sen. Lister Hill, and Cong. Henry Steagall, E Folder, GC 1941, Patterson Papers, TUA. Washington contends that city officials were initially ambivalent over the proposal to establish a military flying field for a black squadron in the area, except to oppose its location in or near the town. He maintains that the news of the Fort Davis selection prompted the town fathers to reassess their attitude toward the project and urge that the field be located closer to the town. Consequently, according to Washington's account, the Chehaw site was selected. Based on the letter cited above and the actual circumstances of the shift of the site from Fort Davis to Chehaw (described below), Washington's interpretation is incorrect (see Washington, *History of Military and Civilian Pilot Training at Tuskegee*, pp. 102–3).

5. "History of Tuskegee Army Air Field," pp. 2–3, 10, 289.28–1, vol. 1 (23 July–6 December 1941), USAF Collection, USAF-HRC. For a copy of the site board report and other documents relating to the selection of the site for TAAF, see Tuskegee 686.1 (Box 1828), Series 295, Correspondence Relating to Airfields, 1939–42, Office of the Commanding General, HQ Army Air Forces, Records of the Army Air Forces, RG 18, NA (hereafter cited as AAF Records, RG 18, NA).

6. "History of Tuskegee Army Air Field," p. 11, 289.28–1, vol. 1 (23 July–6 December 1941), USAF Collection, USAF-HRC. Routing and record sheets, Building and Grounds (B & G) Division to Deputy Chief of Staff (Air), 5 June 1941, and B & G Division to Executive, 12 June 1941, Tuskegee, 600.1 (Box 1828), Series 295, AAF Records, RG 18, NA.

7. "History of Tuskegee Army Air Field," p. 11, 289.28–1, vol. 1 (23 July–6 December 1941), USAF Collection, USAF-HRC.

8. Petition to Sen. John H. Bankhead and Sen. Lister Hill from Tuskegee, Alabama, 21 April 1941, Tuskegee, 350 (Miscellaneous), Series 295, AAF Records, RG 18, NA.

9. Ibid.

10. Telephone conversation transcription, Maj. Gen. George H. Brett with Hill, 28 April 1941, Tuskegee, 350 (Miscellaneous), Series 295, AAF Records, RG 18, NA.

11. Telephone conversation transcription, Brett with Hill's secretary, 28 April 1941, Tuskegee, 350 (Miscellaneous), Series 295, AAF Records, RG 18, NA.

12. Memorandum, Hastie to Robert A. Lovett, Assistant Secretary of War (Air), 23 April 1941, Air Corps Departmental Folder, Civilian Aide Records, RG 107, NA.

13. Telephone conversation transcriptions, Brett with Hill's secretary, 28 April 1941, Tuskegee, 350 (Miscellaneous), Series 295, AAF Records, RG 18, NA; see also Routing and record sheet, Training and Operations Division to B & G Division, 14 March 1941, Tuskegee, 320, Miscellaneous (Box 1826), Series 295, AAF Records, RG 18, NA.

14. Project plan, B & G Division, Office of the Chief of the Air Corps, to the Adjutant General's Office, 13 March 1941, Tuskegee, 600.1 (Box 1828), Series 295, AAF Records, RG 18, NA.

15. Memorandum, Lovett to Hastie, 29 April 1941, Air Corps Departmental Folder, Civilian Aide Records, RG 107, NA.

16. Memorandum, Hastie to Lovett, 2 May 1941, Air Corps Departmental Folder, Civilian Aide Records, RG 107, NA.

17. Osur, *Blacks in the Army Air Forces*, p. 44. The policy of segregated toilet facilities did not, however, originate with Kimble, although it may have first manifested itself under his tenure of command. In September Ellison requested an additional $4,000 to install separate toilets and drinking fountains in four warehouses under construction (Ellison to Chief of the Air Corps, 19 September 1941, Tuskegee, 400, Miscellaneous [Box 1827], Series 295, AAF Records, RG 18, NA).

18. Osur, *Blacks in the Army Air Forces*, pp. 44–45.

19. Routing and record sheet, B & G Division to Deputy Chief of Staff (Air), 5 June 1941, Tuskegee, 600.1 (Box 1828), Series 295, AAF Records, RG 18, NA.

20. "History of Tuskegee Army Air Field," pp. 11–12, 289.28–1, vol. 1 (23 July–6 December 1941), USAF Collection, USAF-HRC.

21. Barnett to Patterson, 10 February 1941, Barnett Correspondence 1936–1942, LC 1937, Patterson Papers, TUA.

22. Telegram, Patterson to Chief of the District Engineers, Mobile, Alabama, 8 March 1941, U.S. War Department, January–July 1941 Folder, Box 2, Series 2, Patterson Papers, TUA.

23. See correspondence with Louis Edwin Fry, architect for the Infantile Paralysis Hospital, November 1940, regarding McKissack and McKissack's work as contractors for the hospital in F Folder, GC 1940, Patterson Papers, TUA. Ironically, the Infantile Paralysis Hospital was dedicated on 16 January 1941, the same day that the War Department released its plans to form a segregated squadron.

24. Patterson to Mr. and Mrs. Calvin McKissack, 12 May 1941, McKissack and McKissack Folder, GC 1941, Patterson Papers, TUA.

25. Patterson to Whom It May Concern, 29 April 1941, McKissack and McKissack Folder, GC 1941, Patterson Papers, TUA.

26. F. D. Patterson to Assistant Secretary of War Robert P. Patterson, 6 May

1941, U.S. War Department, January–July 1941 Folder, Box 2, Series 2, Patterson Papers, TUA.

27. Ibid.

28. Ibid.

29. Memorandum, Assistant Secretary of War Patterson to Roosevelt, 19 May 1941, War Department, Chief of the Air Corps, 1940–1942, OF 25u, FDRL. The story of the March on Washington is told in Herbert Garfinkle, *When Negroes March: The March on Washington Movement in the Organizational Politics for FEPC* (Glencoe, Ill: Free Press, 1959).

30. Memorandum, Assistant Secretary of War Patterson to Eleanor Roosevelt 19 May 1941—with copy of memorandum, Assistant Secretary of War Patterson to Roosevelt, 19 May 1941, attached—War Department, Chief of the Air Corps, 1940–1942, OF 25u, FDRL. The March on Washington movement, spearheaded by A. Philip Randolph, is also described in Sitkoff, *New Deal for Blacks*, pp. 314–21.

31. Memorandum, Assistant Secretary of War Patterson to Eleanor Roosevelt, 13 June 1941, microfilm reel 115, item 2626, Marshall Foundation.

32. Washington to Smith, 22 January 1941, with about ninety "Memoranda on Personnel" forms compiled by G. L. Washington, Tuskegee, 351.28, Series 295, AAF Records, RG 18, NA.

33. Maj. James A. Ellison, Tuskegee Project Officer, to Chief of the Air Corps, 3 March 1941, with about thirty-five "Memoranda on Personnel" forms compiled by G. L. Washington, Tuskegee, 351.28, Series 295, AAF Records, RG 18, NA.

34. Washington, *History of Military and Civilian Pilot Training at Tuskegee*, pp. 118, 123, 128.

35. Ibid., p. 128.

36. *Air Corps News Letter* 24 (15 May 1941): 18, 167.63, USAF Collection, USAF-HRC.

37. Aviation Cadet Nelson S. Brooks to F. D. Patterson, 8 June 1941, Bro–By Folder, GC 1941, Patterson Papers, TUA.

38. Patterson to Marshall, 9 June 1941, with handwritten note, Marshall to Arnold, n.d., 220.765–2, vol. 2 (1941), USAF Collection, USAF-HRC.

39. Memorandum, Arnold to Marshall, 13 June 1941, with handwritten note, Marshall to Arnold, n.d., 220.765–2, vol. 2 (1941), USAF Collection, USAF-HRC.

40. Johnson to Patterson, 13 June 1941, U.S. War Department, January–July 1941 Folder, Box 2, Series 2, Patterson Papers, TUA.

41. Patterson to Johnson, 18 June 1941, U.S. War Department, January–July 1941 Folder, Box 2, Series 2, Patterson Papers, TUA.

42. Patterson to Marshall, 24 June 1941, U.S. War Department, January–July 1941 Folder, Box 2, Series 2, Patterson Papers, TUA.

43. Washington, *History of Military and Civilian Pilot Training at Tuskegee*, pp. 219–20.

44. Patterson to Marshall, 24 June 1941, U.S. War Department, January–July 1941 Folder, Box 2, Series 2, Patterson Papers, TUA.

45. Ibid.

46. Ibid.

47. Ibid.

48. Ibid.

49. Lt. Col. W. B. Smith, Assistant Secretary, General Staff, to Patterson, 19 July 1941, U.S. War Department, January–July 1941 Folder, Box 2, Series 2, Patterson Papers, TUA.

50. Hilyard R. Robinson to Patterson, 9 June 1941, Ro-Ry Folder, GC 1941, Patterson Papers, TUA; Washington, *History of Military and Civilian Pilot Training at Tuskegee*, p. 129.

51. "History of Tuskegee Army Air Field," pp. 11–12, 289.28–1, vol. 1 (23 July–6 December 1941), USAF Collection, USAF-HRC.

52. Ibid., p. 20.

53. Ibid., p. 25.

54. Ibid., p. 26.

55. 1st Indorsement, Col. L. D. Worsham, Division Engineer, South Atlantic Division to the Chief of Engineers, 29 August 1941, Tuskegee, 600.1 (Box 1828), Series 295, AAF Records, RG 18, NA.

56. Gen. George E. Stratemeyer to Commanding General, SEACTC, 4 September 1941, Tuskegee, 600.1 (Box 1828), Series 295, AAF Records, RG 18, NA.

57. Washington to Maj. James A. Ellison, 15 September 1941, U.S. War Department, August–December 1941 Folder, Box 2, Series 2, Patterson Papers, TUA.

58. Washington to Smith, 23 September 1941, U.S. War Department, August–December 1941 Folder, Box 2, Series 2, Patterson Papers, TUA.

59. Smith to Washington, 26 September 1941, U.S. War Department, August–December 1941 Folder, Box 2, Series 2, Patterson Papers, TUA.

60. Telegram, Lt. Col. S. N. Karrick to the Inspector General, 25 September 1941, Tuskegee, 330, Miscellaneous (Box 1827), Series 295, AAF Records, RG 18, NA.

61. Karrick to the Inspector General, 29 September, Tuskegee, 600.1 (Box 1828), Series 295, AAF Records, RG 18, NA.

62. Memorandum, Maj. Gen. Virgil Peterson, Inspector General, to the Chief of Staff, 26 September 1941, with handwritten note, H. H. A.[rnold] to Stratemeyer, 26 September 1941, Tuskegee, 330, Miscellaneous (Box 1827), Series 295, AAF Records, RG 18, NA.

63. Memorandum, Arnold to the Inspector General, 30 September 1941, Tuskegee, 350, Miscellaneous (Box 1827), Series 295, AAF Records, RG 18, NA.

64. Telegram, Weaver to Chief of the Air Corps, 9 October 1941, Tuskegee, 400, Miscellaneous (Box 1827), Series 295, AAF Records, RG 18, NA.

65. Ibid.

66. Telegram, Fairchild to Commanding General, SEACTC, Tuskegee, 400, Miscellaneous (Box 1827), Series 295, AAF Records, RG 18, NA.

67. "History of Tuskegee Army Air Field," pp. 30–31, 49–50, 289.28–1, vol. 1 (23 July–6 December 1941), USAF Collection, USAF-HRC.

68. "History of the [Tuskegee Army Air Field] Corps of Aviation Cadets," p. 2, 289.28–1, vol. 2 (August 1941–March 1944), USAF Collection, USAF-HRC.

69. "History of Tuskegee Army Air Field," p. 31, 289.28–1, vol. 1 (23 July–6 December 1941), USAF Collection, USAF-HRC.

70. "History of Tuskegee Army Air Field," p. 11, 289.28–2 (7 December 1941–31 December 1942), USAF Collection, USAF-HRC.

71. Lt. Col. John R. Hardin, Chief, Construction Section, Office of the Chief of Engineers, to Chief of the Air Corps, 24 September 1941, Tuskegee, 611 (Box 1828), Series 295, AAF Records, RG 18, NA.

72. Telegram, F. D. Patterson to Assistant Secretary of War Robert P. Patterson, 26 September 1941, U.S. War Department, August–December 1941 Folder, Box 2, Series 2, Patterson Papers, TUA.

73. Calvin McKissack to Patterson, 17 October 1941, U.S. War Department, August-December 1941 Folder, Box 2, Series 2, Patterson Papers, TUA.

74. F. D. Patterson to Robert P. Patterson, 20 October 1941, U.S. War Department, August–December 1941 Folder, Box 2, Series 2, Patterson Papers, TUA.

75. Carter to F. D. Patterson, 31 October 1941, U.S. War Department, August–December 1941 Folder, Box 2, Series 2, Patterson Papers, TUA. The original construction estimate (taken from the 18 December 1940 plan the Air Corps submitted to the Adjutant General, subsequently approved by the War Department in January 1941) came to just under $1.1 million; see "History of Tuskegee Army Air Field," appendix 1E, 289.28–1, vol. 1 (23 July–6 December 1941), USAF Collection, USAF-HRC.

76. Carter to F. D. Patterson, 31 October 1941, U.S. War Department,

August–December 1941 Folder, Box 2, Series 2, Patterson Papers, TUA, emphasis added.

77. Telegram, C. G. Kershaw Contracting Company and Daugette–Millican Company, by H. Y. Dempsey, Project Manager to Hill, Bankhead, Steagall, and Starnes, 28 November 1941, McKissack and McKissack Folder, GC 1941, Patterson Papers, TUA. The Kershaw and Daugette–Millican companies are presumed to be the unnamed white contractors cited by G. L. Washington as the firm responsible for grading and excavation operations; see Washington, *History of Military and Civilian Pilot Training at Tuskegee*, p. 239.

78. F. D. Patterson to Robert P. Patterson, 29 November 1941, U.S. War Department, August–December 1941 Folder, Box 2, Series 2, Patterson Papers, TUA.

79. Copies of President Patterson's letters to these individuals, all dated 29 November 1941 and enclosing a copy of his letter to Assistant Secretary of War Patterson, were found in the Patterson Papers, TUA, in the following locations: in GC 1941—Roosevelt (Ro–Ry Folder), Hill (He–Hi Folder), Steagall (Sn–Ste Folder), Wallace (Wa–We Folder), and Bethune (National Youth Administration Folder); in U.S. War Department, August–December 1941 Folder, Box 2, Series 2—Marshall, Arnold, and Gibson; and in the Barnett Correspondence 1936–1942, LC 1937—Barnett.

80. F. D. Patterson to Robert P. Patterson, 29 November 1941, U.S. War Department, August–December 1941 Folder, Box 2, Series 2, Patterson Papers, TUA.

81. Ibid.

82. Ibid.

83. Lt. Col. Ralph E. Stearley, Training Division, to the Secretary of the Air Staff, 6 December 1941, Tuskegee, 353.9 (Box 1827), Series 295, AAF Records, RG 18, NA.

84. Telegram, Barnett to Patterson, 10 December [1941], Barnett Correspondence 1936–1942, LC 1937, Patterson Papers, TUA.

85. Patterson to Barnett, 19 December 1941, Barnett Correspondence 1936–1942, LC 1937, Patterson Papers, TUA.

86. Telegram, Patterson to Weaver, 16 December 1941, U.S. War Department, August–December 1941 Folder, Box 2, Series 2, Patterson Papers, TUA.

87. Patterson to Barnett, 17 December 1941, Barnett Correspondence 1936–1942, LC 1937, Patterson Papers, TUA.

88. J. Henry Smith, Assistant to the President, to Barnett, 18 December 1941, with attached press release, Barnett Correspondence 1936–1942, LC 1937, Patterson Papers, TUA.

89. Ibid.

90. Arnold to Patterson, 3 January 1941, U.S. War Department Folder, Box 1, Series 2, Patterson Papers, TUA.

91. Stratemeyer to Patterson, 1 January 1941, U.S. War Department Folder, Box 1, Series 2, Patterson Papers, TUA.

92. Telegram, Barnett to Patterson, 24 December 1941, Barnett Correspondence 1936–1942, LC 1937, Patterson Papers, TUA.

93. G. L. W.[ashington] to Patterson, 25 November 1941, U.S. War Department, August–December 1941 Folder, Box 2, Series 2, Patterson Papers, TUA.

94. Patterson to Arnold, 26 November 1941, U.S. War Department, August–December 1941 Folder, Box 2, Series 2, Patterson Papers, TUA.

95. Parrish to Chief of the Air Corps, 3 December 1941, 220.765–2, vol. 2 (1941), USAF Collection, USAF-HRC.

96. Ibid.

97. Ibid.

98. Ibid.

99. Washington to John P. Morris, Director, Civil Pilot Training, 11 December 1941, page 1, in U.S. War Department Folder, Box 1, and pages 2 and 3 in U.S. War Department, August–December 1941 Folder, Box 2, Series 2, Patterson Papers, TUA.

100. Washington to Patterson, 22 December 1941, with covering note, n.d., U.S. War Department, August–December 1941 Folder, Box 2, Series 2, Patterson Papers, TUA.

101. Ibid.

102. Ibid.

103. Ibid.

104. At the end of September 1941 the Air Staff told the Richmond (Va.) *News Leader* that there were 255 Negro applicants on the eligible list awaiting appointment as aviation cadets, a backlog of over two years if the quota of ten men every five weeks was maintained (Richmond [Virginia] *News Leader* to Chief of the Air Corps, 25 September 1941, and First Lieutenant William W. Beasley, Military Personnel Division, to Richmond [Virginia] *News Leader*, 27 September 1941, Tuskegee, 351.28 [Box 1827], Series 295, AAF Records, RG 18, NA). The Air Staff readily acknowledged that criticism over the long delay in ordering black aviation cadet selectees to flying training was the primary reason behind the establishment of a second squadron; see memorandum, Lt. Col. C. E. Duncan, Secretary of the Air Staff, to the Chief of Staff, 8 December 1941, Xerox 732, Marshall Foundation.

105. Col. St. Clair Street to Hastie, 21 November 1941, Air Corps Departmental Folder, Civilian Aide Records, RG 107, NA.

106. Routing and record sheet, Chief of the Air Staff to Chief of the Air Corps, 1 November 1941, 220.765–2, vol. 2 (1941), USAF Collection, USAF-HRC.

107. Routing and record sheet, Chief of the Air Corps to Chief of the Air Staff, 21 November 1941, 220.765–2, vol. 2 (1941), USAF Collection, USAF-HRC.

108. Memorandum, Lt. Col. C. E. Duncan, Secretary of the Air Staff, to the Chief of Staff, 8 December 1941, Xerox 732, Marshall Foundation.

109. Ibid.

110. Robert P. Patterson to F. D. Patterson, 17 January 1942, U.S. War Department Folder, Box 1, Series 2, Patterson Papers, TUA.

111. Patterson to Hastie, 21 January 1942, U.S. War Department Folder, Box 1, Series 2, Patterson Papers, TUA.

112. Hastie to Patterson, 28 January 1942, U.S. War Department Folder, Box 1, Series 2, Patterson Papers, TUA.

113. Barnett to Patterson, 15 January 1942, and Patterson to Barnett, 20 January 1941, Barnett Correspondence 1936–1942, LC 1937, Patterson Papers, TUA. See also Hastie to Assistant Secretary of War for Air, 21 April 1942, Air Corps Departmental Folder, Civilian Aide Records, RG 107, NA.

114. Col. W. F. Volandt to Assistant Chief, Materiel Division, Wright Field, 7 February 1942, Tuskegee, 100, Miscellaneous, Series 295, AAF Records, RG 18, NA.

115. The AAF official history of Tuskegee Army Air Field indicates that the first class graduated on 6 March 1942 ("History of Tuskegee Army Air Field," p. 97, 289.28–2 [7 December 1941–31 December 1942], USAF Collection, USAF-HRC), citing graduation orders dated 6 March 1942. Other reliable sources indicate that the graduation ceremenoy was held 7 March. See, for example, Benjamin O. Davis, Jr., *American: An Autobiography* (Washington: Smithsonian Institution Press, 1991), p. 86; "Address by Major General George E. Stratemeyer . . . at the graduation at the Air Corps Advanced Flying School, Tuskegee, Alabama," a press release from Maxwell Field, Alabama, Public Relations Office, with instructions *"For release after delivery scheduled for 10 A.M., Saturday, March 7, 1942,"* in U.S. War Department Folder, Box 1, Series 2, Patterson Papers, TUA; *Pittsburgh Courier* (nat. ed.), 14 March 1942, pp. 1 and ff. Apparently, the orders certifying completion of the training were issued on Friday, 6 March 1942, and the graduation ceremony was held the following day.

116. Ibid.; Washington, *History of Military and Civilian Pilot Training at Tuskegee*, p. 330.

117. Washington, *History of Military and Civilian Pilot Training at Tuskegee*, p. 330.

13. Conclusion

1.   Perhaps the best indication of the general acceptance of the term *Tuskegee Airmen* as the name for the nation's black pilots of World War II came when it appeared in the script of the popular NBC television series "The Bill Cosby Show" during the 1986–87 season. There is also an organization known as The Tuskegee Airmen, Inc., established in 1972, numbering approximately 1,400 members, primarily veterans of the segregated air units of World War II as well as black men and women currently involved in military aviation (see *Encyclopedia of Associations*, 21st ed. [Detroit: Gale Research Co., 1987], entry 16949). The term *Tuskegee Airmen* was first used as early as 1944, when a group of civilian instructor pilots at Tuskegee Institute formed an organization known as the Tuskegee Airmen's Association (see John H. Young, III, Executive Secretary, Tuskegee Airmen's Association to Robert A. Lovett, Assistant Secretary of War for Air, 23 May 1944, 291.2, Negroes, Classified Decimal File, 1940–1946, Office of the Assistant Secretary of War for Air, RG 107, NA). By the mid–1950s the term took on a broader meaning, especially after the publication of Charles E. Francis's *The Tuskegee Airmen: The Story of the Negro in the U.S. Air Force* (Boston: Bruce Humphries, 1955).

2.   For an assessment of the contemporary relationship between black Americans and military service, see Moskos, "Success Story: Blacks in the Army," pp. 64–72.

3.   Richard Wright, *Native Son* (New York: Harper and Brothers, 1940; paperback reprint by the same publisher, Perennial Classics, 1966), p. 20.

4.   Patterson to Barnett, 20 January 1942, Barnett Correspondence 1936–1942, LC 1937, Patterson Papers, TUA.

5.   Harvard Sitkoff, "American Blacks in World War II: Rethinking the Militancy-Watershed Hypothesis," in *The Home Front and War in the Twentieth Century: The American Experience in Comparative Perspective*, edited by James Titus, Proceedings of the Tenth Military History Symposium, October 20–22, 1982 (Washington, D.C.: United States Air Force Academy and Office of Air Force History, 1984), pp. 148–49.

6.   Francis, *Tuskegee Airmen*, p. 30.

7.   Gropman, *The Air Force Integrates*, pp. 86–90; MacGregor, *Integration of the Armed Forces*, pp. 287–90, 338–42, 397–412.

 **Bibliography**

Archival Materials

Henry H. Arnold Papers. Manuscript Division, Library of Congress. Washington, D.C.

Claude A. Barnett Papers. Chicago Historical Society. Chicago, Illinois.

Black Wings Collection. National Air and Space Museum, Smithsonian Institution. Washington, D.C.

Thomas Monroe Campbell Papers. Tuskegee University Archives. Tuskegee, Alabama.

Civil Action Suit no. 9763, 16 January 1941. Civil Docket Book. District Court of the United States for the District of Columbia. Washington, D.C.

Civil Aeronautics Authority. "Wartime History of the Civil Aeronautics Administration." Federal Aeronautics Administration Library. Washington, D.C., n.d.

William H. Hastie Papers. Harvard Law School Library. Cambridge, Massachusetts.

Harold A. McGinnis Papers. United States Air Force Historical Research Center. Maxwell Air Force Base. Montgomery, Alabama.

George C. Marshall Papers. George C. Marshall Foundation. Lexington, Virginia.

Minutes of the Tuskegee Institute Board of Trustees. Tuskegee University Archives. Tuskegee, Alabama.
  Minutes of the Meeting of the Board of Trustees of the Tuskegee Normal and Industrial Institute, October 27, 1933.
  Minutes of the Annual Meeting of the Board of Trustees of the Tuskegee Institute, April 6, 1940.
  Minutes of the Annual Meeting of the Board of Trustees of the Tuskegee Institute, 24 October 1940.
  Minutes of the Spring Meeting of the Board of Trustees of the Tuskegee Institute, April 5, 1941.

Robert Russa Moton Papers. Tuskegee University Archives. Tuskegee, Alabama.

National Association for the Advancement of Colored People Papers. Manuscript Division, Library of Congress. Washington, D.C.

Frederick Douglass Patterson Papers. Tuskegee University Archives. Tuskegee, Alabama.

Records of Civil Action Suit no. 9763, 16 January 1941, of the District Court of the United States for the District of Columbia. File no. CA 9763. Accession no. 21–66–A–1059. Box no. 152. Location no. 16–78–4–5. Washington National Records Center. Suitland, Maryland.

Records of the Army Air Forces. Office of the Commanding General, HQ Army Air Forces. Series 295, Correspondence Relating to Airfields, 1932–42. Record Group 18, National Archives, Washington, D.C.

Records of the Civil Aeronautics Administration. General Files. Record Group 237. National Archives. Washington, D.C.

Records of the Secretary of War. Record Group 107. National Archives. Washington, D. C.
Civilian Aide to the Secretary of War. Subject File, 1940–1947.
Office of Assistant Secretary of War for Air. Classified Decimal File, 1940–1945.

Eleanor Roosevelt Papers. Franklin D. Roosevelt Library. Hyde Park, New York.

Franklin D. Roosevelt Papers. Franklin D. Roosevelt Library. Hyde Park, New York.

United States Air Force Collection. United States Air Force Historical Research Center. Maxwell Air Force Base. Montgomery, Alabama.
101–7B. "Legislation Relating to the Army Air Forces Training Program, 1939–1945." Rev. ed. Army Air Forces Historical Studies no. 7. Headquarters, Army Air Forces, April 1946.
101–90. (1941–1953). "The Preflight Schools in World War II." U.S. Air Force Historical Study no. 90.
220.1645–6 (1941). Office files.
220.765–2, vol. 1 (1940) and vol. 2 (1941). Office files.
220.765–5 (1940). "Reports of Conferences Relative to the Training of Negro Pilots, January 1940."
220.765–6 (1940). Office files.
234.142–1 (28 June 1939–30 March 1940). "Historical Record, Air Corps Training Detachment, Chicago School of Aeronautics, Glenview, Illinois."
234.821 (February–7 December 1941). "History of 66th AAF Flying Training Detachment, Moton Field."
289.28–1, vol. 1 (23 July–6 December 1941). "History of Tuskegee Army Air Field."

289.28–1, vol. 2 (August 1941–March 1944). "History of the [Tuskegee Army Air Field] Corps of Aviation Cadets."

289.28–2 (7 December 1941–31 December 1942). "History of Tuskegee Army Air Field."

Warner, Robert T., and Harold R. Decker. "Auburn–Opelika Airport History." Historical Collection. Auburn University Archives. Auburn, Alabama. 24 May 1974.

Microfilmed Materials

The Claude A. Barnett Papers. Microfilm ed. Frederick, Md.: University Publications of America, 1986.

Kirby, John B., ed. *New Deal Agencies and Black America in the 1930s.* 25 microfilm reels. Frederick, Md.: Univeristy Publications of America, 1983.

Kitchens, John W. and Jonell Chislom Jones, eds. *Microfilm Edition of The Tuskegee Institute News Clippings File.* 252 microfilm reels. Tuskegee, Ala.: Tuskegee Institute, 1978.

Papers of the NAACP. Part 1, *Meetings of the Boards of Directors, Records of Annual Conferences, Major Speeches, and Special Reports, 1909–1950.* 28 microfilm reels. Frederick, Md.: University Publications of America, 1981.

Henry Lewis Stimson Diaries. 9 microfilm reels. Manuscript and Archives, Yale University Library. New Haven, Connecticut.

Interviews and Oral Histories

Allen, Joseph Wren. Interviewed by author. Montgomery, Alabama, 24 October 1985, and by telephone, 26 February 1987.

Anderson, Charles Alfred "Chief." Oral history interview by James C. Hasdorff, 8–9 June 1981. Transcript at United States Air Force Historical Research Center, Maxwell Air Force Base. Montgomery, Alabama.

Parrish, Brigadier General Noel F. Oral history interview by James C. Hasdorff, 10–14 June 1974. Transcript at United States Air Force Historical Research Center, Maxwell Air Force Base. Montgomery, Alabama.

Pitts, Robert G. Interviewed by author. Auburn, Alabama, 17 February 1985.

Shelton, Forrest. Interviewed by author. Atlanta, Georgia, 29 October 1985.

Federal Documents

U.S. Bureau of the Census. *Statistical History of the United States From Colonial Times to the Present.* New York: Basic Books, 1976.

U.S. Congress. *Congressional Record.* 76th Cong., 1st sess., 7 February–21 June 1939. 76th Cong., 3rd sess., 25 January–26 August 1940.

U.S. Congress. House of Representatives. *Supplemental Military Appropriation Bill for 1940: Hearings Before the Subcommittee on Appropriations, House of Representatives, on the Supplemental Military Appropriation Bill for 1940 [H.R. 6791].* 76th Cong., 1st sess., 16 May–5 June 1939.

———. *Military Establishment Appropriation Bill for 1941: Hearings before the Subcommittee of the Committee on Appropriations.* 76th Cong., 3rd sess., 1940.

———. *Second Supplemental National Defense Appropriation Bill for 1941: Hearings before the Subcommittee of the Committee on Approriations.* 76th Cong., 3rd sess., 1940.

———. *Selective Compulsory Military Training and Services: Hearings before the Committee on Military Affairs . . . on H.R. 10132.* 76th Cong., 3rd sess., 1940.

———. *Training of Civil Aircraft Pilots: Hearings Before the Committee on Interstate and Foreign Commerce, House of Representatives, on H.R. 50[9]3.* 76th Cong., 1st sess., 20 and 27 March 1939. (The H.R. number was erroneously printed as 5073 on the cover.)

U.S. Congress. Senate. *Hearings Before the Committee on Military Affairs, United States Senate, Seventy-sixth Congress, First Session, on H.R. 3791, An Act to Provide More Effectively for the National Defense by Carrying Out the Recommendations of the President in His Message of January 12, 1939, to the Congress.* 76th Cong., 1st sess., 1939.

———. *Military Establishment Appropriation Bill for 1940: Hearings Before the Subcommittee of the Committee on Appropriations on H.R. 4630.* 76th Cong., 1st sess., 13–17 March 1939.

———. *Military Establishment Appropriation Bill for 1941: Hearings Before the Subcommittee of the Committee on Appropriations, United States Senate . . . on H.R. 9209.* 76th Cong., 3rd sess., 30 April–17 May 1940.

———. *Training of Civil Aircraft Pilots: Hearing Before a Subcommittee of the Committee on Commerce, United States Senate, on S. 2119.* 76th Cong., 1st sess., 20 April 1939.

U.S. Department of Commerce. Bureau of Foreign and Domestic Commerce. *Negro Aviators* (15 August 1936).

U.S. Department of Commerce. Bureau of the Census. *Negro Aviators.* Bulletin no. 3 (January 1939).

———. *Negro Aviators.* Negro Statistical Bulletin no. 3 (September 1940).

———. *Negro Aviators.* Negro Statistical Release no. 3 (April 1942).

U.S. Patent Office. *Annual Report for the Commissioner of Patents for the Years 1910–1918.* Washington, D.C.: Government Printing Office, 1915–1919.

———. *Official Gazette,* vols. 90–286 (1900–21).

U.S. Supreme Court. Dred Scott v. Sandford. 19 Howard (U.S.), 414–417. 15 L. Ed. 705 (1884).

United States Statutes at Large, vols. 53 (1939) and 54 (1940).

Books, Articles, and Dissertations

Abajian, James de T., comp. *Blacks in Selected Newspapers, Censuses, and Other Sources: An Index to Names and Subjects*. 3 vols. Boston: G. K. Hall and Co., 1977.

Baker, Henry E. "The Negro in the Field of Invention." *Journal of Negro History* 2 (January 1917): 21–36.

Barbour, George Edward. "Early Black Flyers of Western Pennsylvania, 1906–1945," *Western Pennsylvania Historical Magazine* 69 (April 1986): 95–119.

Barbeau, Arthur E., and Florette Henri. *The Unknown Soldiers: Black American Troops in World War I*. Philadelphia: Temple Univ. Press, 1974.

Boatner, Mark M., III. *The Civil War Dictionary*. New York: David McKay Co., 1959.

Bricktop, with James Haskins. *Bricktop*. New York: Atheneum, 1983.

Brooks-Pazmany, Kathleen L. *United States Women in Aviation, 1919–1929*. Washington, D.C.: Smithsonian Institution Press, 1983.

Brown, Richard Carl. "Social Attitudes of American Generals, 1898–1940." Ph.D. dissertation, University of Wisconsin, 1951.

Brown, Walt, Jr., ed. *An American for Layfayette: The Diaries of E. C. C. Genet, Lafayette Escadrille*, with an introduction by Dale L. Walker. Charlottesville: Univ. of Virginia Press, 1981.

Buchanan, A. Russell. *Black Americans in World War II*. Santa Barbara, Calif.: Clio Books, 1977.

Bunch, Lonnie G., III. "In Search of a Dream: The Flight of [James] Herman Banning and Thomas Allen." *Journal of American Culture* 7 (Spring/Summer 1984): 100–103.

Buni, Andrew. *Robert L. Vann of the Pittsburgh Courier: Politics and Black Journalism*. Pittsburgh: Univ. of Pittsburgh Press, 1974.

Butler, Addie Louise Joyner. *The Distinctive Black College: Talladega, Tuskegee and Morehouse*. Mutchen, N.J.: Scarecrow Press, 1977.

Carisella, P. J., and James W. Ryan.*The Black Swallow of Death: The Incredible Story of Eugene Jacques Bullard, the World's First Black Combat Aviator*. Boston: Marlborough House, 1972.

Carroll, John M., ed. *The Black Military Experience in the American West*. New York: Liverwright, 1971.

Carter, Paul. *The Twenties in America*. 2d ed. Arlington Heights, Ill.: AHM Publishing, 1975.

Collings, Kenneth Brown. "America Will Never Fly." *American Mercury* 38 (July 1936): 290–95.

———. "Mr. Collings Replies." *American Mercury* 39 (December 1936): xxx.

Complete Presidential Press Conferences of Franklin D. Roosevelt, with an introduction by Jonathan Daniels. 25 vols. New York: Da Capo Press, 1972.

Corn, Joseph J. *The Winged Gospel: America's Romance with Aviation, 1900-1950*. New York: Oxford Univ. Press, 1983.

Cornish, Dudley Taylor. *The Sable Arm: Negro Troops in the Union Army, 1861–1865*. New York: Longmans, Green, 1956.

Craven, Wesley Frank, and James Lea Cate, eds. *The Army Air Forces in World War II*. 7 vols. Chicago: Univ. of Chicago Press, 1948–58. Reprint. Washington, D.C.: Office of Air Force History, 1983.

Dalfiume, Richard M. *Desegregation of the U.S. Armed Forces: Fighting on Two Fronts, 1939–1953*. Columbia: Univ. of Missouri Press, 1969.

Daniel, Pete. "Black Power in the 1920s: The Case of Tuskegee Veterans Hospital," *Journal of Southern History* 36 (August 1970): 368–88.

Davis, Benjamin O., Jr. *American: An Autobiography*. Washington: Smithsonian Institution Press, 1991.

Davis, Mariaana W., ed. *Contributions of Black Women to America*. 2 vols. Columbia, S.C.: Kenday Press, 1982.

Dictionary of American Biography: Supplement Five, 1951–1955. N.Y.: Charles Scribner's Sons, 1977.

Draper, Theodore. *The Rediscovery of Black Nationalism*. New York: Viking Press, 1970.

"Drive Launched to Honor [Bessie] Coleman with [U.S. Postal] Stamp," *Chicago Defender*, 5 May 1986, p. 13.

Fishel, Leslie H., Jr., and Benjamin Quarles, comps. *The Negro American: A Documentary History*. Glenview, Ill.: Scott, Foresman and Co., 1967.

Flammer, Philip M. *The Vivid Air: The Lafayette Escadrille*. Athens: Univ. of Georgia Press, 1981.

Fletcher, Marvin. *The Black Soldier and Officer in the United States Army, 1891–1917*. Columbia: Univ. of Missouri Press, 1974.

Foner, Jack D. *Blacks and the Military in American History: A New Perspective*. New York: Praeger, 1974.

Forsythe, Albert E. "The Outlook for the Negro in Aviation." *Tuskegee Messenger* 11 (October 1935): 7.

Fowler, Arlen L. *The Black Infantry in the West, 1869–1891*. Westport, Conn.: Greenwood Publishing, 1971.

Francis, Charles E. *The Tuskegee Airmen: The Story of the Negro in the U.S. Air Force.* Boston: Bruce Humphries, 1955.

Franklin, John Hope. *From Slavery to Freedom: A History of Negro Americans.* 5th ed. New York: Alfred A. Knopf, 1980.

Frazier, E. Franklin. *Negro Youth at the Crossways: Their Personality Development in the Middle States.* Washington, D.C.: American Council on Education, 1940.

Garfinkle, Herbert. *When Negroes March: The March on Washington Movement in the Organizational Politics for FEPC.* Glencoe, Ill.: Free Press, 1959.

Garrison, William Lloyd. *The Loyalty and Devotion of Colored Americans in the Revolution and War of 1812.* Boston: R. F. Walcut, 1861. (Reprinted in *The Negro Soldier: A Select Compilation.* New York: Negro Universities Press, 1970.)

Goodrich, J. "Salute to Bessie Coleman." *Negro Digest* 8 (May 1950): 82–83.

Graham, Otis L., Jr., and Meghan Robinson Wander, eds. *Franklin D. Roosevelt: His Life and Times, An Encyclopedic View.* Boston: G. K. Hall, 1985.

Greene, Robert Ewell. *Black Defenders of America, 1775–1973.* Chicago: Johnson Publishing, 1974.

Gropman, Alan L. *The Air Force Integrates, 1945–1964.* Washington, D.C.: Office of Air Force History, 1978.

Gropman, Alan L. "Benjamin O. Davis, Jr.: History on Two Fronts." In *Makers of the United States Air Force.* Edited by John L. Frisbee. Washington, D.C.: Office of Air Force History, 1987.

Hall, James Norman, and Charles Bernard Nordhoff, eds. *The Lafayette Flying Corps.* 2 vols. Boston: Houghton Mifflin, 1920.

Hallion, Richard P. *Legacy of Flight: The Guggenheim Contribution to American Aviation.* Seattle: Univ. of Washington Press, 1977.

Hardesty, Von, and Dominick Pisano. *Black Wings: The American Black in Aviation.* Washington, D.C.: National Air and Space Museum, Smithsonian Institution, 1983.

Harlan, Louis. *Booker T. Washington: The Making of a Black Leader, 1856–1901.* New York: Oxford Univ. Press, 1972.

———. *Booker T. Washington: The Wizard of Tuskegee, 1901–1915.* New York: Oxford Univ. Press, 1983.

Hasdorff, James C. "Reflections on the Tuskegee Experiment: An Interview with Brig. Gen. Noel F. Parrish, USAF (Ret.)." *Aerospace Historian* 24 (Fall, September 1977): 173–80.

Haynes, Elizabeth Ross. *The Black Boy of Atlanta[: A Biography of Richard Robert Wright, Sr.].* Boston: House of Edinboro, 1952.

Haynes, Robert V. "The Houston Mutiny and Riot of 1917." *Southwestern Historical Review* 76 (April 1973): 418–39.

―――. *A Night of Violence: The Houston Riot of 1917*. Baton Rouge: Louisiana State Univ. Press, 1976.

Herr, Allen. "American Pilots in the Spanish Civil War." *American Aviation Historical Society Journal* 22 (Fall 1977): 162–78.

―――. "American Pilots in the Spanish Civil War: Addenda." *American Aviation Historical Society Journal* 23 (Fall 1978): 234–35.

Hill, Robert A., Emory J. Tolbert, and Deborah Forczek, eds. *The Marcus Garvey and Universal Negro Imporvement Association Papers*. Vol. 4, *1 September 1921–2 September 1922*. Berkeley: Univ. of California Press, 1985.

Hinckley, Robert H., and JoAnn J. Wells. *"I'd Rather Be Born Lucky than Rich": The Autobiography of Robert H. Hinckley*. Charles Redd Monographs in Western History no. 7. Provo, Utah: Brigham Young Univ. Press, 1977.

Hopkins, George E. *The Airline Pilots: A Study in Elite Unionization*. Cambridge, Mass.: Harvard Univ. Press, 1971.

Hughes, William H., and Frederick D. Patterson, eds. *Robert Russa Moton of Hampton and Tuskegee*. Chapel Hill: Univ. of North Carolina Press, 1956.

Jones, Allen W. "The Role of Tuskegee Institute in the Education of Black Farmers." *Journal of Negro History* 60 (April 1975): 252–61.

Julian, Hubert, as told to John Bulloch. *Black Eagle*. London: The Adventurers Club, 1965.

King, Anita. "Brave Bessie: First Black Pilot." *Essence* 7 (May 1976): 36.

Kirby, John B. *Black Americans in the Roosevelt Era: Liberalism and Race*. Twentieth-Century America Series. Knoxville: Univ. of Tennessee Press, 1980.

Lane, Ann J. *The Brownsville Affair: National Crisis and Black Reaction*. Port Washington, N.Y.: Kennikat Press, 1971.

Leckie, William H. *The Buffalo Soldiers: A Narrative of the Negro Cavalry in the West*. Norman: Univ. of Oklahoma Press, 1967.

Lee, Ulysses. *United States Army in World War II. Special Studies: The Employment of Negro Troops*. Washington, D.C.: Office of the Chief of Military History, United States Army, 1966.

Lester, Daniel W., Sandra K. Faull, and Lorraine E. Lester, comps. *Cumulative Title Index to United States Public Documents, 1789–1976*. 26 vols. Arlington, Va.: United States Historical Documents Institute, 1979–82.

Lindbergh, Charles A. "Aviation, Geography, and Race." *Reader's Digest* 35 (November 1939): 64–67.

―――. *"We": The Famous Flier's Own Story of His Life and Transatlantic Flight, Together with His Views on the Future of Aviation*. New York: Grosset and Dunlop, 1927.

Logan, Rayford W. *The Betrayal of the Negro from Rutherford B. Hayes to Woodrow Wilson*. New York: Collier Books, 1965.

Louis, Joe. *Joe Louis: My Life.* New York: Harcourt, Brace, Javonovich, 1978.

MacGregor, Morris J., Jr., *Integration of the Armed Forces, 1940–1965.* Defense Studies Series. Washington, D.C.: Center of Military History, United States Army, 1981.

MacGregor, Morris J., Jr., and Bernard C. Nalty, eds. *Blacks in the Military: Basic Documents.* 13 vols. Wilmington, Del.: Scholarly Resources, 1977.

McMurry, Linda O. *George Washington Carver: Scientist and Symbol.* New York: Oxford Univ. Press, 1981.

Manchester, William. *The Glory and the Dream: A Narrative History of America, 1932–1972.* New York: Bantam Books, 1974.

Marable, Manning. "Tuskegee Institute in the 1920's." *Negro History Bulletin* 40 (November–December 1977): 764–68.

Matthews, Carl S. "The Decline of the Tuskegee Machine, 1915–1925: The Abdication of Political Power." *South Atlantic Quarterly* 75 (Autumn 1976): 460–69.

Moore, George Henry. *Historical Notes on the Employment of Negroes in the American Army of the Revolution.* New York: Charles T. Evans, 1862. (Reprinted in *The Negro Soldier: A Select Compilation.* New York: Negro Universities Press, 1970.)

Moskos, Charles. "Success Story: Blacks in the Army." *The Atlantic* 257 (May 1986): 64–72.

Murray, Williamson. *Luftwaffe.* Baltimore, Md.: Nautical and Aviation Publishing Company of America, 1985.

Myrdal, Gunnar. *An American Dilemma: The Negro Problem and Modern Democracy.* New York: Harper and Brothers, 1944.

Nalty, Bernard C. *Strength for the Fight: A History of Black Americans in the Military.* New York: Free Press, 1986.

National Association for the Advancement of Colored People. *NAACP Annual Reports for 1940–1941.*

The Negro Soldier: A Select Compilation. New York: Negro Universities Press, 1970.

Nell, William Cooper. *Services of Colored Americans in the Wars of 1776 and 1812.* Boston: Prentiss and Sawyer, 1851; reprint, New York: AMS Press, 1976. 2d ed., Boston: R. F. Wallcut, 1852; various reprints, 1854 (Canda), 1862, and 1894.

———. *The Colored Patriots of the American Revolution.* Boston: Robert J. Wallcut, 1855. Reprint. New York: Arno Press, 1968.

Nevins, Allan. *The War for the Union: War Becomes Revolution, 1862–1863.* New York: Charles Scribner's Sons, 1960.

Norrell, Robert J. *Reaping the Whirlwind: The Civil Rights Movement in Tuskegee.* New York: Alfred A. Knopf, 1985.

Northrup, Herbert R., Armand J. Thieblot, Jr., and William N. Chernish. *The Negro in the Air Transport Industry.* The Racial Policies of American Industry, Report no. 23. Philadelphia: Univ. of Pennsylvania Press, 1971.

Nugent, John Peer. *The Black Eagle.* New York: Stein and Day, 1971.

Osur, Alan M. *Blacks in the Army Air Forces During World War II.* Washington, D.C.: Office of Air Force History, 1977.

Patterson, Elois. *Memoirs of the Late Bessie Coleman, Aviatrix: Pioneer of the Negro People in Aviation.* N.p., 1969.

Patton, Gerald W. *War and Race: The Black Officer in the American Military, 1915– 1941.* Westport, Conn.: Greenwood Press, 1981.

"Peck, James L(incloln) H(olt)." *Current Biography* 3 (August 1942): 40–42.

Peck, James L. H. "When Do *We* Fly?" *Crisis* 47 (December 1940): 375–78, 388.

Pisano, Dominick A. "A Brief History of the Civilian Pilot Training Program, 1939–1944." *National Air and Space Museum Research Report: 1986.* Washington, D.C.: Smithsonian Institution Press, 1986.

Ploski, Harry A., and Warren Marr, comps. and eds. *The Negro Almanac: A Reference Work on the Afro-American.* 3d ed. New York: Bellwether, 1976.

Powell, William J. *Black Wings.* Los Angeles: Ivan Deach, Jr., 1934.

———. "Negroes in the Air." *American Mercury* 41 (May 1937): 127.

Quarles, Benjamin. *The Negro in the Civil War.* Boston: Little, Brown and Co., 1953.

Ripley, C. Peter, ed. *The Black Abolitionist Papers.* Vol. 1, *The British Isles, 1830– 1865.* Chapel Hill: Univ. of North Carolina Press, 1985.

Roosevelt, Franklin D. *The Public Papers and Addresses of Franklin D. Roosevelt.* Edited by Samuel Rosenman. 13 vols. New York: Macmillan, 1938–50.

Ross, B. Joyce *J. E. Spingarn and the Rise of the NAACP, 1911–1939.* New York: Atheneum, 1972.

Ruby, Barbara C. "General Patrick Cleburne's Proposal to Arm Southern Slaves." *Arkansas Historical Quarterly* 30 (Autumn 1971): 193–212.

Schreiner, Herm. "The Waco Story: Part II—Expansion with the 'F' Series." *American Aviation Historical Society Journal* 29 (Fall 1984): 214–27.

Schuyler, George S. "Negroes in the Air." *American Mercury* 39 (December 1936): xxviii–xxx.

Scott, William Randolph. "Colonel John C. Robinson: The Condor of Ethiopia." *Pan-African Journal* 5 (Spring 1972): 59–69.

————."A Study of Afro-American and Ethiopian Relations, 1896–1941." Ph.D. dissertation, Princeton University, 1971.

Sherry Michael S. *The Rise of American Air Power: The Creation of Armageddon.* New Haven: Yale Univ. Press, 1987.

Sitkoff, Harvard. "American Blacks in World War II: Rethinking the Militancy-Watershed Hypothesis." In *The Home Front and War in the Twentieth Century: The American Experience in Comparative Perspective,* edited by James Titus. Proceedings of the Tenth Military History Symposium, October 20–22, 1982. Washington, D.C.: United States Air Force Academy and Office of Air Force History, 1984.

————. *A New Deal for Blacks. The Emergence of Civil Rights as a National Issue: The Depression Decade.* New York: Oxford Univ. Press, 1978.

Smith, Raymond A., Jr. "'He Opened Holes Like Mountain Tunnels.'" *Palimpsest* 66 (May/June 1985): 87–100.

Spencer, Chauncey E. *Who is Chauncey Spencer?* Detroit: Broadside Press, 1975.

Strickland, Patricia. *The Putt-Putt Air Force: The Story of the Civilian Pilot Training Program and the War Training Service (1939–1944).* Washington, D.C.: U.S. Dept. of Transportation, Federal Aviation Administration, [1971].

"Success Story: Blacks in Uniform." *Wilson Quarterly* 8 (Spring 1984): 80–81.

"They Take to the Sky: Group of Midwest women follow path blazed by Bessie Coleman." *Ebony,* May 1977, pp. 88–96.

Ware, Gilbert. *William Hastie: Grace Under Pressure.* New York: Oxford Univ. Press, 1984.

Washington, George L. *The History of Military and Civilian Pilot Training of Negroes at Tuskegee, Alabama, 1939–1945.* Washington, D.C.: George L. Washington, 1972.

Washington, Mary J. "A Race Soars Upward." *Opportunity* 12 (October 1934): 300.

Waters, Enoch P. *American Diary: A Personal History of the Black Press.* Chicago: Path Press, 1987.

Weaver, John D. *The Brownsville Raid.* New York: W. W. Norton, 1970.

Weisbord, Robert G. "Black America and the Italian-Ethiopian Crisis: An Episode in Pan-Negroism." *Historian* 34 (February 1972): 230–41.

Weiss, Nancy J. *Farewell to the Party of Lincoln: Black Politics in the Age of FDR.* Princeton: Princeton Univ. Press, 1983.

Whittier, John Greenleaf. "The Black Men of the Revolution and War of 1812." *National Era,* 22 July 1847.

Wiener, Willard. *Two Hundred Thousand Flyers: The Story of the Civilian–AAF Pilot Training Program*. Washington, D.C.: The Infantry Journal, 1945.

Wilkins, Roy, and Tom Matthews. *Standing Fast: The Autobiography of Roy Wilkins*. New York: Viking Press, 1982.

Williams, Charles H. *Negro Soldiers in World War I: The Human Side*. With an introduction by Benjamin Brawley. New York: AMS Press, 1970. (Reprint of 1923 ed. published under the title *Sidelights on Negro Soldiers*.)

Wilson, John R. M. *Turbulence Aloft: The Civil Aeronautics Administration amid Wars and Rumors of Wars*. Washington, D.C.: U.S. Dept. of Transportation, Federal Aviation Administration, 1979.

Wolf, William. "U.S.A.A.F. Pilot Training in World War II," pt. 1. *The Historical Aviation Album* 16 (1980): 60.

Wolters, Raymond. *Negroes and the Great Depression: The Problem of Economic Recovery*. Contributions in American History no. 6. Westport, Conn.: Greenwood Publishing, 1970.

Woodson, Carter G., and Charles H. Wesley, *The Negro in Our History*. 12th ed., rev. and enl. Washington, D.C.: Associated Publishers, 1972.

Wynes, Charles E. "'Alexander the Great,' Bridge Builder," *Palimpsest* 66 (May/June 1985): 78–86.

Wynn, Neil A. *The Afro-American and the Second World War*. New York: Holmes and Meier Publishers, 1975.

Young, David, and Neal Callahan. *Fill the Heavens with Commerce: Chicago Aviation, 1855–1926*. Chicago: Chicago Review Press, 1981.

Index

A

AAF. *See* United States Army Air Forces
Abbott, Robert S., 58, 67, 104
Adams, Maj. Gen. E. S., 223
Aerial Observers School, 56
Aeronautical University of Chicago, 195
African Americans: interest in aviation among, ix, 7, 34, 35, 52, 53-68 passim, 69, 79, 306-7; interest in military careers among, ix, 33-52 passim, 69, 81, 104, 306-7; migration to northern cities, 71-72, 75; political affiliations of, 35; race heroes for, 27-29; voting rights of, 70-71, 72, 74, 75, 176, 307
African Methodist Episcopal Church, 191
Agricultural Adjustment Administration (AAA), 73-74
Air Commerce Bulletin, 113
Air Corps Expansion Act. *See* Public Law 18
Air Corps Technical School (Rantoul, Ill.), 195, 197, 281. *See also* Chanute Field
Air Corps Training Center (San Antonio), 92-93, 98, 104, 162, 165, 193, 340-41 (n. 23)

Air Line Pilots Association, 63
Airmail pilots, black, 79, 80-81, 89, 97, 106, 338 (n. 44)
Air Medals, viii
Air Site Board, 272-73
Alabama Air Service, 118, 119, 120, 124, 125-26, 137, 139, 145, 151, 152, 153, 265
Alabama Aviation Commission, 127, 133, 134, 154, 157, 248
Alabama Institute of Aeronautics, 152
Alabama Polytechnic Institute (API), 119, 125, 126, 133, 137, 146, 153, 265
Alexander, Archie A., 248, 254-55, 265-67, 277
Alexander, Will W., 35, 76
Allen, George W., 264, 357 (n. 76)
Allen, Joseph Wren, 118, 119, 121, 125-26, 130, 131, 133, 137, 146, 151, 153, 308, 352 (n. 10)
Allen, Thomas, 60, *65*, 65-66, 79, 319 (n. 18)
Altman, Dale E., 127
American Dilemma, An (Myrdal), 77
American Federation of Labor (AFL), 75
American Mercury, 53-54
American Revolution: blacks'

military service during, 38, 39-40, 41; exclusion of blacks from military service during, 37
"Amos 'n Andy" (radio show), 60
Anderson, Alexander S., 129, 237-39, 245
Anderson, Charles Alfred: as black role model, 79, 114, 158; and christening of *Booker T. Washington*, 10, *12*; and CPT Program at Tuskegee, 114-15, 117, 118, 131, 144-45, 147-49, 150, 154, 241, 246, 260, 264-65, 310, 357 (n. 76), 365 (n. 56); illustrations of, *246-47, 268*; and military flight training at Tuskegee, 260, 264-65; and Pan-American Goodwill Flight, 7, 8, 10, 17-20; transcontinental flights by, 66
Anderson, Grady P., 95
Anderson, John C., Jr., 381-82 (n. 116)
Anderson, Marian, 46
Andrews, Maj. Gen. Frank M., 200
ANP. *See* Associated Negro Press
API. *See* Alabama Polytechnic Institute
Armies With Wings (Peck), 191
Armstrong, Daniel W., 9
Armstrong, Henry, 175
Armstrong, Samuel C., 2
Army Air Forces (AAF). *See* United States Army Air Forces
Army Reorganization Bill of 1920, 78
Arnold, Maj. Gen. Henry H. ("Hap"), 89, 92, 93, 100-101, 105, 165, 166, 171-72, 178, 184, 189, 194, 196, 200, 201, 205, 211-12, 215, 216, 221, 223, 233-34, 250, 258, 259, 282, 287-89, 292, 296, 297, 301, 309, 314
Ashburn Field (Chicago), 163
Associated Negro Press (ANP), 22, 87, 165, 236, 278, 308, 322 (n. 61)
Atlanta Compromise, 3
AT-6 trainer, *302*

Attucks, Crispus, 34
Auburn University. *See* Alabama Polytechnic Institute
Austin, Warren R., 181
Aviation mechanics: Air Corps training of, 185, 192, 193-94, 195, 198, 261; training of at Tuskegee, 116-17, 120, 135-37, 141, 142, 152, 156, 261-62, 263

B

Baker, George ("Father Divine"), 59
Baker, Newton D., 49
Baltimore *Afro-American*, 104, 219, 220
Bankhead, John H., 156, 208-9, 210, 273, 291
Banning, James Herman, 60, 63, *65,* 65-66, 79, 158, 319 (n. 18), 320 (n. 19)
Barbour, W. Warren, 177, 178
Barnett, Claude A., 22, 23-29, 30, 119-20, 220, 225-26, 229, 231-32, 263, 264, 278, 292, 294, 295, 303, 308, 313
Barnstormers, black, 55, 58
Battle of New Orleans, 39
Bayen, Malaku E., 22
Beachey, Lincoln, 62
Belknap, Jeremy, 37
Bessie Coleman Aero Clubs, Inc., 62-63, 64
Bessie Coleman Aero News, 62-63, 64
Bethune, Mary McLeod, 20, 117, 174, 261-62, 292
Bishop, Billy, 58
Black Cabinet, 34, 74, 76, 80, 174
Black Eagle Flying Corps, 59
Black Renaissance, 72
Blacks. *See* African Americans; Free blacks; Slaves
Bluford, F. D., 223
Boatner, Maj. Mark M., Jr., 272
Boland, Jesse, 62, 64
Booker T. Washington (airplane), 7, 9, 10, *12-13*, 17-18, 20, 66

Booker T. Washington Monument, 258

Boston Chronicle, 59-60

Bousefield, Dr. M. O., 161, 163, 167

Boyack, James Edmund, 9, 11, 14, 15, 25

Boyd, J. Erroll, 11

Branche, Dr. George C., 124

Brett, Maj. Gen. George H., 251, 274-76

Bridges, H. Styles, 98, 99, 100, 101, 164-65, 170-71

Brimhall, Dean, 140

Brotherhood of Sleeping Car Porters, 6, 75

Brown, Charles D., 381 (n. 116)

Brown, Edgar G., 34, 35, 97, 98, 99-100, 103-5, 106, 110, 120-21, 156, 161, 162-63, 168, 181

Brown, Theodore E., 381 (n. 116)

Brown, Willa B., 66, 67, 142, 143, 161, 164, 165, 202, 225, 232, 255

Brown Eagle Aero Club, 63, 64

Brownsville Riot, 46-47, 181

Brown v. Board of Education, 70

BT-13 trainer, *302*

Buffalo soldiers, 44, 328 (n. 44)

Bullard, Eugene Jacques, 56-58, *57*, 64, 303

Bunche, Ralph, 46

Bureau of Air Commerce: black pilots licensed by, 8, 34, 35, 63, 68, 80, 100, 334 (n. 49); replaced by Civilian Aeronautics Authority, 90

Bureau of Colored Troops, 41

Burgin, Emile H., 11

Burke, Edmund H., 117

Burke-Wadsworth Bill. *See* Selective Training and Service Act

Butler, Gen. Benjamin F., 41

Byrnes, James F., 181

C

C. G. Kershaw Contracting Company, 291

CAA. *See* Civil Aeronautics Authority

Cable, Dr. Theodore, 80, 97, 106

California Eagle, 61

"Call of the Wings" (Levette), 64-65

Camilleri, Joseph T., 264, 357 (n. 76)

Campbell, T. M., 116

Carnegie Corporation, 245

Carnera, Primo, 22

Carter, Brig. Gen. A. H., 290-91, 294

Carter, Marion A. B., 381 (n. 116)

Carver, George Washington, 31

Challenger Air Pilot Association, 63-64

Chanute Field (Rantoul, Ill.), 192, 195, 197, 198, 200, 211, 221, 226, 229, 231, 281-85, 301, 312

Chattanooga News-Free Press, 129

Checkerboard Field (Chicago), 58, 62

Chenault, Dr. John W., 124

Chicago: annual tribute to Bessie Coleman in, 64; black aviation in, ix, 10, 27, 32, 58, 62, 63, 66-68, 109, 116, 183-84; and CAA training of black pilots, 103, 225; rejected as home base of Ninety-ninth Pursuit Squadron, 200-203, 212, 228, 309-10; rivalry with Tuskegee for CPT Program, 143

Chicago Bee, 165, 176

Chicago Defender, 6, 27, 58, 67, 72, 79, 104, 109, 161, 165, 225, 231, 234, 264

Chicago School of Aeronautics: changing names of, 344 (n. 65); training of black aviation mechanics at, 195, 261; training of black pilots at, 142, 143, 145, 147, 160, 162, 163, 164, 165-68, 184, 185, 189-90, 200, 243-44, 255, 358-59 (n. 5)

Chicago Tribune, 175-76

Chicago Tuskegee Club, 155

Chicago World, 165

Civil Aeronautics Authority (CAA): establishment of, 90; Private

Flying Development Division, 118;
and training of civilian pilots, ix,
89, 90-91, 96-97, 99, 100, 102-3,
108, 184, 191
Civilian Pilot Training Act of 1939,
ix, 91, 94, 108-10, 117, 309
Civilian Pilot Training (CPT) Program: confusion of relationship to
Air Corps training, 93-94, 96-97,
99, 100-101, 109, 119-20, 340-41
(n. 23); costs of and scholarships
for, 122, 123-24, 140, 147, 150-51,
154, 348 (n. 44); eligibility for,
122; establishment of in 1939, ix,
32, 90-91; parental consent for
participation in, 122, 123; and
training of black pilots, 108-11,
112-31 passim, 158-60, 183-84,
191-92
Civil Rights Act of 1875, 70
Civil rights movement, and campaign for black pilots, ix, 34, 35-36,
68, 69-87 passim, 306, 307. See also
Racism; Segregation
Civil War, 36, 37, 40-42
Cleburne, Gen. Patrick, 42
Cockburn, Rear Adm. George, 39
Coffey, Cornelius R., 66-67, 112,
142, 161, 201, 228, 255, 318 (n. 1)
Coffey School of Aeronautics
(Chicago), 142, 143, 145, 149,
162, 163, 167, 200, 229, 255, 315,
335-36 (n. 70), 368 (n. 4)
Col. J. C. Robinson School of
Aviation (proposed), 30, 31
Coleman, Bessie ("Brave Bessie"), 58,
60, 62, 64, 67
Collings, Kenneth Brown, 53-54
Colored Patriots of the American
Revolution, The (Nell), 40
Commission on Interracial Coopera-
tion (CIC), 174
Committee on the Participation of
Negroes in the National Defense
Program, 169-70, 172, 176, 181

Congressional Medals of Honor, 44,
47
Congress of Industrial Organiza-
tions, 75
Connally, Tom, 177, 179-81
Connor, John, 127, 350 (n. 65)
Connor, Maj. Gen. William D., 256
Continental Army: blacks in, 38;
exclusion of blacks from, 37
Cornell, Bloomfield M., 125, 129,
131, 139, 146, 148, 308, 352 (n.
10)
CPT Program. See Civilian Pilot
Training Program
Craftsmen Aero News, 84
Craftsmen of Black Wings, 80, 84
Crenshaw, Milton P., 265
Crisis, The, 47, 67, 191, 224, 227,
329 (n. 58)
Croix de guerre, 34, 50, 56
Curtis, William, 145
Curtis JN-4 "Jenny," 61
Curtiss Aeroplane Company, 58
Curtiss Airport (Glenview, Ill.),
103, 104, 105, 161, 244, 255,
358-59 (nn. 5-6)
Curtiss-Wright Flying School
(Chicago), 10, 63, 160, 344 (n.
65)
Custis, Lemuel R., 269, 304, 305

D

Daniel Guggenheim Fund for the
Promotion of Aeronautics, 10
Darden, Charles S., 78
Darr, Harold S., 160-61, 163, 243-
44, 255, 358-59 (nn. 5-6)
Darr Aero Tech (near Albany, Ga.),
243-44, 248
Darty, Warren G., 265
Daugette-Millican Company, 291
Davis, Dr. A. Porter, 62
Davis, Brig. Gen. Benjamin O., Sr.,
10, 13, 107

Davis, Capt. Benjamin Oliver, Jr., 105, 107, *186*, 256, *257*, 258, 269, 304, *305*, 315
Davis, John P., 129-30
Dawson, Robert A., 265
DeBow, Charles H., Jr., 269, 304, *305*
Delaware State College for Colored Students, 184
Democratic party, 35, 71, 72-73, 76, 86, 97, 98, 176
Department of Commerce: black pilots licensed by, 8, 34, 35, 63, 68, 80, 100; Division of Negro Affairs, 67, 80
De Priest, Oscar, 8, 63, 71
Dibble, Dr. Eugene, 260
Dirksen, Everett M., 106-7, 108, 109, 110, 169
Distinguished Flying Cross, viii
Dole, James, 62
Dothan Advanced Flying School, 286, 287
Doughtry, Mary, 54
Douglass, Frederick, 3, 42
Draper, Owen, 134
Dred Scott case, 40
Dryden, Charles W., 236
Du Bois, W. E. B., 47-48
Duncan, L. N., 146, 308, 352 (n. 10)
Dunham, Mr. (masonry foreman), 266
Dunham, Royal B., 128
Dunmore, Lord, 38

E
Eaker, Gen. Ira C., 200
Eich, S. M., 249
Eighth Illinois Regiment, 45
Ellender, Allen J., 177, 179
Ellison, Maj. James A., 258, 272, 277, 285, 286
Emancipation Proclamation, 41
Emergency Education Program, 79
Employment of Negro Troops, The (Lee), 36

Engel, Albert J., 105
Ethiopia, 2, 21-24, 59, 88, 304, 307
Ethiopian Air Force, 2, 21-23, 59, 66, 318 (n. 5)
Ethiopian Regiment (American Revolution), 38
Evans, James C., 117-18

F
Federation Aeronautique Internationale, 58
Fifteenth Amendment, 70
Finnish Air Force, 59
Firestone Tire and Rubber Company, 27
First Louisiana Native Guards, 41
Fish, Hamilton, 85-86, 105-6
Flight instructor training, 137-38, 150, 222
Florida Agricultural and Mechanical College (Tallahassee), 104
Flying clubs, black, 62-63
Foraker, Joseph B., 47
Foreman, Clark, 35, 76
Foreman, Joel ("Ace"), 61, 64
Forsythe, Dr. Albert E., 7-8, 10-20, *12*, 66, 79, 158, 308
Fort Davis (Alabama), 272, 382 (n. 4)
Fort Des Moines (Iowa), 49, 50, 235
Fortieth Infantry Regiment, 43
Forty-first Infantry Regiment, 43
Fourteenth Amendment, 70
Foxx, Charles R., 129, 130-31, 265
Franco, Gen. Francisco, 88
Free blacks: excluded from colonial militia, 37; excluded from state militia, 39, 40; military service by, 37, 38, 40
French and Indian War, 38
French Flying Service, 56
French Foreign Legion, 56

G
Garrison, William Lloyd, 41
Garvey, Marcus, 58-59, 62

Gayden, Earnest L., 338 (n. 44)
Gayle, W. A., 121
General Education Board, 138, 245
Gerni Aeroplane Company, 58
Gibson, Truman K., 220, 227, 229, 292
Goens, Julia, 9
Göring, Hermann, 59
Gould, Howard, 201-3, 225-26, 232
Grant, Maj. Gen. Walter S., 223
Gray, G. N. T., 80
Green, Alfred M., 41
Greene, John W., Jr., 335 (n. 60)
Guido, Dominick J., 264
Gulf Coast Air Corps Training Center, 250

H

Haile Selassie (emperor of Ethiopia), 21, 23, 29, 59, 66, 87, 304
Haitian Coffee and Products Trading Company, 114
Hall, James Norman, 56
Hampton Institute: Aviation Club at, 95-96; CPT Program at, 110, 147, 159, 184
Harlem Airport (Chicago), 142, 163, 167, 192, 201, 225
Harlem Renaissance, 72, 75
Hastie, William H., 173, 201-3, 211, 213-14, 215, 216, 219, 220, 225, 226, 227, 229-30, 235, 270, 275-77, 284, 301, 303, 312, 316
Hayden, Lucian Arthur, 55, 56, 332-33 (n. 25)
Headen, L. Arthur, 331 (n. 14)
Headin, Lucian, 331 (n. 14)
Hill, Lister, 156, 208-9, 210, 273-75, 276, 291, 292
Hill, T. Arnold, 72, 185, 188
Hinckley, Robert H., 90, 91, 96, 103, 113, 120, 122, 140
Hines, Percy L., 80
Hitler, Adolf, 58, 83, 88, 89, 95, 172
Holland, E. H., 260

Hooter, H. E., 54
Hoover, Herbert, 72, 73
Houston, Charles H., 82, 176
Houston Riot, 51, 181
Howard University, CPT Program at, 110, 144, 147, 159
Howe, Arthur, 95, 96
Humbles, Austin P., 265
Hunter, Gen. David, 41

I

Ickes, Harold L., 35, 76
IGAC. *See* Interracial Goodwill Aviation Committee
Imes, G. Lake, 7, 9, 10-11, 14-16, 17, 20
Improved, Benevolent and Protected Order of Elks of the World, 103-4
Indian Wars, 43-44
Institute of Aeronautical Sciences, 130
International Colored Aeronautical Association, 64
International Council of Friends of Ethiopia, 21
Interracial Goodwill Aviation Committee (IGAC), 7, 8-9, 15, 20
Inventors, black, 54-55, 58
Isaacs, Lloyd, 115

J

Jackson (Ajax Montmorency), 54
Jackson, Andrew, 41
Jackson, Lewis A., 145, 149, 151, 153, 260, 264-65, 357 (n. 76), 365 (n. 56)
James, Wilson, 322 (n. 61)
Jefferson (of Northwestern), 175
John C. Robinson Aviation Fund, 29
John C. Robinson National Air College (Chicago), 113, 115-16, 119
John C. Robinson School of Aviation (Chicago), 32

Johnson, Gen. Davenport, 196, 197-
    200, 252, 282-83
Johnson, Sgt. Henry, 34
Johnson, James Weldon, 334 (n. 46)
Jones, Casey, 11
Jones, Eugene Kinckle, 8, 80
Jones, Willie ("Suicide"), 87
Julian, Hubert Fauntleroy ("Black
    Eagle of Harlem"), 22, 58-60, 62,
    64
Julius Rosenwald Fund, 245, 248,
    259, 293

K

Karmin (NYU professor of aeronauti-
    cal engineering), 11
Karrick, Lt. Col. S. N., 286-87
Kelly, Rev. C. W., 258
Kelly Field (San Antonio), 92, 191
Kennedy, Stanley, 127, 131, 152
Kennedy Field (Tuskegee, Ala.), 127-
    28, 133, 134, 135, 139, 146, 148,
    152, 153, 241, 245, 249, 265, 266-
    67, 345 (n. 3), 350 (n. 65), 351 (n.
    6)
Kimble, Col. Frederick V. H., 277
King George's War, 38
Kleindienst, Richard, 329 (n. 58)
Knox, Frank, 185-87
Kountze, Mabe, 87
Ku Klux Klan, 72, 176

L

Labor unions: blacks excluded from,
    71; support for black workers
    from, 74, 75, 307
Lafayette Escadrille, 332 (n. 22)
Lafayette Flying Corps, 56, 332 (n.
    22)
Lafayette Flying Corps (Hall and
    Nordhoff), 56
LaMar, J. H., 266
Lane, Gen. James H., 41
Langley Field, 96
Lansing Airport (near Chicago), 212

Lautier, Louis, 170
Lea, Clarence F., 109, 340 (n. 12)
League of Nations, 21, 22
Lee, Robert E., 42
Lee, Ulysses, viii, 33, 36, 43
Levette, Harry, 64
Lewis, Alfred Baker, 204
Lewis, John L., 75
Liberia, 59
Lincoln, Abraham, 41
Lindbergh, Charles A., 9, 29, 35, 53,
    60-61, 64
Logan, Marvel Mills, 100-101
Logan, Rayford W., 46, 169-71, 172,
    176
Long, R. M., *305*
Louis, Joe ("Brown Bomber"), 22,
    29, 46, 175
Lovell, John, Jr., 235
Lovett, Robert A., 215, 230, 275,
    276-77, 301
Lowry Field (Denver), 192
Ludlow, Louis, 106, 169, 174, 175
Lyon, Robert G., 11, 14

M

MacArthur, Gen. Douglas, 78
McCarran, Pat, 340 (n. 12)
McConnell, Wendell L., 223
McCormick, Col. Robert R., 175
McIntyre, M. H., 95, 96
McKeough, Raymond S., 106
McKinley, William, 45
McKissack, Calvin, 278, 290
McKissack and McKissack (contrac-
    tors), 278-80, 291
MacLean, Malcolm S., 217-18
McWhorter, John E., 54
Madison, Walter G., 54
Maine (battleship), 44
Manumissions, 37-38, 40, 42
March on Washington movement,
    280, 384 (n. 30)
Marshall, Gen. George C., 89, 166,
    174-75, 176, 184, 185, 189, 195,

200, 223, 232, 233-34, 251, 258, 259, 282, 284, 286-87, 292, 301, 312
Marshall, James E., 54
Marshall, Thurgood, 204-5, 219, 223, 224, 227, 235
Martin, Clarence E., 62, 64
Martin, J. Sella, 39
Maxwell Field (Montgomery, Ala.), 197, 198, 205, 206-8, *207*, 209, 211, 215, 218, 221-22, 230-31, 240, 242, 249, 251-54, 258, 271, 310
Menelik II (emperor of Ethiopia), 21
Metropolitan Post, 165
Mexican War, 39, 326 (n. 24)
Militia, American: black military service in, 39; exclusion of blacks from, 39, 40
Militia, colonial, 37
Militia Act of 1792, 37
Ming, W. Robert, Jr., 205, 219, 223
Moore, Frederick H., Jr., 269
Moore, George H., 41
Morgan, Robert O., 155, 156
Morris, Elvatus C., 129
Moseley, Maj. Gen. George Van Horn, 78
Moton, Robert Russa: and christening of *Booker T. Washington*, 8-9, 10, *12*; Moton Field named after, 249, 376 (n. 37); and National Aviation Fund (proposed), 68; and Pan-American Goodwill Flight, 7, 8-9, 14, 15, 17, 20, 32; as second president (principal) of Tuskegee, 3-4, 6, 7, 8-9, 23, 285
Moton, Mrs. Robert Russa, 10, *12*, 20
Moton Field, 243-49, 251, 254-55, 265-67
Murphy, Doris, 10
Murphy, George B., 227
Murray, John (earl of Dunmore), 38
Mussolini, Benito, 21, 23, 95
Myrdal, Gunnar, 35, 77

N
NAAA. *See* National Airmen's Association of America
NAACP. *See* National Association for the Advancement of Colored People
NAACP Annual Report for 1941, 223
Nash, Grover, 80, 318 (n. 1)
National Airmen's Association of America (NAAA): and campaign for training of black pilots, 104, 106, 109-10, 161-63, 164, 165-67, 181, 202; establishment of, 67; opposition to establishment of segregated air squadron, x, 170, 201, 225-30, 231, 264, 294, 310, 311
National Association for the Advancement of Aviation Amongst Colored Races, 62
National Association for the Advancement of Colored People (NAACP): and campaigns for black military participation, 52, 79, 81-82, 83, 87, 159, 172-73, 181, 193, 202, 203-5, 218-20, 223-24, 307, 316; establishment and growth of, 6, 74; and the New Deal, 74-75; opposition to establishment of segregated air squadron, x, 188-89, 191, 201, 203, 210-11, 222, 225, 226-27, 231, 234-38, 294, 310, 311-12; sponsorship of black aviators by, 61; during World War I, 47
National Aviation Fund (proposed), 68
National Baptist Convention, 191
National Era, 39, 326-27 (n. 25)
National Negro Business League, 14
National Negro Congress, 129
National Negro Council, 190
National Postal Alliance, 80
National Recovery Administration (NRA), 74

National Urban League (NUL), 6, 61, 72, 74-75
National Youth Administration (NYA), 117, 120, 134, 136, 189, 191, 192, 193-94, 261
Neely, Alvin J., 1, 123, 135
Negro Aviators (U.S. Dept. of Commerce), 80, 235, 344-45 (n. 74)
Negro World, 59
Nell, William Cooper, 40, 41
New Deal, 35, 69, 74, 76-77, 80-81
New Negro Movement, 72
New York Amsterdam News, 104
New York Times, 26, 189, 191, 226
Nilson, Edward C., 127, 128, 136, 142-43, 145, 146, 150
Ninety-ninth Pursuit Squadron: activation of, 282; establishment of, ix-x, 32, 36, 81, 111, 172, 182, 183-215 passim; home station at Tuskegee Army Air Field, x, 195-215 passim, 270-305 passim; opposition to, x, 170, 188-89, 191, 201, 203, 210-11, 216-39 passim
Ninety-second Division, 50-51
Ninety-third Division (Provisional), 50
Ninth Cavalry Regiment, 43, 45, 46
Ninth U.S. Volunteer Infantry Regiment, 45
Nordhoff, Charles Bernard, 56
North Carolina Agricultural and Technical College, CPT Program at, 118, 147, 159
North Suburban Flying Corporation (Glenview, Ill.), 103, 104, 105, 344 (n. 65)
NRA. *See* National Recovery Administration
NUL. *See* National Urban League

O

Oliver, Oscar, 236-37
100th Pursuit Squadron , 300-303
O'Neill, Col. R. E., 283

Opportunity, 62
Oscar De Priest (airplane), 63
Overton, John H., 177-78, 179
Owens, Jesse, 29
Owsley, Mr. (sheet metal work foreman), 266

P

Pan-American Interracial Goodwill Aviation Flight, 7-20, 66, 320 (n. 20)
Pannell, Ulysses S., 381 (n. 116)
Parachutists, black, 54, 87, 238
Paris, Leon, 64
Parks Air College (East St. Louis), 358-59 (n. 5)
Parrish, Capt. Noel, 167-68, *207*, 255, 258, 277, 296, 297-98, 315, 316
Patterson, Frederick Douglass: and advanced pilot training at Tuskegee, 139, 141-42, 144, 146, 148, 150, 202; and aviation mechanics training at Tuskegee, 116-17, 120, 136-37; and campaign to establish an airport at Tuskegee, 133, 154-56, 208-11, 272, 278-79, 280, 290-93; and christening of *Booker T. Washington*, 10, *13*; and expansion of Tuskegee's CPT Program, 150-53; and flight instructor training at Tuskegee, 137-38; illustration of, *26*; interest in establishing aviation training at Tuskegee, 24-25, 29, 31, 32, 112-14, 116, 118, 121, 123, 124, 239; and military flight training at Tuskegee, 208-11, 217-18, 220-21, 223, 228, 230, 232-34, 240, 244, 249, 251-52, 254-55, 259, 271, 281, 282-85, 293-97, 298-300, 303, 311, 312, 314; as third president of Tuskegee, 5-6, 21
Patterson, Robert P., 187, 194-95,

213, 279-80, 290, 291, 294, 303, 312

Patzolcl, J. L., 147

Peck, James L. H., 191-93, 227

Peters, Charles Wesley, 55, 56

P-40 trainer, *302*

Phelps, Gen. John W., 41

Philippine Insurrection, 44, 45-46

Pickering, John F., 54

Pilot training, civilian: elementary (primary), 93-94, 101, 102-3, 104, 105, 137, 152-53, 200, 222, 340-41 (n. 23); secondary (advanced), 139-49, 152-53, 157, 166, 202, 212, 218, 221, 222, 226, 228-29, 240-42, 254, 314, 340-41 (n. 23)

Pilot training, military: advanced, 92, 104, 198, 221, 287, 303, 340-41 (n. 23); basic, 92, 104, 198, 212, 221, 228, 240, 251, 254, 267, 287, 340-41 (n. 23); ceremony inaugurating at Tuskegee, 256-60; costs of, 243; and preflight training, 250-51, 253, 254, 255, 267; primary, 92, 93-94, 140, 162, 167, 198, 212, 228-31, 240-41, 242-50, 267-69, 270, 340-41 (n. 23); transition, 92, 340-41 (n. 23)

Piper J-3 "Cubs," 145, 151, 153

Pitts, Robert G., 125, 129, 131, 139, 308, 352 (n. 10)

Pittsburgh Courier, 23, 60, 62, 64, 66, 68, 81, 82-86, 104, 158, 160, 165, 169, 170, 175-76, 190, 193-94, 307

Plattsburg movement, 49

Plessy v. Ferguson, 70, 71

Porter, Don S., 126

Powell, Dr. C. B., 104

Powell, William J., 63, 64, 79-80, *84*, 320 (n. 19), 323 (n. 75)

Powers, D. Lane, 105, 106

PT-17 trainer, *302*

Public Law 18: and campaign for black pilots, ix, 97-110 passim, 113, 159, 160, 166-68, 171, 190, 309; and expansion of pilot training programs, 91-92; and expansion of the Army Air Corps, 91, 271; failure of, 181; passage of, 91, 101-2

Q

Quarles, Benjamin, 40

Quiet Birdmen, 11

R

Racism: against black aviators, 53-54, 60-61, 148; black military participation, 33. *See also* Civil rights movement; Segregation

Randolph, A. Philip, 48, 75, 185, 188, 280, 384 (n. 30)

Randolph Field (San Antonio), 92, 191, 215, 239

Reconstruction, 70

Reconstruction Finance Corporation (RFC), 243, 244

Renfroe, Earl, 161

Repass, M. A., 248

Republican party, 35, 71, 72-73, 86, 98, 176

Reserve Officers Training Corps (ROTC): black units for, 52, 56, 105, 107; and training of civilian pilots, 91, 100, 101, 115

Revised Statutes of 1878, 43

RFC. *See* Reconstruction Finance Corporation

Richmond (Va.) *News Leader*, 388 (n. 104)

Roberts, George S., 205, 254, 269, 304, *305*

Robinson, Hilyard R., 258, 285

Robinson, Jackie, 46

Robinson, John C. ("Brown Condor"): as aviation instructor (proposed) at Tuskegee, 2, 23-25, 27, 29-32, 113-14, 115-16, 118-19, 144, 145, 261-64; as aviator in

Chicago, 10, 27, 32, 63, 66, 160-61, 163; as black role model, 79, 87, 158, 308; as correspondent for Associated Negro Press, 322 (n. 61); criticism of War Department training initiative, 190; denied opportunity in American armed forces, 87; in Ethiopian Air Force, 2, 21-24, 59, 66, 303-4; flight to Tuskegee reunion by, 1, 6-7; homecoming from Ethiopia for, 25-29, *28*

Rockefeller, John D., 138

Rogers, Joel, 23, 25

Roosevelt, Eleanor, 35, 76, 183, 185, 245-47, *246-47*, 259, 280, 310, 312, 375 (n. 28)

Roosevelt, Franklin D., 15, 34, 35, 68, 69-70, 72, 73-74, 76-77, 79, 82, 83, 86, 89-90, 94, 95, 97, 102, 112, 116, 159, 165, 172, 174, 176, 183, 184, 185-87, 211, 259, 280, 299, 307, 309, 310

Roosevelt, Theodore, 46

Roosevelt Field (Long Island), 65

Rosenberg, Frank, 264, 357 (n. 76)

Rosenwald Foundation, 161, 245. *See also* Julius Rosenwald Fund

Ross, Mac, 269, 304, *305*

Rough Riders, 45

Rountree, Asa, Jr., 133-34

S

Sanders, Ed, 68, 335 (n. 60)

San Juan (Puerto Rico) Tuskegee Club, 18, *19*

Sauzereseteo, Samuel V. B., 62, 64

Savory, Dr. P. M., 26

Scarborough, W. S., 55-56

Schiefflin, Dr. William Jay, 26, 27, 244

Schroeder, Maj. R. W., 359 (n. 6)

Schwartz, H. H., 98-99, 100-101, 105, 164, 165, 170, 177, 178

Scott, Emmett J., 8, 85, 97

SEACTC. *See* Maxwell Field; Southeast Air Corps Training Center

Segregated opportunity, 5-6, 239

Segregation: black acceptance of, 3, 5-6; upheld by Supreme Court, 70; white justification of, 51. *See also* Civil rights movement; Racism

Selective Training and Service Act (Burke-Wadsworth Bill), 159, 172, 176-82, 183, 185, 224, 309

Services of Colored Americans in the Wars of 1776 and 1812 (Nell), 40

Sharecroppers, 71, 72, 74

Shell Intercollegiate Aviation Scholarship, 130

Shelton, Forrest, 127, 264-65, 357 (n. 76)

Siegel, Alexander B., 244

Simmons, Ozzie, 175

Sitkoff, Harvard, 35

Sixth Massachusetts Regiment, 328 (n. 52)

Slade, William H., 381 (n. 116)

Slater Fund, 31

Slaves, military service by, 37-39, 42

Smith, James, 54

Smith, Maj. Luke S., 206, *207*, 221-22, 230, 242-43, 249, 251, 252, 255, 260-61, 272, 280, 281-82, 286

Snyder, J. Buell, 169, 171

Sorrell, Mr. (painting foreman), 266

Southeast Air Corps Training Center (Montgomery, Ala.), 197, 200, 218, 231, 240, 242, 249, 250, 251-54, 258, 260, 271, 287, 310

Souther Field (near Americus, Ga.), 248

Southern Aviation School (Camden, S.C.), 248

Southern Conference on Human Welfare, 76

Spanish-American War, 44-46

Spencer, Chauncey, 109-10
Spingarn, Joel, 49
Starnes, Joe, 291
Steagall, Henry B., 208-9, 210, 291, 292
Stimson, Henry L., 176, 183, 184, 185-87, *186*, 196, 223, 259
Stratemeyer, Brig. Gen. George E., 286, 296, 304
Street, Col. St. Clair, 301
Swaby, Ralph W., 115, 119
Swagerty, Walter E., 54, 62, 64

T

TAAF. *See* Tuskegee Army Air Field
Taney, Roger B., 40
Tenth Cavalry Regiment, 43, 44, 45, 46
Terry, David D., 107-8
Thair, Reginald, 337 (n. 36)
Thirty-eighth Infantry Regiment, 43
Thirty-ninth Infantry Regiment, 43
Thomas, Elbert D., 181
Thomason, R. Ewing, 107-8
Time magazine, 75, 219
Training aircraft, 145, 151, 153, 242, *302*, 354-55 (n. 46)
Trent, William J., Jr., 95, 96
Truman, Harry S., 110, 316
Tuskegee Airmen: definition of, viii, 306; historical interest in, viii, 390 (n. 1)
Tuskegee Airmen, Inc., 390 (n. 1)
Tuskegee Airmen's Association, 390 (n. 1)
Tuskegee Army Air Field (TAAF): construction of, x, 94, 111, 270-80, 285-92, *288*, 300; cost of, 290-92, 386 (n. 75); Henry L. Stimson at, *186*; segregated facilities at, 275-77; site selection, 271-75; training of black pilots at, vii, 270-305 passim
Tuskegee Choir, 27
Tuskegee Executive Council, 14
Tuskegee Institute: administrators'

early interest in establishing aviation program at, 1-32 passim; aviation conference at, 217-18, 220, 221-23; black veterans hospital at, 4, 105, 124, 223, 279; campaign to establish an airport at, 132-57 passim; christening of *Booker T. Washington* at, 9-10, *12-13*; college-level courses offered at, 4, 31; CPT Program at, ix, viii, 110, 112-31 passim, 132-57 passim, 184, 198, 200, 240-42, 307-9; early presidents (principals) of, 2-6, 21, 319 (n. 15); establishment of, 2-3; ground school for CPT Program at, 122, 124-25, 128-29, 139, 145-46; military flight training program established at, x, 195-215 passim, 240-69 passim, 309-11; suitability as aviation training center, 2, 104, 105, 107, 112, 113, 114, 115, 307-11
Tuskegee Institute Airport Fund Campaign, 244
Tuskegee Institute Alumni Association, 135
Tuskegee Institute Board of Trustees, 21, 26, 135, 153, 154, 155, 156, 245, 248
Tuskegee Institute Quintet, 131
Twenty-fifth Infantry Regiment, 43, 46
Twenty-fourth Infantry Regiment, 43, 45, 46, 51, 258
Twenty-third Kansas Regiment, 45
Tydings, Millard E., 181

U

UNIA. *See* Universal Negro Improvement Association
United Aid for Ethiopia, 26, 27
United Government Employees, 103, 191
United States Air Force (USAF), establishment of, vii (n.)

United States Army: black combat troops in, 50-51, 82, 85; black officers in, 49-50, 51, 56, 85, 235; black regiments in, 34, 35, 41, 43-44, 45-47, 49, 50-52, 78, 82, 171; desegregation following World War II, 43; dishonorable discharge of blacks from, 46-47; exclusion of blacks from, 39; "immune" regiments organized in, 45

United States Army Air Corps: black aviation mechanics in, 185, 192, 193-94, 195, 198; black pursuit squadron established in, ix-x, 32, 36, 81, 111, 172, 182, 183-215 passim; campaign to admit blacks to, viii-ix, 33-36, 52, 53, 69, 73, 77-87 passim, 110-11, 158-82 passim, 217, 224, 228, 306-7; exclusion of blacks from, vii, 25, 33, 78-79, 85, 87, 97, 98, 99, 102-3, 110, 119, 158, 161, 166-67, 168-69, 171-72, 172-75, 176-82, 183; expansions of, ix, 33-34, 78, 86, 88-111 passim, 160, 194; renaming of in 1926, vii (n.); second black pursuit squadron established in, 300-303; training centers for, 92-93, 98, 104, 162, 165, 193, 197, 200, 218, 231, 240, 242, 249, 250, 251-54, 258, 260, 271, 310. *See also* Ninety-ninth Pursuit Squadron; 100th Pursuit Squadron

United States Army Air Forces (AAF), vii (n.)

United States Army Air Service: black military pilots qualified by, vii, 55-56; establishment of, vii (n.); exclusion of blacks from, 78

United States Army Ground Forces, vii (n.)

United States Army Service Forces, vii (n.)

United States Army Signal Corps, aviation in, vii (n.)

United States Colored Troops, 41

United States House of Representatives Committee on Interstate and Foreign Commerce, 91

United States National Guard: black units in, 45, 49, 50, 52, 82; Chicago regiment of, 161

United States Navy: black pilots in, vii, 191; blacks excluded from, vii, 85

United States Senate Military Affairs Committee, 34, 86, 91, 97, 98-99, 164

United States Supreme Court: and black military service, 40; and civil rights movement, 70

Universal Aviation Association, 62

Universal Negro Improvement Association (UNIA), 58-59, 62

"Use of Negro Manpower in War, The" (U.S. Army War College), 51

V

Vandenberg, Arthur H., 177, 179

Van Leer, Blake R., 96

Vann, Robert L., 82-86, 97, 104, 106

Vaughn, Mr., 152, 249

Vocational Education of Defense Workers program, 262

Vocational training at Tuskegee, 2, 3-4, 5

W

Waco Aircraft Company, 145, 147

Waco UPF-7 (YPT-14), 145, 242, 354-55 (n. 46)

Wadsworth, James W., 108

Wagener, Mr. (welding and machine work foreman), 266

Wagner, Robert F., 177-81, 182

Walker, D. Ormande, 95, 96

Walker, Leslie A., 121, 126, 127, 131, 139, 140-41, 308

Wallace, Henry A., 292

Ward, Artis, 61

War of 1812, 38-39, 41

Washington, Booker T., 2-3, 4, 31,

70, 74, 85, 121, 124, 239, 258, 259, 285, 310, 312

Washington, G. L.: as director of pilot training programs at Tuskegee, 122-29, 131, 132-57 passim, 218, 220, 221-23, 225, 229, 230, 240-45, 248-49, 252, 254-67, 271, 278, 280-85, 296, 298-300, 308, 310, 311, 312, 313; illustrations of, *24*, *246*; interest in establishing aviation training at Tuskegee, 1-2, 23, 29, 30-32, 115, 116, 117-18, 120, 121, 131

Washington, George, 38, 41

Washington, Nettie H., 10

Washington Tribune, 104, 176

Waters, Enoch P., 66, 67, 109, 161, 163, 201, 203, 225-26, 228, 232, 310

Watson, Col. Edwin M., 96, 191

Watson, Thomas J., 27

We (Lindbergh), 60-61

Weaver, Gen. Walter, 197-99, 205-8, *207*, 231, 242, 243, 249, 251, 258, 259-60, 273, 281, 286, 287-89

Webster, Grove, 118, 120-21, 128, 145, 147, 152, 156, 161-62, 164, 198, 230, 232, 258, 259

Webster, Maj. R. M., 161-62, 164

Weissman, Helene, 94-95

West Coast Air Corps Training Center, 250

West Point Military Academy: appointment of blacks to, 86, 104-5, 107, 173, 256

West Virginia State College for Negroes: aviation mechanics training at, 136, 346-47 (n. 19); CPT Program at, 118, 147, 159, 184

Wheeler, Gen. Joseph, 45

White, Dale, 110

White, Walter, 78, 83, 160, 161, 173, 177, 185-87, 188, 191, 202-3, 204, 210-11, 225, 226-27, 235, 237-38, 284, 312

Whittier, John Greenleaf, 39-40, 41

Whooter, J. E., 54

Wiggs, George A., 128-29

Wilberforce University, CPT Program at, 115-16

Wilkerson, Joe Wright, 127

Wilkie, Wendell, 176

Wilkins, Roy, 83, 86, 203, 224

Williams, Aubrey, 76

Williams, Howard, 183-84

Williams, Roderick C., 381 (n. 116)

Williams, Roger Q., 11

Williams, Yancey, 204-5, 210, 218-19, 220, 223-24

Wilson, J. Finley, 103-4

Wilson, Woodrow, 48, 49, 73

Women, black: as applicants for CPT Program, 123; as pilots, 54, 58, 60, 64, 142

Woodring, Harry H., 89, 101, 106, 107, 113, 120

Works Progress Administration (WPA), 134, 155-56, 165

Works Progress Administration Act, 109

World War I: role of black aviators during, 55-56; role of black soldiers during, 34, 47-51, 82

World War II: black pilots in, vii-viii, x, 315, 316; black soldiers in, 36, 187; impact on Tuskegee aviation program, 271, 294-96

Wright, E. C., 114

Wright, Orville, 54

Wright, Maj. R. R., 55, 114, 346 (n. 9)

Wright, Bshp. R. R., Jr., 346 (n. 9)

Wright, Richard, 311

Wright, Wilbur, 54

Wright Field (Dayton, Ohio), 115, 253

Y

Yount, Gen. Barton K., 105, 166-67, 184, *186*